STUDENT'S SOLUTIONS MANUAL

James W. Hall's

BEGINNING
ALGEBRA

S E C O N D   E D I T I O N

Prepared by Vicki L. Beitler

PWS Publishing Company

I(T)P   An International Thomson Company

Boston • Albany • Bonn • Cincinnati • Detroit • London • Madrid • Melbourne • Mexico City •
New York • Paris • San Francisco • Singapore • Tokyo • Toronto • Washington

PWS PUBLISHING COMPANY
20 Park Plaza, Boston, Massachusetts 02116-4324

International Thomson Publishing
The trademark ITP is used under license

For more information, contact::

PWS Publishing Co.
20 Park Plaza
Boston, MA 02116

International Thomson Publishing Europe
Berkshire House I68-I73
High Holborn
London WC1V 7AA
England

Thomas Nelson Australia
102 Dodds Street
South Melbourne, 3205
Victoria, Australia

Nelson Canada
1120 Birchmont Road
Scarborough, Ontario
Canada M1K 5G4

International Thomson Editores
Campos Eliseos 385, Piso 7
Col. Polanco
11560 Mexico D.F., Mexico

International Thomson Publishing GmbH
Konigswinterer Strasse 418
53227 Bonn, Germany

International Thomson Publishing Asia
221 Henderson Road
#05-10 Henderson Building
Singapore 0315

International Thomson Publishing Japan
Hirakawacho Kyowa Building, 31
2-2-1 Hirakawacho
Chiyoda Ku, Tokyo 102
Japan

ISBN: 0-534-94039-0

Printed and bound in the United States of America
95 96 97 98 99---10 9 8 7 6 5 4 3 2 1

Problem Set 1.1

1. See answer section in the text.

5. See answer section in the text. Remember that "and" is used to indicate the
    presence of a decimal point.

9. See answer section in the text. Be sure to use the word "and" to replace the decimal point.

13. See answer section in the text. Be sure to use "and" between the whole number and the
    fractional portion.

17. See answer section in the text.

21. See answer section in the text.

25.
$$
\begin{array}{r}
713 \\
11.9\cancel{8}\cancel{8}7 \\
-8.0050 \\
\hline
3.9787
\end{array}
$$

When adding or subtracting decimals,
align the decimal points.

29.
$$
\begin{array}{r}
1.07 \\
23.4\,_\wedge\overline{)25\,.\,0\,_\wedge 38} \\
234 \\
\hline
1638 \\
1638 \\
\hline
0
\end{array}
$$

The decimal point is moved over one place.

Since 234 goes into 163 zero times we bring down
the eight and try again.

33a. Multiplication by 10 has the effect of moving the decimal point to the right one place: 98,765.4
33b. Multiplication by 100 moves the decimal point to the right two places: 987,654
33c. Multiplication by 1000 moves the decimal point to the right three places: 9,876,540
33d. Any real number times zero is zero. Answer: **0**

37. $973.000 + 0.409 = 973.409$

41. $71.705 + 8.910 + 312.000 = 392.615$

45.
$$
\begin{array}{r}
0.053 \\
\times572 \\
\hline
106 \\
371 \\
265 \\
\hline
30.316
\end{array}
$$

The number of places to the right of the decimal in the answer is found by adding the number of
places to the right of the decimal points in the numbers that were multiplied.

49. This translates to $453.7 + 95.08$. See answer section in the text.

53. This translates to 800 ÷ 0.05. (After you set up the division, don't forget to move the decimal points over two places.) See answer section in the text.

57. This translates to $(10.04)^2$ which is equivalent to 10.04×10.04. See answer section in the text.

61. B. This sum is close to being $75 + 25$, which is 100.

65. D. This is close to 20^2 or 20×20, which is 400.

69. terms, or addends. Reread the text to locate this definition.

73. factors. Reread the text to locate this definition.

77. Terms are added, factors are multiplied. Notice that this is just restating what was said in exercises 69 and 73.

81. This translates to $47.03 + 3.047 + 734$. If we write each number with three places to the right of the decimal point, we have $47.030 + 3.047 + 734.000$. See answer section in the text.

85. This translates to $47.75 \div 25$. See answer section in the text.

Problem Set 1.2

1. The even numbers are those divisible by 2. Look for numbers with a last digit of either 0, 2, 4, 6 or 8. See answer section in the text.

5. Any integer ending in 0 or 5 is divisible by 5. See answer section in the text.

9. 20, 22, 24, 26 and 28 are divisible by 2; 21, 27 (and 24) are divisible by 3; 25 (and 20) are divisible by 5. The only prime numbers between 18 and 30 are 19, 23 and 29.

13. 105 is not divisible by 2 but is divisible by 3, so 3 is the smallest prime factor.

17. 1729 is not divisible by 2, 3 or 5, so the next number to test is 7.
$1729 = 7 \cdot 247$, so 7 is the smallest prime factor.

21. $44 = 4 \cdot 11 = 2^2 \cdot 11$

25. $165 = 5 \cdot 33 = 5 \cdot 3 \cdot 11 \text{ or } 3 \cdot 5 \cdot 11$

29. $\dfrac{99}{121} = \dfrac{9(11)}{11(11)} = \dfrac{9}{11}$

33. $\dfrac{582}{873} = \dfrac{2(291)}{3(291)} = \dfrac{2}{3}$

37. We need to compare the part to the whole:

$\dfrac{14}{49} = \dfrac{7(2)}{7(7)} = \dfrac{2}{7}$ This says that 14 is two-sevenths of 49.

41. $\dfrac{80}{165} = \dfrac{5(16)}{5(33)} = \dfrac{16}{33}$

45. $\dfrac{\text{diamonds}}{\text{all cards}} = \dfrac{13}{52} = \dfrac{13(1)}{13(4)} = \dfrac{1}{4}$

The ratio is 1 : 4, read "one to four".

49. 127 is prime.

53. 361 is not divisible by 2, 3 or 5. (It will therefore not be divisible by any larger number that has 2, 3 or 5 as a factor.) Checking each of 7, 11, 13, 17 and 19 finally gives $361 = 19^2$.

57. $\dfrac{1155}{1617} = \dfrac{5(231)}{3(539)} = \dfrac{5(3)(77)}{3(7)(77)} = \dfrac{5}{7}$

While it is necessary to have the numerator and denominator *factored before doing any canceling*, they need not be completely factored into primes. In this problem we did not need to factor 77, as it was a common factor to both numerator and denominator and dropped out.

61. $\dfrac{\text{defective chips}}{\text{total chips}} = \dfrac{132}{1056} = \dfrac{2(66)}{2(528)} = \dfrac{66}{528} = \dfrac{6(11)}{6(88)} = \dfrac{11}{88} = \dfrac{11(1)}{11(8)} = \dfrac{1}{8}$

This says that one-eighth of all the computer chips were defective.

65. $\dfrac{\text{banded ducks}}{\text{total captured}} = \dfrac{17}{85} = \dfrac{1}{5}$ The ratio is 1 : 5 , read "one to five".

$\dfrac{\text{banded ducks}}{\text{unbanded ducks}} = \dfrac{17}{85 - 17} = \dfrac{17}{68} = \dfrac{1}{4}$ The ratio is 1 : 4 , read "one to four".

69. A. Two times 117 is 234; this suggests that this fraction is approximately $\frac{1}{2}$. Or, round to the nearest 10 and the fraction is nearly $\dfrac{120}{240}$ which is $\frac{1}{2}$.

73. $\dfrac{\text{banded rabbits}}{\text{total rabbits}} = \dfrac{26}{117} = \dfrac{2(13)}{9(13)} = \dfrac{2}{9}$ Therefore, this ratio is 2 : 9.

Out of the 117 rabbits, 26 were banded. This means that the number of non-banded rabbits was $117 - 26 = 91$. We use this figure to construct the next ratio:

$\dfrac{\text{banded rabbits}}{\text{non-banded rabbits}} = \dfrac{26}{91} = \dfrac{2(13)}{7(13)} = \dfrac{2}{7}$ which can be expressed as the ratio 2 : 7.

77. The number of right-handed players is $770 - 77 = 693$. Therefore:

$\dfrac{\text{left-handed players}}{\text{right-handed players}} = \dfrac{77}{693} = \dfrac{77(1)}{77(9)} = \dfrac{1}{9}$

This says that the ratio of left-handed players to right-handed players is 1 : 9 ; or, there are 9 times as many right-handed players as left-handed players.

81. $(419)^2 = 419 \cdot 419$, and so $(419)^2$ is composite.

1. $\dfrac{5}{19} \cdot \dfrac{11}{7} = \dfrac{5(11)}{19(7)} = \dfrac{55}{133}$

5. $\dfrac{18}{24} \cdot \dfrac{16}{27} = \dfrac{18(16)}{24(27)} = \dfrac{9(2)(8)(2)}{8(3)(9)(3)} = \dfrac{4}{9}$

(In #5, notice that we only *set up* the multiplication of the numerators (and denominators) but did not actually carry out the calculations. To do so would only hinder our next step, which is to determine common factors in the numerator and denominator. We want the head start.)

9. $\dfrac{7}{8} \div \dfrac{4}{5} = \dfrac{7}{8} \cdot \dfrac{5}{4} = \dfrac{7(5)}{8(4)} = \dfrac{35}{32}$ or $1\dfrac{3}{32}$

13. $\dfrac{143}{429} \div \dfrac{156}{385} = \dfrac{143(385)}{429(156)} = \dfrac{143(5)(77)}{3(143)3(52)} = \dfrac{5(77)}{3(3)(52)} = \dfrac{5(7)(11)}{3(3)(4)(13)} = \dfrac{385}{468}$

17. $\dfrac{7}{33} + \dfrac{22}{21}$ $\qquad \left.\begin{array}{l} 33 = 3 \cdot 11 \\ 21 = 3 \cdot 7 \end{array}\right\}$ The LCD is $3 \cdot 7 \cdot 11$.

$\dfrac{7(7)}{3(11)(7)} + \dfrac{22(11)}{3(7)(11)} = \dfrac{49 + 242}{3(7)(11)} = \dfrac{291}{3(7)(11)} = \dfrac{3(97)}{3(7)(11)} = \dfrac{97}{7(11)} = \dfrac{97}{77} = 1\dfrac{20}{77}$

$\boxed{IMPORTANT}$ *Notice that the factors in the denominator are not multiplied until the fraction is completely reduced.*

21. $\dfrac{9}{35} - \dfrac{2}{35} = \dfrac{7}{35} = \dfrac{7(1)}{7(5)} = \dfrac{1}{5}$

25. $\dfrac{3}{238} - \dfrac{2}{255}$ $\qquad \left.\begin{array}{l} 238 = 2 \cdot 119 = 2 \cdot 7 \cdot 17 \\ 255 = 5 \cdot 51 = 5 \cdot 3 \cdot 17 \end{array}\right\}$ LCD $= 2 \cdot 3 \cdot 5 \cdot 7 \cdot 17$

$\dfrac{3(3)(5)}{2(7)(17)(3)(5)} - \dfrac{2(2)(7)}{5(3)(17)(2)(7)} = \dfrac{45 - 28}{2(3)(5)(7)(17)} = \dfrac{17}{2(3)(5)(7)(17)} = \dfrac{1}{210}$

29. $\dfrac{5}{14} + \dfrac{3}{10} - \dfrac{3}{35}$ $\qquad \left.\begin{array}{l} 14 = 2 \cdot 7 \\ 10 = 2 \cdot 5 \\ 35 = 5 \cdot 7 \end{array}\right\}$ LCD $= 2 \cdot 5 \cdot 7$

$\dfrac{5(5)}{2(7)(5)} + \dfrac{3(7)}{2(5)(7)} - \dfrac{3(2)}{5(7)(2)} = \dfrac{25 + 21 - 6}{2(5)(7)} = \dfrac{40}{2(5)(7)} = \dfrac{2(5)(4)}{2(5)(7)} = \dfrac{4}{7}$

33. $\dfrac{14}{15} \div 6 = \dfrac{14}{15} \cdot \dfrac{1}{6} = \dfrac{14}{15(6)} = \dfrac{2(7)}{15(2)(3)} = \dfrac{7}{45}$

37. $3\dfrac{2}{9} \div 2\dfrac{1}{3}$ \qquad Notice that $3\dfrac{2}{9} = 3 + \dfrac{2}{9} = \dfrac{3(9)}{9} + \dfrac{2}{9} = \dfrac{29}{9}$.

$$\boxed{\text{A mixed number } A\tfrac{b}{c} \text{ is equal to the fraction } \dfrac{A(c)+b}{c}.}$$

$= \dfrac{29}{9} \div \dfrac{7}{3} = \dfrac{29(3)}{9(7)} = \dfrac{29(3)}{3(3)(7)} = \dfrac{29}{21}$ or $1\dfrac{8}{21}$

41. Approach #1

$6\dfrac{5}{21} = 6\dfrac{25}{3(7)(5)}$ \qquad\qquad $\left.\begin{array}{l} 21 = 3 \cdot 7 \\ 15 = 3 \cdot 5 \end{array}\right\}$ LCD $= 3 \cdot 5 \cdot 7$ \quad (105)

$2\dfrac{4}{15} = 2\dfrac{28}{3(7)(5)}$

Borrowing is required. We borrow 1 from 6 in the form $\dfrac{105}{105}$ and add this to the fractional part of $6\dfrac{25}{3(7)(5)}$, giving us $5\dfrac{130}{3(7)(5)}$. Remember, these fractions are equivalent, they just look different.

$$\begin{array}{r} 5\dfrac{130}{3(5)(7)} \\ - \ 2\dfrac{28}{3(5)(7)} \\ \hline = 3\dfrac{102}{3(5)(7)} \end{array} \qquad = 3\dfrac{3(34)}{3(5)(7)} = 3\dfrac{34}{35}$$

Approach #2: Begin by changing both mixed numbers to improper fractions:

$6\dfrac{5}{21} - 2\dfrac{4}{15} = \dfrac{131}{21} - \dfrac{34}{15} =$

$\dfrac{131(5)}{3(7)(5)} - \dfrac{34(7)}{3(5)(7)} = \dfrac{655 - 238}{3(5)(7)} = \dfrac{417}{3(5)(7)} = \dfrac{3(139)}{3(5)(7)} = \dfrac{139}{5(7)} = \dfrac{139}{35} = 3\dfrac{34}{35}$

45. Zero divided by any *nonzero number* is zero.

49. $\dfrac{1(2)(3)(4)(5)}{2(3)(4)(5)(6)} = \dfrac{1}{6}$ \qquad\qquad 53. $\left(\dfrac{7}{30}\right)^2 = \dfrac{49}{900}$

57. $\dfrac{4}{23} \div \dfrac{8}{69} = \dfrac{4(69)}{23(8)} = \dfrac{4(23)(3)}{23(4)(2)} = \dfrac{3}{2}$

61. Zero times any number of factors is zero.

65. Notice that 112 is almost 110, which is 10(11), so the fraction $\dfrac{11}{112}$ is approximately $\dfrac{1}{10}$. $\dfrac{1}{10}(1008) \approx \dfrac{1}{10}(1000) = 100.$ The closest answer is 99, which is response D.

69. $\left(\dfrac{4}{21} + \dfrac{1}{42}\right)^2 = \left(\dfrac{4(2) + 1}{42}\right)^2 = \left(\dfrac{9}{42}\right)^2 = \left(\dfrac{3}{14}\right)^2 = \dfrac{9}{196}$

73. $\dfrac{85}{86} \cdot \dfrac{86}{115} = \dfrac{85}{115} = \dfrac{5(17)}{5(23)} = \dfrac{17}{23}$

$\dfrac{17}{23} \div \dfrac{34}{46} = \dfrac{17(46)}{23(34)} = \dfrac{17(2)(23)}{23(2)(17)} = 1$

77. $\left(4\dfrac{2}{15}\right)\left(1\dfrac{1}{4}\right) = \dfrac{62}{15} \cdot \dfrac{5}{4}$

$= \dfrac{2(31)(5)}{3(5)(2)(2)} = \dfrac{31}{6}$ or $5\dfrac{1}{6}$

Problem Set 1.4

1a. $0^n = 0$ where n is any natural number.

1b. $1^n = 1$ where n is any natural number.

1c. Answer: 18. $n^1 = n$ for any number n.

1d. $10^3 = 10 \cdot 10 \cdot 10 = 1000$

5a. $5 + 2 \cdot 8$
$= 5 + 16$
$= 21$

5b. $(5 + 2) \cdot 8$
$= 7 \cdot 8$
$= 56$

9a. $10^2 = 100$

9b. $36 + 16 = 52$

13. $6 \cdot 7 - 5 \cdot 3$
$= 42 - 15$
$= 27$

17. $15 - 3^2 + 6$
$= 15 - 9 + 6$
$= 6 + 6$
$= 12$

21. $182 - 5(3^3)$
$= 182 - 5 \cdot 27$
$= 182 - 135$
$= 47$

25. $16 \div 4 \cdot 3 + 5$
$= 4 \cdot 3 + 5$
$= 12 + 5$
$= 17$

29. $(3 + 4)^2 - (3^2 + 4^2)$
$= 7^2 - (9 + 16)$
$= 49 - 25$
$= 24$

33. $(2 + 3)^3 - (2^3 + 3^3)$
$= 5^3 - (8 + 27)$
$= 125 - 35$
$= 90$

37. $12 + 2[8 - 3(2)]$
$= 12 + 2(8 - 6)$
$= 12 + 2(2)$
$= 16$

41. $18 + 2[(41 - 8) - (5 + 9)]$
$= 18 + 2[33 - 14]$
$= 18 + 2(19)$
$= 18 + 38$
$= 56$

45. This can be approximated by:
$(5 + 4)^2 = 9^2 = 81$
We choose C as the best answer.

49. $(7.018)^2 + (8.956)^2$
$\approx 7^2 + 9^2$
$= 49 + 81$
$= 130$
Answer: C

53. associative
57. distributive

61. $\left(\frac{5}{7} + \frac{2}{7} + \frac{1}{7}\right) + \left(\frac{3}{8} + \frac{5}{8}\right)$
$= \frac{8}{7} + \frac{8}{8} = \frac{8}{7} + 1$
$= \frac{15}{7}$ or $2\frac{1}{7}$

65. $\left(1\frac{1}{2}\right)^2 - \left(1\frac{1}{3}\right)^2 = \left(\frac{3}{2}\right)^2 - \left(\frac{4}{3}\right)^2$
$= \frac{9}{4} - \frac{16}{9} = \frac{9(9)}{4(9)} - \frac{16(4)}{4(9)}$
$= \frac{81 - 64}{4(9)} = \frac{17}{4(9)} = \frac{17}{36}$

69. Seventeen is going to be subtracted from a sum, so first determine the sum.
The two numbers to be added are 5^2 and 6^2; their sum is $5^2 + 6^2$.

$(5^2 + 6^2) - 17$ Remember the order of operations: exponentiation first.
$= (25 + 36) - 17$ Now add what is in parentheses.
$= 61 - 17$
$= 44$

73. The order of operations for arithmetic is followed: seven is first squared, then we multiply by two, and finally five is added. Answer: B

77. $[8 - 2(5 - 4)][11 - 3(7 - 4)]$

Problem Set 1.5

1. The check is for $1033.14 so the balance will be
$1033.14 + 135.47$
$= \$1168.61$

5. cost $=$ (number of pmts)(amount of pmt)
$= 32(7.85)$
$= \$251.20$

9. $\frac{7}{8} - \frac{3}{16}$
$= \frac{14}{16} - \frac{3}{16}$
$= \frac{11}{16}$ inch

13. The perimeter is the length of the border:
$110.7 + 85.6 + 48.6 + 79.3 = 324.2$ m

17. The sum of the kwh used over the seven days is 150.5 kwh. Dividing by seven gives 21.5 kwh per day.

21. The sales tax is 6% of 499.99. To compute this, write 6% as 0.06.
0.06(499.99) = 29.9994
The sales tax would be $30.00.

25. Compute regular pay and overtime pay separately. In other words, work on the problem in manageable pieces. We will add the results to get total pay.

regular pay = ($4.20 per hour)(40 hours) = $168.00
overtime pay = ($6.30 per hour)(12 hours) = $75.60 } total earnings: $243.60

$\boxed{COMMENT}$ *It seems a good time to make a point. Regular pay can also be written:*

$$\frac{4.20 \text{ dollars}}{1 \text{ hour}} \cdot 40 \text{ hours} \quad \text{or} \quad \frac{4.20 \text{ dollars}}{1 \text{ hour}} \cdot \frac{40 \text{ hours}}{1} = 168.00 \text{ dollars}$$

The rate of pay is written as a fraction, $\frac{4.20 \text{ dollars}}{1 \text{ hour}}$, where the numerator is equivalent to the denominator. This is sometimes called a conversion factor, since we multiply 40 hours by it in order to convert the 40 hours of work to the equivalent amount of dollars of pay.
Notice although "hours" is a word, it seems to cancel out of the problem because it appears in the numerator and denominator of our calculations, and the final answer has only the word "dollars" in it. This is what we want, of course, because we are computing the worker's pay. The fact that the hours cancels out and only dollars remains tells us that the problem is set up correctly.
Suppose we were not sure how to set up the problem and so we had considered division instead of multiplication. By setting up the problem with the proper units, we can see if what we have makes sense or not:

$$40 \div 4.20 = \frac{40 \text{ hours}}{1} \cdot \frac{1 \text{ hour}}{4.20 \text{ dollars}} \qquad ???$$

Notice that the units don't work. The "hours" do not cancel out, and even so, "dollars" ends up in the denominator, which doesn't make sense. This calculation would not give the correct answer.

29. The mileage was 20,073.4 − 17,598.3 = 2,475.1 (miles).

$$\frac{\text{miles}}{\text{gallon}} = \frac{2475.1}{81.2} \approx 30 \text{ miles per gallon}$$

33. The child's dosage is $\frac{3}{4}$ of 12 cc. $\frac{3}{4}(12) = \frac{3}{4} \cdot \frac{12}{1} = 9$ cc.

37. D. The mean (which is the same thing as an average) is found by adding the numbers and dividing by how many there are in the list. Rounding to the nearest 10 gives:
50 + 60 + 70 + 70 + 60 = 310
Dividing by five, we have 62 as an approximation for the mean.

41. $2\frac{1}{2} + 3\frac{1}{4} + 2\frac{3}{4} + 4 + 3\frac{3}{8}$

$= (2 + 3 + 2 + 4 + 3)$

$\qquad\qquad + (\frac{1}{2} + \frac{1}{4} + \frac{3}{4} + \frac{3}{8})$

$= 14 + (\frac{1}{2} + 1 + \frac{3}{8})$

$= 15\frac{7}{8}$ or $\frac{127}{8}$

Now divide by 5: $\quad \frac{127}{8} \div 5$

$= \frac{127}{8} \div 5$

$= \frac{127}{8} \cdot \frac{1}{5}$

$= \frac{127}{40}$ or $3\frac{7}{40}$ boxcar loads per day

45. cost for the front
(122.5 m)($9.40 per meter)
= $1151.50

> or: $\dfrac{122.5 \text{ m}}{1} \cdot \dfrac{\$9.40}{1 \text{ m}} = \$1151.50$
>
> See discussion with #25.

cost for the sides
(126 m)($8.75 per meter)
= $1102.5

total cost = $2254.00

Solutions To All Review Exercises — Chapter One

1. See answer section in the text.

2. The test for divisibility by 3 is to add the digits in the numbers; if the resulting number is divisible by 3, then so is the original number. We test each one:

   $7 + 8 = 15$ $\qquad$ so 78 is divisible by 3
   $8 + 7 = 15$ $\qquad$ so 87 is divisible by 3
   $2 + 2 + 2 = 6$ so 222 is divisible by 3
   $3 + 1 + 3 = 7$ so 313 is not divisible by 3

3. If a number ends in 5 or 0, it is divisible by 5; otherwise it is not.
   Answer: 45, 450.

4. A prime number is divisible only by itself and 1.
   70, 72, 74, 76, 78 and 80 are divisible by 2; 75 is divisible by 5. 71, 73 and 79 are prime.

5. 84, 86 and 88 are divisible by 2; 85 is divisible by 5; 87 (and 84) are divisible by 3.
   All the integers between 83 and 89 are composite.

6. $\dfrac{77}{187} = \dfrac{7(11)}{17(11)} = \dfrac{7}{17}$ $\quad$ Once you notice that $77 = 7(11)$, try 7 and 11 as factors of 187.

7. $\dfrac{195}{255} = \dfrac{5(39)}{5(51)} = \dfrac{39}{51} = \dfrac{3(13)}{3(17)} = \dfrac{13}{17}$

8. $\dfrac{1200}{3300} = \dfrac{12(100)}{33(100)} = \dfrac{12}{33} = \dfrac{3(4)}{3(11)} = \dfrac{4}{11}$

9. 167.366

10. 3697.883

11. 316.214

12. 8.5

13. $16 \div 8 \cdot 2 = 2 \cdot 2 = 4$

14. $(8 - 7)^9 = 1^9 = 1$

15. $(48 - 3 \cdot 4^2)^7 = (48 - 3 \cdot 16)^7 = (48 - 48)^7 = 0^7 = 0$

16. $10^5 = 100{,}000$ Remember: $10^5 = 10 \cdot 10 \cdot 10 \cdot 10 \cdot 10$

17. $(0.1)^4 = 0.0001$ or $(0.1)^4 = \left(\dfrac{1}{10}\right)^4 = \dfrac{1}{10{,}000} = 0.0001$

18. $7^2 - 3^2 + (7 - 3)^2 = 7^2 - 3^2 + 4^2 = 49 - 9 + 16 = 40 + 16 = 56$

19. $56 - 3 \cdot 8 = 56 - 24 = 32$

20. $73 + 5 \cdot 2^3 = 73 + 5 \cdot 8 = 73 + 40 = 113$

21. $[3 + 7(6)][11 - 3(17 - 15)] = [3 + 42][11 - 3(2)] = (45)(11 - 6) = 45(5) = 225$

22. $\dfrac{8 + 3 \cdot 19}{3 + 2 \cdot 5} = \dfrac{8 + 57}{3 + 10} = \dfrac{65}{13} = 5$

23. $\dfrac{75}{98} \cdot \dfrac{42}{45} = \dfrac{75(42)}{98(45)} = \dfrac{5(15)(6)(7)}{7(14)9(5)} = \dfrac{15(6)}{14(9)} = \dfrac{3(5)(2)(3)}{2(7)(3)(3)} = \dfrac{5}{7}$

24. $\dfrac{88}{63} \div \dfrac{66}{45} = \dfrac{88}{63} \cdot \dfrac{45}{66} = \dfrac{88(45)}{63(66)} = \dfrac{8(11)(5)(9)}{7(9)(6)(11)} = \dfrac{8(5)}{7(6)} = \dfrac{4(5)}{7(3)} = \dfrac{20}{21}$

25. $2\dfrac{4}{7} \cdot 5\dfrac{4}{9} = \dfrac{18}{7} \cdot \dfrac{49}{9} = \dfrac{18(49)}{7(9)} = \dfrac{9(2)(7)(7)}{7(9)} = 14$

26. $4\dfrac{1}{8} \div 3\dfrac{3}{4} = \dfrac{33}{8} \cdot \dfrac{15}{4} = \dfrac{33}{8} \cdot \dfrac{4}{15} = \dfrac{33(4)}{8(15)} = \dfrac{3(11)(4)}{2(4)(3)(5)} = \dfrac{11}{10}$ or $1\dfrac{1}{10}$

27. $\dfrac{25}{87} + \dfrac{4}{87} = \dfrac{25 + 4}{87} = \dfrac{29}{87} = \dfrac{29}{(29)(3)} = \dfrac{1}{3}$

28. $\dfrac{31}{69} - \dfrac{8}{69} = \dfrac{31 - 8}{69} = \dfrac{23}{69} = \dfrac{23}{3(23)} = \dfrac{1}{3}$

29. $\dfrac{49}{143} \div \dfrac{11}{35} \cdot \dfrac{39}{105} = \left(\dfrac{49(35)}{143(11)}\right)\left(\dfrac{39}{105}\right) = \dfrac{7(7)(5)(7)(3)(13)}{143(11)(5)(3)(7)} = \dfrac{7(7)(13)}{143(11)} = \dfrac{7(7)(13)}{11(13)(11)} = \dfrac{49}{121}$

Chapter One Review

30. $23 - 4(11 - 6) = 23 - 4(5) = 23 - 20 = 3$

31. $36 \div 4 \cdot 9 \div 3 = 9 \cdot 9 \div 3 = 81 \div 3 = 27$

32. $4\frac{1}{9} + 5\frac{1}{3} = \frac{37}{9} + \frac{16}{3} = \frac{37}{9} + \frac{48}{9} = \frac{37 + 48}{9} = \frac{85}{9}$ or $9\frac{4}{9}$

33. $33 - 3[18 - 4 \cdot 2] = 33 - 3(18 - 8) = 33 - 3(10) = 33 - 30 = 3$

34. $\frac{8}{21} + \frac{13}{35} - \frac{2}{105}$

$$\left. \begin{array}{l} 21 = 3 \cdot 7 \\ 35 = 5 \cdot 7 \\ 105 = 3 \cdot 5 \cdot 7 \end{array} \right\} \quad LCD = 3 \cdot 5 \cdot 7$$

$$\frac{8(5)}{3(7)(5)} + \frac{13(3)}{5(7)(3)} - \frac{2}{3(5)(7)} = \frac{40 + 39 - 2}{3(5)(7)} = \frac{77}{3(5)(7)} = \frac{11}{15}$$

35. $\left(\frac{1}{2} + \frac{1}{3}\right)^2 - \left[\left(\frac{1}{2}\right)^2 + \left(\frac{1}{3}\right)^2\right] = \left(\frac{5}{6}\right)^2 - \frac{1}{4} - \frac{1}{9}$

$\quad = \frac{25}{36} - \frac{9}{36} - \frac{4}{36} = \frac{12}{36} = \frac{1}{3}$

36. $\left(\frac{1}{2} + \frac{1}{5}\right)^2 = \left(\frac{7}{10}\right)^2 = \frac{49}{100}$

37. $\frac{151}{401} \cdot \frac{401}{151} = \frac{151(401)}{401(151)} = 1$ Any number times its reciprocal is one.

38. $\frac{67}{103} \div \frac{67}{103} = 1$ Any number divided by itself is 1.

39. $0 \div 5.6 = 0$ Zero divided by any nonzero number is zero.

40.   Division by zero is undefined.   41. $101.1 - 45.007 = 56.093$

42. $(33.08)(40.7) = 1346.356$.   Rounding gives 1346.4.

43. $\frac{325.78}{204.5} \approx 1.59$ 44. $\frac{6(\text{ways a 7 could occur})}{36(\text{possible pairs})} = \frac{6}{36} = \frac{1}{6}$

45.   The team's losses are $24 - 18 = 6$.   The ratio is $\frac{\text{losses}}{\text{total games}} = \frac{6}{24} = \frac{1}{4}$ or $1 : 4$.

46.   First we need to know how many meters make up the trench. The unlabeled side has the same measure as the sum of the two sides opposite it, and so it is 30 meters.
$20 + 20 + 10 + 10 + 10 + 30 = 100$ meters of trench
The cost will be ($2.60 per meter)(100 meters) = $260.00

47. $217.85 - 10 - 10 - 10 - 38.19 + 45 = 194.66$
The new balance is \$194.66.

48.   The ratio is $\dfrac{\text{cost}}{\text{pounds of flour}} = $ cost per pound. We want to see which costs less per pound,

so we will compute this ratio for each.

$\dfrac{0.75}{5} = 0.15$, or \$0.15 per pound

$\dfrac{1.56}{10} = 0.156$, or \$0.156 per pound

The five-pound sack costs a bit less per pound than the larger sack, and so is the better buy.

49.   Solution #1

(amount of gas)(rate of usage) = distance
amount of gas = distance ÷ rate of range
amount of gas = 370 miles ÷ 27 mi/g
$\approx$ 14 gallons

Solution #2

$\dfrac{\text{number of gallons}}{370 \text{ miles}} = \dfrac{1 \text{ gallon}}{27 \text{ miles}}$

(number of gallons)27 = 370(1)

number of gallons = $\dfrac{370}{27} \approx$ 14 gallons

Solution #3

$\dfrac{1 \text{ gallon}}{27 \text{ miles}} \cdot \dfrac{370 \text{ miles}}{1} = \dfrac{370}{27}$ gallons $\approx$ 14 gallons

Notice in solution #3, we express the rate 27 mpg in an alternate way, as a fraction equivalent to one. We choose 1 gallon in the numerator and 27 miles in the denominator (instead of the other way around), since this makes sense with the units. We note that the "miles" units will drop out, and we are left with only gallons, which is what is needed for the units in the answer. *See also in this manual, section 1.5, #25, for a similar problem involving an employee's pay.*

50.   We will figure the tax first: $(0.04)(49.95) = 1.998$.   Add this to the price, and since this is money, we must round to the nearest hundredth: \$51.95.

51. $\dfrac{79 + 83 + 91 + 76}{4} = \dfrac{329}{4} = 82.25$

52.   We can estimate as follows: $9 + 7 + 13 = 29$ Answer: D

53.   We may estimate as follows: $1050 - 250 = 800$ Answer: A

54.   We may estimate as follows: $(2360)(100) = 236{,}000$ Answer: B

55.   We may estimate as follows $(300)(0.03) = 300\left(\dfrac{3}{100}\right) = 9$ Answer: C

56.   We may estimate as follows: $42 \div 7 = 6$ Answer: C

57. $300 \div 0.03 = 300 \div \left(\dfrac{3}{100}\right) = 300\left(\dfrac{100}{3}\right) = 10{,}000$ Answer: B

58. Estimate using $5^3 = 125$.   Answer: A

59. Use $5(8 + 12) = 5(20) = 100$.   Answer: D

60. Remember that division is done before addition.   Answer: B

61. $(7 + 3^2) \div (7 + 3)^2 = \dfrac{7 + 3^2}{(7 + 3)^2}$

    The parentheses are very important here.   Answer: D

62. $50 - \dfrac{36}{4} = 50 - 9 = 41$

63. The range is the largest value minus the smallest value: $43.81 - 20.08 = 23.73$

    The mean is the average. The sum of the bills is 380.68; then divide by 12: $\dfrac{380.68}{12} \approx 31.72$

64. commutative, multiplication   65. associative, addition

66. associative, multiplication   67. commutative, addition

68. 1   69. distributive

70. 0   71. 0

72. $22 - 7(3) = 22 - 21 = 1$

73. This is explained in the answer section in the text.

Solutions To All Mastery Test Problems — Chapter One

1a. 3011.002   1b. nineteen thousand one and five hundredths

1c. Looking at 73.6784, the digit after the hundredth place is 8 which is greater than 5, so we add one to the hundredths position. Answer: 73.68

2a.
```
  12.409
+  7.230
 -------
  19.639
```

2b.
```
      3 10
  1 2 . 4̸ 0̸ 9
 -  7 . 2 3 0
 -----------
    5 . 1 7 9
```

2c.
```
      2.1
  ×   9.45
  -------
      105
       84
      189
  -------
   19.845
```

2d.
```
              4.5
  2.1∧ | 9.4∧5
         8 4
         -----
         1 0 5
         1 0 5
         -----
             0
```

page 13

Chapter One Mastery Test

3a. $104.38 + 35.12 = 139.5$

3b. $0.001 \times 35.12 = 0.03512$

3c. $35.12 - 12.5 = 22.62$

3d. $\dfrac{35.12}{0.04} = 878$

3e. $(3.2)^2 = 10.24$

4a. We may estimate this with 350×0.02 or $(350)(\frac{2}{100}) = \frac{700}{100} = 7$. The answer is (v).

4b. We may estimate this with $480 \div 240 = 2$. Answer: (i)

5a. The even numbers are those divisible by 2. Answer: 16, 42, 50 and 252

5b. A number is divisible by three if the number that results by adding the digits is divisible by 3. (Since this new number is relatively small, it's easier to determine if it is divisible by 3.) From the list, 42, 45 and 252 have this property and so are divisible by three.

5c. A number is divisible by 5 if it ends in 0 or 5. Answer: 45, 50 and 55

5d. A number is divisible by 10 if it ends in 0. Answer: 50

6a. $66 = 2(3)(11)$

6b. $115 = 5(23)$

6c. $210 = 10(21)$
$= 2(5)(3)(7)$
$= 2(3)(5)(7)$

6d. $195 = 5(39)$
$= 5(3)(13)$
$= 3(5)(13)$

7a. $\dfrac{30}{55} = \dfrac{5(6)}{5(11)} = \dfrac{6}{11}$

7b. $\dfrac{28}{34} = \dfrac{2(14)}{2(17)} = \dfrac{14}{17}$

7c. $\dfrac{22}{77} = \dfrac{11(2)}{11(7)} = \dfrac{2}{7}$

7d. $\dfrac{260}{390} = \dfrac{10(26)}{10(39)} = \dfrac{26}{39} = \dfrac{13(2)}{13(3)} = \dfrac{2}{3}$

8a. $\dfrac{2}{3} + \dfrac{3}{5} = \dfrac{10+9}{15} = \dfrac{19}{15}$ or $1\dfrac{4}{15}$

8b. $\dfrac{2}{3} \cdot \dfrac{3}{5} = \dfrac{2(3)}{3(5)} = \dfrac{2}{5}$

8c. $\dfrac{11}{12} - \dfrac{5}{8}$ $\begin{cases} 12 = 2^2 \cdot 3 \\ 8 = 2^3 \\ \text{LCD} = 2^3 \cdot 3 \end{cases}$

$\dfrac{11(2)}{12(2)} - \dfrac{5(3)}{8(3)}$

$= \dfrac{22-15}{2^3 \cdot 3} = \dfrac{7}{24}$

8d. $\dfrac{5}{18} \div \dfrac{11}{12} = \dfrac{5(12)}{18(11)}$

$= \dfrac{5(2)(6)}{(3)(6)11}$

$= \dfrac{5(2)}{(3)(11)} = \dfrac{10}{33}$

Chapter One Mastery Test

9a.　$4^3 = (4)(4)(4) = 64$　　　　9b.　$3^4 = (3)(3)(3)(3) = (9)(9) = 81$

9c.　$\left(\dfrac{5}{7}\right)^2 = \dfrac{5^2}{7^2} = \dfrac{25}{49}$　　　　9d.　$10^5 = (10)(10)(10)(10)(10) = 100{,}000$

10a.　$18 - 5(9 - 6)$
$= 18 - 5(3)$
$= 18 - 15$
$= 3$

10b.　$3^2 + 7^2 + (3 + 7)^2$
$= 9 + 49 + 10^2$
$= 9 + 49 + 100$
$= 158$

10c.　$57.68 - 47.9(4^2 - 2^4)$
$= 57.68 - 47.9(16 - 16)$
$= 57.68 - 47.9(0)$
$= 57.68$

10d.　$8 \cdot 8 \div 8 \cdot 8$
$= 64 \div 8 \cdot 8$
$= 8 \cdot 8$
$= 64$

11a.　$350 + 5.25 + 175 - 125$
$= 530.25 - 125 = 405.25$

11b.　$\dfrac{82 + 71 + 95 + 87 + 78}{5}$
$= \dfrac{413}{5}$
$= 82.6$

Chapter One Mastery Test

Problem Set 2.1

1. $5 + 8 = 13$

5. $\dfrac{5 + 4}{6} = \dfrac{9}{6} = \dfrac{3}{2}$

9. $\dfrac{5^2 - 3}{2(5) + 1} = \dfrac{25 - 3}{10 + 1} = \dfrac{22}{11} = 2$

13. $5(2) - 9 = 10 - 9 = 1$

17. $\dfrac{6}{2} + \dfrac{9}{3} = 3 + 3 = 6$

21. $2^2 + 4(2)(9) + 9^2$
     $= 4 + 72 + 81$
     $= 157$

25. "Times" indicates multiplication. Seven times n would be written 7n.

29. "Two more than" means that you will add 2 to a quantity. That quantity is 5n, so the expression is 5n + 2.

33. Break up the sentence into parts:

nine times a number	9m	
is	=	$9m = 369$
three hundred sixty-nine	369	

37.
forty-eight	48	
twice a number	2m	
their quotient	$\dfrac{48}{2m}$	$\dfrac{48}{2m} = 8$
is equal to eight	$= 8$	

41.
four more than a number	m + 4	
the ratio of five to that quantity	$\dfrac{5}{m + 4}$	$\dfrac{5}{m + 4} = \dfrac{3}{5}$
is three-fifths	$= \dfrac{3}{5}$	

45. $2(8) - 7 = 16 - 7 = 9$ (not 11)   The number 8 is not a solution.

49. $\left[3\left(\dfrac{2}{3}\right) - 2 \right]\left[2\left(\dfrac{2}{3}\right) + 3 \right] = \left[2 - 2 \right]\left[2\left(\dfrac{2}{3}\right) + 3 \right] = 0\left[2\left(\dfrac{2}{3}\right) + 3 \right] = 0$

The number $\dfrac{2}{3}$ is a solution.

53a. $C = 2\pi r = 2\pi(5.6) = 11.2\pi$ cm
Using $\pi \approx 3.1415927$, the circumference
is approximately 35.2 cm.

53b. $A = \pi r^2 = \pi(5.6)^2$
$= \pi(31.36)$ cm^2 or 31.36π cm^2
Using $\pi \approx 3.1415927$, the area is
approximately 98.5 cm^2.

57. You can estimate by using 12 feet and 15 feet. $12 \times 15 = 180$, so the best estimate is C.

61. The old retail price is R dollars; if we increase it by \$2 we are adding \$2. The new retail price will be R + 2 dollars.

65. Use the formula provided; Profit = R − 5.

69.
$5(\frac{3}{5}) + 2 \stackrel{?}{=} 10(\frac{3}{5}) - 1$

$3 + 2 \stackrel{?}{=} 6 - 1$

$5 = 5$ ✓

$5(\frac{7}{10}) + 2 \stackrel{?}{=} 10(\frac{7}{10}) - 1$

$\frac{7}{2} + 2 \stackrel{?}{=} 7 - 1$

$\frac{11}{2} \neq 6$

$\frac{3}{5}$ is a solution of the equation, but $\frac{7}{10}$ is not.

73. If 2 is substituted in for x, you have $0(-1) = 0$.
If 3 is substituted in for x, you have $1(0) = 0$. Both 2 and 3 satisfy the equation.

77. See answer section in the text. See also #81 below.

81. Suggestion :

(1) Write down a very general idea of what your statement will be. For example:
Two quantities will be subtracted to get a third.

(2) Write phrases to describe the quantities in the problem:
    x   the number we wish to find
    3x   three times the number
    5   five
    2x   two times the number

(3) Write a sentence that describes what the equation says:
    3x   −   5   =   2x

| Three times a number | minus five | equals | two times a number. |

Problem Set 2.2

1a. See answer section in the text. Remember:
The commutative property has to do with *reversing order.*
The associative property has to do with *regrouping.*
In this problem, the order of multiplication between x and (y+z) was reversed.

1b.   See answer section in the text.   The multiplication is carried out across addition.

1c.   See answer section in the text.   The order of addition was reversed between z and y.

5.   See answer section in the text. The additive inverse is found by multiplying the number by −1.

> $\boxed{\textit{COMMENT}}$ *To find the absolute value:*
> *The absolute value of a number is its distance from the origin.*
> *If the number is positive or zero, then the absolute value of the number is the number itself.*
> *If the number is negative, then its absolute value is its opposite (additive inverse).*

9a. $17 + 8 = 25$ 9b. $|\,9\,| = 9$ 9c. $17 - 8 = 9$ 9d. $-\,|\,9\,| = -9$

13a.   Remember, the notation $\sqrt{}$ means only the principle square root, which is either positive or zero.   Although 9 has two square roots, -3 and 3, the radical notation means only 3, not both. See answer section in the text.

13b.   See answer section in the text, and comment in 13a.   The positive square root of 16 is 4.

13c.   See answer section in the text.   Be careful here; we have to compute $\sqrt{9}$ and $\sqrt{16}$ separately and *then* add.

13d. $\sqrt{9 + 16} = \sqrt{25} = 5$

Contrast this problem with 13c.   Here, the sum of 9 and 16 is under the radical, so we must find this sum <u>before</u> taking the square root.

> NOTE: BE CAREFUL WHEN RADICAL NOTATION ($\sqrt{}$) IS USED.
>
> 25 HAS TWO SQUARE ROOTS, 5 AND -5 BECAUSE $5^2 = 25$ AND $(-5)^2 = 25$.
>
> HOWEVER, THE NOTATION $\sqrt{}$ MEANS ONLY THE POSITIVE SQUARE ROOT, NOT BOTH.
>
>   EXAMPLES: $\sqrt{25} = 5$; $\sqrt{36} = 6$; $\sqrt{100} = 10$.
>
> WE CALL THIS ONE ROOT THE PRINCIPAL SQUARE ROOT OF THE NUMBER.

17. $x \leq 3$ represents all numbers either equal to 3 or to the left of 3 on the number line.
$x \geq 3$ represents all numbers either equal to 3 or to the right of 3 on the number line.
The only number that can satisfy both descriptions is the number 3.

21a.   All negative numbers are less than 0.   21b.   0 is smaller than all positive numbers.

25a.   They are equal. The decimal form of $\frac{1}{2}$ is 0.5.

25b.   They are equal. Absolute value can be described as distance from the origin. Both 21 and -21 are the same distance from the origin, 21 units. Therefore $|\,21\,|$ and $|-21\,|$ are both equal to 21.   *The absolute value of a number and the absolute value of its opposite are always equal. Absolute value measures size, or magnitude.*

29a. Work inside absolute value bars first: $|-1.5| = 1.5$; $|1.3| = 1.3$
We now have: -1.5 _____ -1.3 .

(Notice that we still retain the minus signs that were outside of the absolute values. The absolute value operation has no effect on them.)
The answer is $<$, because -1.5 is further to the left on the number line than -1.3, and this is because 1.5 is larger in <u>size</u> than 1.3.

29b. Again, work inside absolute value bars first: $|7-4| = |3| = 3$; $|9-7| = |2| = 2$.
We now have: -3 _____ -2 .   The answer is $<$, because -3 is to the left of -2 on the number line.

33. First go back to your text and review the definitions of *natural numbers, whole numbers, integers, rational numbers,* and *irrational numbers.* These definitions are very specific and will tell you which numbers to choose.

    a. The natural numbers: { 1, 2, 3, 4, 5, ...} These are the counting numbers.   Answer: 15

    b. The whole numbers are just the natural numbers with 0 included.   Answer: 0, 15

    c. The integers: { ..., -3, -2, -1, 0, 1, 2, 3, 4, ...}
       Answer: -11, $-\sqrt{9}$, 0, 15. Notice that because $-\sqrt{9}$ is equivalent to -3, it is an integer.

    d. The rational numbers are all those numbers that can be written as a fraction, where the numerator and denominator are integers.   This also includes integers themselves (for example, $-8 = \frac{-8}{1}$) and mixed numbers (for example, $1\frac{3}{5} = \frac{8}{5}$).
       A decimal number may or may not be rational. The test is: if it has a finite number of decimal places, or *repeats* indefinitely, then the number is rational. Examples: -7.231, and 0.3333.... are rational numbers.
       Roots are irrational *unless* they reduce to a rational number; *BE CAREFUL.*
       For example, $\sqrt{\frac{9}{4}}$ reduces to $\frac{3}{2}$ and so is rational but $\sqrt{5}$ is irrational.
       Answer: all the numbers in the list except $\sqrt{5}$ are rational.

    e. See #33d for discussion.   Answer: only $\sqrt{5}$ is irrational.

37. See answer section in the text. You may find the discussion above for #33 helpful.

41. See answer section in the text. You may find the discussion above for #33 helpful.

Also notice that $5\frac{3}{7}$ is equivalent to $\frac{38}{7}$.

45. See answer section in the text. Suggestion: Start by evaluating those numbers with absolute values. For example, $|-3| = 3$ and $|4| = 4$. Notice that $-|4| = -4$. The absolute value notation has no effect on that minus sign on the outside.

49. $|-25| = 25$

53. $|-144| - |25|$
$= 144 - 25$
$= 119$

57. See the discussion in this manual, Section 2.2, #13d. The root symbol $\sqrt{}$ refers to only the positive square root, not both. Answer: 5

61. B. When estimating a square root, it is helpful to think of some perfect squares. The perfect square 64 is close to 62.41, therefore we can estimate $\sqrt{62.41}$ as $\sqrt{64}$ which is 8.

65. For a simple calculator, key in 54.6 and press the square root key (and perhaps the equals key.) Not all calculators operate in the same way. Ask your instructor if you are having trouble.

> $\boxed{COMMENT}$ *It can be helpful to notice whether or not an answer is reasonable. Since 54.6 is in the vicinity of 49, the answer should be at least in the vicinity of 7. You might say that 54.6 isn't at all close to 49, so this is pretty rough estimation. You are correct — but the idea is to point out that an answer on the calculator display that is nowhere near 7 is clearly wrong and we should check our steps. Always keep an eye out for ridiculous results.*

69. (a) When 2 is divided by 3 using long division, the sixes obtained in the quotient go on without end, so that $\frac{2}{3} = 0.666\cdots$

> $\boxed{WARNING}$ *A calculator can display only a finite number of digits. What you see on your calculator display is your calculator's best approximation of $\frac{2}{3}$, not its exact value.*

If you perform the long division by hand, after awhile you will be convinced that the sixes go on without end.

(b) Since $\frac{2}{3} = 0.66666....$ (without end), we may write

$$\frac{2}{3} = 0.66 + 0.006 + 0.0006 + 0.00006 + \cdots \text{ (without end)}$$

which is certainly larger than 0.66. Consequently $\frac{2}{3} > 0.66$.

73. See answer section in the text. The number x must be negative.

77. Don't try to write the whole equation at once. Start with with one expression:
two more than a number: n + 2
Since this quantity is 11, we write n + 2 = 11.

81. Don't try to write the whole inequality at once. Start with with one expression:

the sum of a number and eight: n + 8 $\left.\begin{array}{l}\\\\\end{array}\right\}$ $n + 8 \leq 2n$
twice the number: 2n

85. The set of whole numbers is almost the same as the set of natural numbers. Review the definition given in the text. The only whole number that is not a natural number is 0.

89. There are many possible answers, just be sure that the fractions do not reduce to integers. Example: $-\frac{3}{2}$ and $-\frac{27}{123}$ but not $-\frac{36}{12}$ (it equals -3). There are many others.

Problem Set 2.3

1. See answer section in the text.

5. Starting at zero, -37 takes you *left* 37 units; adding plus 35 then takes you *right* 35 units. You end up at 2 units to the left of zero, at -2.

9. A number added to its additive inverse equals zero.

13. $-8 + (-7) + (-6)$
$= -15 + (-6) = -21$

17. $[(-7) + (-6)] + (-13)$
$= -13 + (-13) = -26$

21. $[-5 + (-6) + (-7)]$
$\qquad + [2 + 3 + 4]$
$= [-11 + (-7)] + [5 + 4]$
$= -18 + 9 = -9$

25. This is the sum of a number and its additive inverse, and so the result is zero.

29. $-5 + 7 = 2$

33. $-3 + (-5 + 7) = -3 + 2 = -1$

37. $|-3 + 7| = |4| = 4$

41. $-15.53 + 27.68 = 12.15$ (yes)

45. $-7 + 15 + 23 = 8 + 23 = 31$

49. $-\pi + (+\pi) = 0$

53. $-\dfrac{2}{5} + \left(-\dfrac{5}{8}\right)$
$= -\dfrac{16}{40} + \left(-\dfrac{25}{40}\right)$
$= \dfrac{-16 + (-25)}{40}$
$= -\dfrac{41}{40}$ or $-1\dfrac{1}{40}$

57. $-12.45 + (-8.92) + 4.06$
$= -21.37 + 4.06$
$= -17.31$

61. These are equal by the associative property of addition. The sum has been regrouped.

65. left side: $|-19 + 23| = |4| = 4$
right side: $|19 + (-23)| = |-4| = 4$
They are equal.

69. Rounding to the nearest unit, the sum is approximately
$-5 + (-1) + (-2) = -8$
Answer: B

73. $\approx -40 + 70 = 30$
Answer: D

77.   keystrokes:

basic calculator: $\boxed{3}\,\boxed{.}\,\boxed{1}\,\boxed{4}\,\boxed{\pm}\,\boxed{+}\,\boxed{\pi}\,\boxed{=}$

graphics calculator: $\boxed{(-)}\,\boxed{3}\,\boxed{.}\,\boxed{1}\,\boxed{4}\,\boxed{+}\,\boxed{\pi}\,\boxed{\text{Enter}}$

To the nearest thousandth, we have: 0.002

NOTE. On many calculators, π is found directly above another key. You may first have to press Shift or 2nd to access it.

---

81. $(+20) + (+30) = 20 + 30 = 50$
We have a gain, or profit, of \$50.

85. $-32 + 30 = -2$
The thermometer registers $-2°$ Fahrenheit, or $2°$ below zero.

Problem Set 2.4

1. $7 + (-13) = -6$

5. $-15 + 8 = -7$

9. $-11 + 17 = 6$

13. $-143 + 143 = 0$

17. $[-18 + (-18)] + (-21)$
$= -36 + (-21)$
$= -57$

21. $-7 + (-8) + (-9) + (-10)$
$= -15 + (-9) + (-10)$
$= -24 + (-10) = -34$

25. $46 - \left([17 + (-35)] - [41 + (-56)]\right)$
$= 46 - (-18 + 15)$
$= 46 - (-3)$
$= 46 + 3$
$= 49$

29. $-7 + 11 + (-15) + (-16) + 8 + (-4)$
$= 4 + (-15) + (-16) + 8 + (-4)$
$= -11 + (-16) + 8 + (-4)$
$= -27 + 8 + (-4)$
$= -19 + (-4) = -23$

33. $14 - [(9 + (-12)) - (-5 + 11)]$
$= 14 - [-3 - 6]$
$= 14 - (-9)$
$= 14 + 9 = 23$

37. $-9 - (-2) = -9 + 2 = -7$ (yes)

41.   Both numbers are approximately 13, but we are subtracting the larger from the smaller, so the result will be negative. The difference in quantity between 12.9 and 13.0 is less than one unit, so the answer is B.

45.   This can be estimated as $-20 - 80$ by rounding to the nearest 10. This gives an estimate of -100.   Answer: A

49. $-\dfrac{5}{6} - \left(-\dfrac{7}{15}\right) = -\dfrac{5}{6} + \dfrac{7}{15} = -\dfrac{5 \cdot 5}{6 \cdot 5} + \dfrac{7 \cdot 2}{15 \cdot 2} = -\dfrac{25}{30} + \dfrac{14}{30} = \dfrac{-25 + 14}{30} = \dfrac{-11}{30} = -\dfrac{11}{30}$

53. $-7.25 - [8.49 - (-4.26)] = -7.25 - [8.49 + 4.26] = -7.25 - 12.75 = -20$

57. $(-5) - (-7) - (-8) = -5 + 7 + 8 = 2 + 8 = 10$

61. $-5 - (-7) + (-8) - (-11)$
       $= -5 + 7 + (-8) + 11$
       $= 2 + (-8) + 11$
       $= -6 + 11 = 5$

65. -148.497.   Suggestion: be sure you have -47.08 on the display before subtracting 93.017. Note: Not all calculators work in the same way. If you are having trouble with yours, ask your instructor for help. If your school has a tutoring room or other place where students go for help, ask there. The math tutors there can help you with the keystrokes.   Consider also asking another student in the class for advice.

69. $-\sqrt{25} - \sqrt{64} = -5 - 8 = -13$

73.   The change from a to b is b $-$ a.
       $-9 - (-7) = -9 + 7 = -2$ The temperature fell 2°.

77.   Read the problem carefully. Since we want to decrease -14, we <u>begin</u> with -14 and subtract 6:
       $(-14) - 6 = -20$

       Notice that -20 *is* less than -14, which was expected in the words "6 less than...."

81. $(-7) + 10 + 2\frac{1}{2} = 3 + 2\frac{1}{2} = 5\frac{1}{2}$ The net yardage is $5\frac{1}{2}$ yards.

85.   We can calculate the net change in the price and then add that to the original price.
       net change in price, <u>in cents</u>: $7 + (-5) + (-4) + (-9) = 7 + (-18) = -11$

       new price of silver $=$ original price $+$ net change

$$= 6.49 + (-0.11)$$
$$= 6.38$$

new price: $6.38 per ounce

Problem Set 2.5

1. See answer section in the text.

    [REMEMBER] *When two numbers have the same sign, their product is positive. When they have different signs, their product is negative.*

5a. Be careful here. It is 5 that is raised to the second power, not -5. So we have:
    $$-5^2 = -(5^2) = -25$$

5b. Now -5 is raised to the second power. That is the whole reason for the parentheses, to control the order of operations. $(-5)^2 = (-5)(-5) = 25$

5c. $-1^6 = -(1^6) = -1$
    $(-1)^4 = 1^4 = 1$ $\Big\}$ $-1^6 + (-1)^4 = -1 + 1 = 0$

9. See answer section in the text.
    Remember: If two numbers are of the same sign and they are divided, the result will be positive. If they are of different signs, the result will be negative.

13a. $3^2 - 5^2 = 9 - 25 = -16$ 13b. $(3 - 5)^2 = (-2)^2 = 4$

13c. $3 - 5^2 = 3 - 25 = -22$

17. $19 + 11(-7 - 3)$
     $= 19 + 11(-10)$
     $= 19 + (-110) = -91$

21. $(-16 + 11)[-4 - 4]$
     $= (-5)(-8)$
     $= 40$

25. $-7 + 6 \cdot 8^2$
     $= -7 + 6 \cdot 64$
     $= -7 + 384 = 377$

29. $\dfrac{32 - 25}{8 - 9} = \dfrac{7}{-1} = -7$

33. $\dfrac{3(2) - 7}{4(2) - 11} = \dfrac{6 - 7}{8 - 11} = \dfrac{-1}{-3} = \dfrac{1}{3}$ (no)

37. We may estimate this with $(-4)^3$ which is -64. The closest answer is B.

41. $-\dfrac{2}{3}\left(-\dfrac{3}{4}\right) = \dfrac{2 \cdot 3}{3 \cdot 4} = \dfrac{1}{2}$

45. $\dfrac{7}{12} \div \left(-\dfrac{5}{6}\right) = \dfrac{7}{12} \cdot -\dfrac{6}{5} = -\dfrac{7 \cdot 6}{12 \cdot 5} = -\dfrac{7}{10}$

49. $-0.039 \div (-1.3)$ This can be done by a calculator, giving 0.03.

Without a calculator, notice that we have $\frac{0.039}{1.3}$. If we multiply the numerator and denominator by 1000, the decimals will be removed:

$$\frac{0.039(1000)}{1.3(1000)} = \frac{39}{1300} = \frac{3(13)}{100(13)} = \frac{3}{100} \text{ which is } 0.03.$$

53. Before doing any multiplying, notice what is really being multiplied together. The numbers are $(-5)(-4)(-3) \text{ .... } (2)(3)(4)$
This pattern of numbers *also contains zero*. If zero is a factor in any product, the result is zero.

57. See answer section in the text.

61. $18.3^2 - 5.7^2 - (5.7 - 18.3)^2$
$= 334.89 - 32.49 - (-12.6)^2$
$= 334.89 - 32.49 - 158.76$
$= 143.64$

65. $[-3 - (-4)][-5 - (-6)]$
$= (-3 + 4)(-5 + 6)$
$= (1)(1)$
$= 1$

69. $[-3 + (-4) + (-5)]^2$
$= (-12)^2$
$= 144$

73. $[2(-4) - 5(-5)][3(-3) - (-6)]$
$= (-8 + 25)(-9 + 6)$
$= (17)(-3)$
$= -51$

77. $\frac{-36}{-9} = 4$

Solutions To All Review Exercises — Chapter Two

1. -12

2. -20

3. -20

4. $-16 - (-4) - 16 + 4 = -12$

5. -64

6. 64

7. -4

8. 4

9. 0.71

10. $14 + (-10) = 4$

11. $-7 - (-11) = -7 + 11 = 4$

12. $12 - (-2) = 12 + 2 = 14$

13. $-(-9) - (-30) = 9 + 30 = 39$

14. -4170

15. 0.00417

16. $(-10)(6) = -60$

17. $-35 - (63) = -35 - 63 = -98$

18. 0 Zero divided by any nonzero is zero.

19. Undefined

20. $(-3)(-3)(-3)(-3) = 81$

21. $-(3^4) = -81$

22. $\dfrac{4}{-2} = -2$

23. $\dfrac{14 + 35}{-27 + 20} = \dfrac{49}{-7} = -7$

24. $\dfrac{5(12)}{6(11)} = \dfrac{5(2)(6)}{6(11)} = \dfrac{10}{11}$

25. $-\dfrac{14}{15} \div \left(-\dfrac{21}{25}\right) = -\dfrac{14}{15} \cdot \left(-\dfrac{25}{21}\right) = \dfrac{14(25)}{15(21)} = \dfrac{2(7)(5)(5)}{3(5)(3)(7)} = \dfrac{10}{9}$ or $1\dfrac{1}{9}$

26. $-\dfrac{1}{2} + \dfrac{2}{3} = -\dfrac{3}{6} + \dfrac{4}{6} = \dfrac{-3 + 4}{6} = \dfrac{1}{6}$

27. $-\dfrac{4}{5} - \dfrac{3}{8} = -\dfrac{4(8)}{5(8)} - \dfrac{3(5)}{8(5)} = -\dfrac{32 + 15}{5(8)} = -\dfrac{47}{40}$ or $-1\dfrac{7}{40}$

You may find other correct ways to deal with the minus signs.

28. $(-1)(-2)(-5)(-8)(-3) = (2)(40)(-3) = 80(-3) = -240$

29. See answer section in the text.

30. Since 0 is one of the factors, the product is 0.

31. $(417 - 418)^{419} = (-1)^{419} = -1$

32. $(3 - 4)^2 - (3^2 - 4^2) = (-1)^2 - (9 - 16) = 1 - (-7) = 8$

33. $(5 + 6)^2 - (5^2 + 6^2) = 11^2 - (25 + 36) = 121 - 61 = 60$

34. $|3 - 8| = |-5| = 5$

35. $|5 - 6(9)| = |5 - 54| = |-49| = 49$

36. $|3| - |8| = 3 - 8 = -5$

37. $|-8 - 9| - |-9 + 8| = |-17| - |-1| = 17 - 1 = 16$

38. $\sqrt{64 + 36} = \sqrt{100} = 10$

39. $\sqrt{64} + \sqrt{36} = 8 + 6 = 14$

40. $\left(\dfrac{-2}{3}\right)^2 \div \left(-\dfrac{2}{9}\right) = \dfrac{4}{9} \div -\dfrac{2}{9} = \dfrac{4}{9} \cdot -\dfrac{9}{2} = -\dfrac{4(9)}{9(2)} = -2$

41. $-6[5 - 8(-1) + 11] = -6[5 + 8 + 11] = -6(24) = -144$

42. $-2 + 3 - (-4) + 5 = -2 + 3 + 4 + 5 = 1 + 4 + 5 = 10$

43. $-2 - 3 - (-4 - 5) = -2 - 3 - (-9) = -2 - 3 + 9 = -5 + 9 = 4$

44. $-2 - 3(-4)(5) = -2 - 3(-20) = -2 - (-60) = -2 + 60 + 58$

45. $\dfrac{2(5) - 5(-4)}{(-2)(3) + 1} = \dfrac{10 + 20}{-6 + 1} = \dfrac{30}{-5} = -6$

46. Replace x with -3.
$-17 - 2(-3) \overset{?}{=} 3(-3) - 2$
$-17 + 6 \overset{?}{=} -9 - 2$
$-11 = -11$
-3 is a solution.

47. Replace x with 10.
$3(30 - 7) \overset{?}{=} 4(40 + 2)$
$3(23) \overset{?}{=} 4(42)$
$69 \neq 168$
10 is not a solution.

48. $\dfrac{5(5) + 1}{-3(5) + 2} = \dfrac{26}{-13} = -2$

5 is a solution.

49. $\dfrac{-7(-2) + 4}{-11(-2) + 2} = \dfrac{18}{24} = \dfrac{3}{4}$

-2 is a solution.

50. $-4.17 < 7.14$

51. $-4.17 > -7.14$

52. Answer: $<$ This is because $|-2| = 2$ and $|-3| = 3$.

53. Answer: $>$ This is because $|-9| = 9$ and $|-8| = 8$.

54. The area of a triangle is one-half the product of the base and height.
area $= \frac{1}{2}(7.2)(3.1) = 11.16$ cm^2

55. The formula for the area of a circle is $A = \pi r^2$.
$A = \pi(5.6)^2 = 31.36\pi$ cm$^2 \approx 98.52034562$
Rounded to the nearest tenth of a square centimeter, we have 98.5 cm^2.

56. The perimeter is the sum of lengths of the sides:
$5 + 2.2 + 2 + 2 + 3.2 + 2 + 6.2 + 6.2 = 28.8$ cm

57. $I = 650(0.08)(1) = \$52.00$

58. The average is $\dfrac{-17 - 8 + 5 + 3 + 12 + 11 + 20}{7} = \dfrac{26}{7} \approx 3.7°$

The range is $20 - (-17) = 20 + 17 = 37°$

59. $2(n + (-7)) = n - 4$

60. $5(n + 2) = 16$

61. $\dfrac{n}{12} = \dfrac{1}{2}$

62. $n^2 = n + 3$

63. $2n \leq 13$

64. $n + 6 > -5$

65. xy will have to be positive, being a product of negatives. For the product wxy to be negative (which is must be, since z is negative), the factor w must be negative.

66. Any nonzero number raised to an even power will be positive. For example,
$$(-2)^{10} = (-2)(-2)(-2)(-2)(-2)(-2)(-2)(-2)(-2)(-2) = 1024$$

67. A sum of negative numbers is a negative number.

68. xy will be positive, so 5xy is positive.

69. 7   B. a prime natural number

70. $\sqrt{36}$ C. a composite natural number, since $\sqrt{36} = 6 = 2 \cdot 3$

71. $\sqrt{2}$ A. an irrational number

72. -17 E. a negative integer

73. 0   F. a number that is neither negative nor positive

74. $\frac{3}{5}$ D. a rational number that is not an integer

75. a. natural numbers: 2
    b. whole numbers: 0,2
    c. integers: $-1,0,2$
    d. rational numbers: $-4.6, -\frac{5}{4}, -1, 0, 2$
    e. irrational numbers: $-\sqrt{5}, \pi$

76. We are regrouping the addition, and so this is the associative property of addition.

77. The order of multiplication has been reversed. Commutative property of multiplication.

78. Distributive property

79. Commutative property of addition. The order of addition has been reversed.

80. $\approx (-50)(10) = -500$ Answer: B

81. $\approx (-2)^2 = 4$ Answer: E

82. $\approx \frac{-45}{-5} = 9$ Answer: D

83. $\approx -900 - 400 - 200 = -1500$ Answer: A

84. $\approx |-17 + 13| = |-4| = 4$ Answer: D

85. The correct sequence is A.
    The $\boxed{+/-}$ key only changes the sign of 3.73.

86. The correct sequence is C. In the order of operations, division is done before subtraction.

87. This is equivalent to $-5^2 - \frac{15}{3} - 4 = -25 - 5 - 4 = -34$.

88. $-5^2 - 3^2 = -25 - 9 = -34$
    The error here is confusing -5^2 with $(-5)^2$. Notice that $-5^2 = -25$, but $(-5)^2 = 25$.

89. The correct steps are shown in the answer section in the text.
The error in the problem is that Student B did not follow the proper procedure to subtract fractions. The common denominator needed is found by using the smallest number that both 5 and 3 will divide into; and that is 15. We cannot subtract fractions by subtracting their denominators.

Solutions To All Mastery Test Problems – Chapter Two

1a. $2(-2) - 3(-5) = -4 + 15 = 11$

1b. $(-2)^2 + (-5)^2 = 4 + 25 = 29$

1c. $(-2 + (-5))^2$
$= (-7)^2$
$= 49$

1d. $\dfrac{2(-5)-(-2)}{3(-2)+(-5)-1}$
$= \dfrac{-10 + 2}{-6 + (-5) + (-1)}$
$= \dfrac{-8}{-12}$
$= \dfrac{2}{3}$

1e. $\dfrac{2}{-2} - \dfrac{3+2}{-5}$

$= -1 - \dfrac{5}{-5}$

$= -1 - (-1) = -1 + 1 = 0$

1f. $\dfrac{2(-5) - 3(-2) + 4}{7(-2) + 8(-5) - 21}$

$= \dfrac{-10 + 6 + 4}{-14 - 40 - 21}$

$= \dfrac{0}{-75} = 0$

2a. $3n = n - 5$

2b. $3n < 5n$

2c. $3n > n + 5$

2d. $|n - 8| = 11$

2e. $\dfrac{n}{n-4} = \dfrac{6}{5}$

3a. The perimeter P is the length of the boundary, so we need the lengths of all the segments. The unlabeled horizontal segment is 10 cm − 6 cm = 4 cm; the unlabeled vertical segment is 9 cm − 6 cm = 3 cm.

$P = 9 + 10 + 3 + 4 + 6 + 6 = 38$ cm

Chapter Two Mastery Test

3b. The area of a rectangle equals length × width. Divide the region into two rectangles as shown:

$A_1 = 6(9) = 54 \text{ cm}^2$
$A_2 = 3(4) = 12 \text{ cm}^2$

The sum of these areas equals the
total area of the region. Answer: 66 cm^2

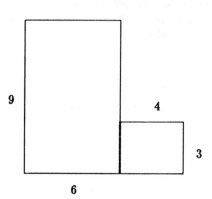

4a. B The set of whole numbers is { 0, 1, 2, 3, ... } , but the set of integers is the set { ..., -3, -2, -1, 0, 1, 2, 3, ... }. Every whole number is also integer, but not every integer is a whole number.

4b. C A rational number is by definition a quotient of two integers, and $-\frac{2}{3}$ can be thought of as $\frac{-2}{3}$ or $\frac{2}{-3}$. The set of integers is the set { ..., -3, -2, -1, 0, 1, 2, 3, ... }, which does not contain $-\frac{2}{3}$.

4c. A The whole numbers is the set {0, 1, 2, 3, ... } while the natural numbers is the set {1, 2, 3, ... }.

4d. E $\sqrt{9}$ can also be written as 3, so this is a natural number in disguise.

4e. D $\sqrt{11}$ is clearly positive.
In this course, it is safe to say that roots that still contain radical notation (or π) after simplifying are irrational. Contrast this with $\sqrt{9}$ and $\sqrt{\frac{25}{4}}$ which are rational because they can be simplified as 3 and $\frac{5}{2}$, respectively.

5a. commutative property of multiplication   **5b.** associative property of addition

5c. commutative property of addition   **5d.** distributive property

6a. $-(-17) = 17$ **6b.** 17   **6c.** -17

6d. $|-40| = 40$ **6e.** $17 - 23 = -6$ **7a.** >   **7b.** <

7c. <   **7d.** >   **8a.** -16 **8b.** -5.2

Chapter Two Mastery Test

8c. $-\dfrac{2}{7} + \dfrac{3}{8} = \dfrac{-2(8) + 3(7)}{56}$

$= \dfrac{-16 + 21}{56}$

$= \dfrac{5}{56}$

8d. $-7 + (-9) + (-25) + 11 + 17$

$= -41 + 28$

$= -13$

9a. -40

9b. $\dfrac{11}{14} - \dfrac{6}{49}$ $\quad \begin{cases} 14 = 2(7) \\ 49 = 7(7) \\ \text{LCD} = 2(7^2) \end{cases}$

9c. $-5 - [-15] = -5 + 15 = 10$

$\dfrac{11(7) - 6(2)}{2(7)(7)} = \dfrac{65}{2(7)(7)} = \dfrac{65}{98}$

9d. $-(-3) - (-22) = 3 + 22 = 25$

10a. -105 10b. -160 10c. 0 10d. 64

11a. 26 11b. 0 11c. undefined 11d. -120

12a. $-16 + 3(-3) = -16 - 9 = -25$

12b. $-24 + 6 - 2 = -18 - 2 = -20$

12c. $(-17)[-13 - 4(-2)]$

$= (-17)[-13 + 8]$

$= -17(-5)$

$= 85$

12d. $\dfrac{-3(-7) + 3}{6 - 2(-3)} = \dfrac{21 + 3}{6 + 6} = \dfrac{24}{12} = 2$

Chapter Two Mastery Test

Problem Set 3.1

1. $42\left(\dfrac{5}{6}\right) - 42\left(\dfrac{3}{14}\right)$

$= \dfrac{42(5)}{6} - \dfrac{42(3)}{14}$

$= \dfrac{6(7)(5)}{6} - \dfrac{3(14)(3)}{14}$

$= 7(5) - 3(3)$

$= 35 - 9$

$= 26$

5. $-7(2x) - 7(3y) - 7(-8)$

$= -14x - 21y + 56$

9. $-2a - (-3b) - 9c$

$= -2a + 3b - 9c$

13. $8x + 12$

$= 4(2x) + 4(3) = 4(\underline{2x+3})$

17. $6a + 10b - 14c$

$= 2(3a) + 2(5b) + 2(-7c)$

$= 2(\underline{3a + 5b - 7c})$

21. Terms are items added or subtracted. See answer section in the text.

25. $4x + 11x = (4 + 11)x = 15x$

29. $3a - 5a + 11a = (3 - 5 + 11)a = 9a$

33. $5v + 8w - 3 - 9v + 7$

$= 5v - 9v + 8w - 3 + 7$

$= (5 - 9)v + 8w + (-3 + 7)$

$= -4v + 8w + 4$

37. $18m - 5(2m - 3)$

$= 18m - 5(2m) - 5(-3)$

$= 18m - 10m + 15$

$= 8m + 15$

41. $-(2x - 3y) - 5x$

$= -2x - (-3y) - 5x$

$= -2x + 3y - 5x$

$= -2x - 5x + 3y$

$= -7x + 3y$

45. $6(2x - 7) - 7(4x + 6)$

$= 6(2x) - 6(7) - 7(4x) - 7(6)$

$= 12x - 42 - 28x - 42$

$= 12x - 28x - 42 - 42$

$= -16x - 84$

49. $3\pi r + 2\pi r = (3+2)\pi r = 5\pi r$

Since $\pi \approx 3.14$, a rough estimate for $5\pi r$ would be $5(3)r$ or $15r$. We choose A because π actually is a little larger than 3, making an estimate of $16r$ the best choice.

53. $-x(3x) - x(7) - 2(x^2) - 2(-x) - 2(5)$

$= -3x^2 - 7x - 2x^2 + 2x - 10$

$= -3x^2 - 2x^2 - 7x + 2x - 10$

$= (-3 - 2)x^2 + (-7 + 2)x - 10$

$= -5x^2 - 5x - 10$

57. $\frac{1}{2}x - \frac{1}{3} + \frac{1}{4}x + \frac{1}{5}$

$= \left(\frac{1}{2} + \frac{1}{4}\right)x - \frac{1}{3} + \frac{1}{5}$

$= \left(\frac{2+1}{4}\right)x + \frac{-5+3}{15}$

$= \frac{3}{4}x - \frac{2}{15}$

61. $2.3(2x) + 2.3(y) - 1.9(3x) - 1.9(-y)$
$= 4.6x + 2.3y - 5.7x + 1.9y$
$= (4.6 - 5.7)x + (2.3 + 1.9)y$
$= -1.1x + 4.2y$

65. The perimeter is the length of the boundary, so add up the lengths of all sides:

$$2 + (x + 2) + (x + 2) + 1 + x + (x + 1) = 4x + 8$$

To find the area, consider this region as two adjoined rectangles as shown.
The area of a rectangle = length · width.
For the rectangle on the left, area = (x+2)(2).
For the rectangle on the right, area = (x)(1).
The total area is :

$(x + 2)(2) + (x)(1)$

$= 2x + 4 + x$

$= 3x + 4$

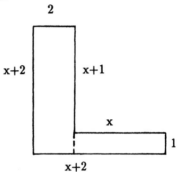

69. $(-3)[\,-4 - (-5)\,]$
$= (-3)[\,-4 + 5\,]$
$= (-3)(1)$
$= -3$

73. $[\,-(-5) - (-4)\,]^2 - (-3)^2$
$= [\,5 + 4\,]^2 - 9$
$= 9^2 - 9$
$= 81 - 9$
$= 72$

77. $(-2.39)^2 + 4.87^2 - (-2.39 + 4.87)^2$ Answer: 23.2786
It saves a step to notice that we get the same result if we square 2.39 instead of -2.39.

The calculator key sequence for a basic calculator would be:

$\boxed{2}\,\boxed{.}\,\boxed{3}\,\boxed{9}\,\boxed{x^2}\,\boxed{+}\,\boxed{4}\,\boxed{.}\,\boxed{8}\,\boxed{7}\,\boxed{x^2}\,\boxed{-}\,\boxed{(}\,\boxed{2}\,\boxed{.}\,\boxed{3}\,\boxed{9}\,\boxed{\pm}\,\boxed{+}\,\boxed{4}\,\boxed{.}\,\boxed{8}\,\boxed{7}\,\boxed{)}\,\boxed{x^2}\,\boxed{=}$

For a graphing calculator:

$\boxed{2}\,\boxed{.}\,\boxed{3}\,\boxed{9}\,\boxed{x^2}\,\boxed{+}\,\boxed{4}\,\boxed{.}\,\boxed{8}\,\boxed{7}\,\boxed{x^2}\,\boxed{-}\,\boxed{(}\,\boxed{(-)}\,\boxed{2}\,\boxed{.}\,\boxed{3}\,\boxed{9}\,\boxed{+}\,\boxed{4}\,\boxed{.}\,\boxed{8}\,\boxed{7}\,\boxed{)}\,\boxed{x^2}\,\boxed{\text{Enter}}$

Not all calculators operate in the same way. Ask your instructor if you need help.
If your school has a tutoring room or lab where students go for help with math, ask the people working there. It is also helpful to talk to other students in your class.

81. We substitute in -3 for x:

left side	right side
$3[2(-3) + 5]$	$-3 + 1$
$3(-6 + 5)$	-2
$3(-1)$	-2
$-3 \neq$	-2

Since the left side and right side are not equal, -3 is not a solution.

85. Be careful: the word 'quantity' indicates that 4 will be multiplied by something more than m. It will be multiplied by m + 7.

$$4(m + 7) - 2(m - 3)$$
$$= 4m + 28 - 2m + 6$$
$$= 2m + 34$$

Problem Set 3.2

1. $x - 11 = 13$
    $x - 11 + 11 = 13 + 11$
    $x = 24$

5. $2y = y - 1$
    $2y - y = y - y - 1$
    $y = -1$

9. $3m = 4m - 9$
    $3m + 9 = 4m - 9 + 9$
    $3m + 9 = 4m$
    $3m + 9 - 3m = 4m - 3m$
    $m = 9$

13. $6x + 2 = 5x + 18$
    $6x + 2 - 2 = 5x + 18 - 2$
    $6x = 5x + 16$
    $6x - 5x = 5x - 5x + 16$
    $x = 16$

17. $-5v + 7 = 9 - 4v$
    $-5v + 7 + 5v = 9 - 4v + 5v$
    $7 = 9 + v$
    $7 - 9 = 9 + v - 9$
    $v = -2$

21. $5m - 7 + 3m = 19 + 7m$
    $8m - 7 = 19 + 7m$
    $8m - 7 + 7 = 19 + 7m + 7$
    $8m = 26 + 7m$
    $8m - 7m = 26 + 7m - 7m$
    $m = 26$

25. $9t = 10t$
    $9t - 9t = 10t - 9t$
    $t = 0$

29. $3x - 4 + 5x = 6x + 3 + x$
    $8x - 4 = 7x + 3$
    $8x - 4 + 4 = 7x + 3 + 4$
    $8x = 7x + 7$
    $8x - 7x = 7x + 7 - 7x$
    $x = 7$

33. $-4(w + 3) = -5(w - 2)$
    $-4w - 4(3) = -5w - 5(-2)$
    $-4w - 12 = -5w + 10$
    $-4w - 12 + 5w = -5w + 10 + 5w$
    $w - 12 = 10$
    $w - 12 + 12 = 10 + 12$
    $w = 22$

37. $3(5 - n) = 4(6 - n)$
    $15 - 3n = 24 - 4n$
    $15 - 3n + 4n = 24 - 4n + 4n$
    $15 + n = 24$
    $15 + n - 15 = 24 - 15$
    $n = 9$

41. $x = 2x$
$x - x = 2x - x$
$x = 0$ (conditional)

45. $w + w = 2w$
$2w = 2w$
$2w - 2w = 2w - 2w$
$0 = 0$ (identity)
solution: all real numbers

49. $1 + 2z = 2z + 1$
$1 + 2z - 2z = 2z + 1 - 2z$
$1 = 1$ (identity)
solution: all real numbers

53. $-3.8v - 4.6 = -4.8v - 2.6$
$-3.8v - 4.6 + 4.8v = -4.8v -2.6 + 4.8v$
$v - 4.6 = -2.6$
$v - 4.6 + 4.6 = -2.6 + 4.6$
$v = 2$

57. $\frac{1}{4}\left(y - 8\right) = \frac{3}{4}\left(4 - y\right)$

$\frac{1}{4}y - \frac{1}{4}(8) = \frac{3}{4}(4) - \frac{3}{4}y$

$\frac{1}{4}y - 2 = 3 - \frac{3}{4}y$

$\frac{1}{4}y + \frac{3}{4}y - 2 = 3 - \frac{3}{4}y + \frac{3}{4}y$

$y - 2 = 3$
$y = 5$

61. $2(4v - 5) - 3(3v + 4) = -27$
$8v - 10 - 9v - 12 = -27$
$-v - 22 = -27$
$-v - 22 + v = -27 + v$
$-22 = -27 + v$
$-22 + 27 = -27 + v + 27$
$v = 5$

65. D. Rounding to the nearest half unit gives the best results. An approximation to the original
problem would be:
$x - 1 = 0.5$ (0.819 rounds up to 1; 0.687 rounds down to 0.5)
$x = 1.5$

69. Don't try to write the equation all at once.

three times a number	3m
three times a number plus seven	3m + 7
four times the number	4m
four times the number minus 8	4m − 8

$3m + 7 = 4m - 8$
$3m + 15 = 4m$
$m = 15$

73.

three times a number minus 9	3m − 9
twice that quantity	2(3m − 9)
the sum of the number and thirteen	m + 13
five times that sum	5(m + 13)

$$2(3m - 9) = 5(m + 13)$$
$$6m - 18 = 5m + 65$$
$$6m = 5m + 83$$
$$m = 83$$

77. We substitute in 5.5 for x. Calculate the value of the left side, and then the right side and compare. If they are equal, then 5.5 is a solution; if not, then 5.5 is not a solution.

left side : $35.83(5.5+12.3) = 637.774$
right side : $28.7(5.5) + 51.515 = 209.365$

Since the left side is not equal to the right side, 5.5 is not a solution to the equation.

81. The perimeter of a figure is the distance around it, or the length of the boundary. The rectangle has two long sides of 17 units each, and two short sides of 5 units each, so the perimeter is $2(17) + 2(5) = 34 + 10 = 44$ units.

We are told that the perimeter is also $x + 35$ units. This must be equivalent to 44, so:

$$x + 35 = 44$$
$$x = 44 - 35$$
$$x = 9$$

Problem Set 3.3

1. See answer section in the text.

5. $45y = 5$

$$\frac{45y}{45} = \frac{5}{45}$$

$$y = \frac{1}{9}$$

9. $-3m = -39$

$$\frac{-3m}{-3} = \frac{-39}{-3}$$

$$m = 13$$

13. $-\frac{2}{3}m = 48$

$$(-\tfrac{3}{2})(-\tfrac{2}{3}m) = (-\tfrac{3}{2})48$$

$$m = \left(-\frac{3}{2}\right)\left(\frac{48}{1}\right)$$

$$m = -\frac{3(2)(24)}{2}$$

$$m = -72$$

17. $24 = -18z$

$$\frac{24}{-18} = \frac{-18z}{-18}$$

$$\frac{4(6)}{-3(6)} = z$$

$$z = -\frac{4}{3}$$

21. $-47t = 0$

$$\frac{-47t}{-47} = \frac{0}{-47}$$

$t = 0$

25. $\frac{-5m}{9} = -45$ **Multiply both sides by** $-\frac{9}{5}$.

$$\left(-\frac{9}{5}\right)\left(\frac{-5m}{9}\right) = \left(-\frac{9}{5}\right)(-45)$$

$m = 81$

29. $\frac{x}{13} = -2$

$$\left(\frac{13}{1}\right)\left(\frac{x}{13}\right) = \left(\frac{13}{1}\right)(-2)$$

$x = -26$

33. $33 = 9 - 8t$
$33 - 9 = 9 - 8t - 9$
$24 = -8t$
$\left(-\frac{1}{8}\right)(24) = \left(-\frac{1}{8}\right)(-8t)$

$t = -3$

37. $8x - 1 = 13x - 1$
$8x - 1 + 1 = 13x - 1 + 1$
$8x = 13x$
$8x - 8x = 13x - 8x$
$0 = 5x$
$\left(\frac{1}{5}\right)(0) = \left(\frac{1}{5}\right)5x$

$x = 0$

41. $4\frac{3}{7}\, v = 2\frac{20}{21}$ **Rewrite the mixed numbers.**

$\frac{31}{7}v = \frac{62}{21}$ **Multiply both sides by** $\frac{7}{31}$.

$$\left(\frac{7}{31}\right)\left(\frac{31}{7}\right)v = \left(\frac{7}{31}\right)\left(\frac{62}{21}\right)$$

$v = \frac{2}{3}$

45. $0.001t = 13.7$

$$\frac{0.001t}{0.001} = \frac{13.7}{0.001}$$

$t = 13,700$

49. $-4.25 = 1000w$

$$\frac{-4.25}{1000} = \frac{1000w}{1000}$$

$-0.00425 = w$

53.   D. Using $5x = 30$ to approximate, we have $\frac{5x}{5} = \frac{30}{5}$ which gives $x = 6$.

57.   E. Using $-6z = -60$ to approximate, we have $z = 10$.

61. $m + (m + 2) + (m + 4) = 99$
$3m + 6 = 99$
$3m = 93$
$m = 31$

65. $\frac{x}{10.5} = \frac{20}{1}$ Multiply both sides by 10.5, or cross-multiply.

$x = (10.5)(20) = 210$

69. Do not expect to immediately write down the whole equation. Begin with key phrases. Let the number be represented by n.

three-eighths of a number: $\quad\quad\quad\quad\quad \frac{3}{8}$ n

negative nine-sixteenths: $\quad\quad\quad\quad -\frac{9}{16}$

Now we reread the problem and write the equation: $\frac{3}{8}$ n $= -\frac{9}{16}$

$(\frac{8}{3})(\frac{3}{8}$ n$) = (\frac{8}{3})(-\frac{9}{16})$

n $= -\frac{3}{2}$

73. $\quad \frac{n}{8} = 16$

$(8)\frac{n}{8} = 8(16)$

n $= 128$

77. Area of a triangle $= \frac{1}{2}$(base)(height).

$24 = \frac{1}{2}(5v)(4)$

$24 = 10v$

$\frac{24}{10} = \frac{10v}{10}$

v $= \frac{12}{5} = 2\frac{2}{5}$ or 2.4

81. $\quad 14.732w = 736.6$

$\frac{14.732w}{14.732} = \frac{736.6}{14.732}$

w $= 50$

Problem Set 3.4

1. $\quad 2(m + 1) = 18$

$2m + 2 = 18$

$2m + 2 - 2 = 18 - 2$

$2m = 16$

$\frac{2m}{2} = \frac{16}{2}$

m $= 8$

5. $\quad -3(4x - 1) = -(x - 14)$

$-12x + 3 = -x + 14$

$-12x + 3 + 12x = -x + 14 + 12x$

$3 = 11x + 14$

$3 - 14 = 11x + 14 - 14$

$-11 = 11x$

$\frac{-11}{11} = \frac{11x}{11}$

x $= -1$

9. $\dfrac{y}{3} + \dfrac{2}{3} = \dfrac{y}{2} + \dfrac{3}{2}$

$6\left(\dfrac{y}{3} + \dfrac{2}{3}\right) = 6\left(\dfrac{y}{2} + \dfrac{3}{2}\right)$

$6\left(\dfrac{y}{3}\right) + 6\left(\dfrac{2}{3}\right) = 6\left(\dfrac{y}{2}\right) + 6\left(\dfrac{3}{2}\right)$

$2y + 4 = 3y + 9$
$2y + 4 - 2y = 3y + 9 - 2y$
$4 = y + 9$
$4 - 9 = y + 9 - 9$
$y = -5$

13. $3w - 7 + w + 1 = 11w + 1$
$4w - 6 = 11w + 1$
$4w - 6 - 4w = 11w + 1 - 4w$
$-6 = 7w + 1$
$-6 - 1 = 7w + 1 - 1$
$-7 = 7w$
$w = -1$

17. $6 - 3(2v + 2) = 4(8 - v)$
$6 - 6v - 6 = 32 - 4v$
$-6v = 32 - 4v$
$-6v + 4v = 32 - 4v + 4v$
$-2v = 32$
$\dfrac{-2v}{-2} = \dfrac{32}{-2}$

$v = -16$

21. $x - 2 - 5x + 7 = 5 - 9x$
$-4x + 5 = 5 - 9x$
$-4x + 5 - 5 = 5 - 9x - 5$
$-4x = -9x$
$-4x + 9x = -9x + 9x$
$5x = 0$
$\dfrac{5x}{5} = \dfrac{0}{5}$; $x = 0$

25. $\dfrac{v}{6} + \dfrac{1}{3} = \dfrac{v}{2} + \dfrac{1}{4}$

$12\left(\dfrac{v}{6} + \dfrac{1}{3}\right) = 12\left(\dfrac{v}{2} + \dfrac{1}{4}\right)$

$12\left(\dfrac{v}{6}\right) + 12\left(\dfrac{1}{3}\right) = 12\left(\dfrac{v}{2}\right) + 12\left(\dfrac{1}{4}\right)$

$2v + 4 = 6v + 3$
$2v + 4 - 2v = 6v + 3 - 2v$
$4 = 4v + 3$
$4 - 3 = 4v + 3 - 3$
$1 = 4v$

$v = \dfrac{1}{4}$

29. $\dfrac{a}{3} - \dfrac{1}{6} = \dfrac{a + 1}{9}$

$18\left(\dfrac{a}{3} - \dfrac{1}{6}\right) = 18\left(\dfrac{a + 1}{9}\right)$

$18\left(\dfrac{a}{3}\right) - 18\left(\dfrac{1}{6}\right) = 2(a + 1)$

$6a - 3 = 2a + 2$
$6a - 3 + 3 = 2a + 2 + 3$
$6a = 2a + 5$
$6a - 2a = 2a + 5 - 2a$
$4a = 5$

$a = \dfrac{5}{4}$

33. $\dfrac{x + 2}{3} - 4 = -\dfrac{x}{2}$

$6\left(\dfrac{x + 2}{3} - 4\right) = 6\left(-\dfrac{x}{2}\right)$

$6\left(\dfrac{x + 2}{3}\right) - 24 = -3x$

$2(x + 2) - 24 = -3x$
$2x - 20 = -3x$
$2x - 20 - 2x = -3x - 2x$
$-20 = -5x$

$\dfrac{-20}{-5} = \dfrac{-5x}{-5}$

$x = 4$

37. $\dfrac{4x + 2}{5} + 2 = 3 - \dfrac{3 - 7x}{2}$

$10\left(\dfrac{4x + 2}{5} + 2\right) = 10\left(3 - \dfrac{3 - 7x}{2}\right)$

$2(4x + 2) + 20 = 30 - 5(3 - 7x)$
$8x + 4 + 20 = 30 - 15 + 35x$
$8x + 24 = 15 + 35x$
$24 = 15 + 27x$
$24 - 15 = 15 + 27x - 15$
$9 = 27x$

$\dfrac{9}{27} = \dfrac{27x}{27}$

$x = \dfrac{1}{3}$

41. $2x = \frac{1}{3}(x - 10)$

$3(2x) = 3(\frac{1}{3})(x - 10)$

$6x = x - 10$
$6x - x = x - 10 - x$
$5x = -10$
$x = -2$

45. $1.5(3 - y) - 41.5 = 2.5(3y - 4)$

$4.5 - 1.5y - 41.5 = 7.5y - 10$
$-37 - 1.5y = 7.5y - 10$
$-37 - 1.5y + 1.5y = 7.5y - 10 + 1.5y$
$-37 = 9y - 10$
$-27 = 9y$
$y = -3$

49.   This is supposed to be a mental exercise, so please refrain from setting up an equation or using a calculator. We admit that it is easy to find the answer that way, but the purpose of this problem is to exercise your mental estimating skills.

You can estimate the tax as $7.50. Since $7.50 is the tax on the room charges, the question is: 7.50 is 10% of what number? In other words, what did we start out with (the room charge) so that 10% of it is 7.50? Notice that taking 10% of something is a fairly easy computation, since 10% is equivalent to one-tenth, and to find one-tenth of a quantity you can just divide by 10. So we ask the question again, rephrasing it: what number can we divide by 10 and get 7.50? Think a minute, and you will have 75.00. We conclude that the room charge is approximately $75.00.

(You can perform similar reasoning with the original figure of $7.58 and just as easily conclude that the exact room charges are $75.80. We really didn't have to estimate the tax as $7.50.)

53.

the quantity three v minus 2	$3v - 2$
twice that quantity	$2(3v - 2)$
the quantity v plus 9	$v + 9$
four times that quantity	$4(v + 9)$

$2(3v - 2) = 4(v + 9)$
$6v - 4 = 4v + 36$ Add 4 to both sides.
$6v = 4v + 40$ Subtract 4v from both sides.
$2v = 40$ Divide both sides by 2.
$v = 20$

57.

the quantity two x plus five	$2x + 5$
one-third of that	$\frac{1}{3}(2x + 5)$
four x plus 5	$4x + 5$

$\frac{1}{3}(2x + 5) = 4x + 5$

$3(\frac{1}{3})(2x + 5) = 3(4x + 5)$ Multiply both sides by 3 to remove fractions.

$2x + 5 = 12x + 15$ Subtract 2x from both sides.
$5 = 10x + 15$ Subtract 15 from both sides.
$-10 = 10x$ Divide both sides by 10.
$x = -1$

61. The perimeter of a figure is the length of the boundary. A square has four equal sides, so the perimeter equals four times the length of one side.

perimeter = (4)(length of one side)

$36 = 4(4x + 3)$ Divide both sides by 4.
$9 = 4x + 3$ Subtract 3 from both sides.
$6 = 4x$ Divide both sides by 4 and reduce.

$x = \frac{3}{2}$ or 1.5

65. The area of a triangle is one-half the product of the base and height.

$51 = \frac{1}{2}(5x + 7)(6)$ We may take advantage of the fact that $\frac{1}{2} \cdot 6 = 3$.

$51 = 3(5x + 7)$ Since 51 is divisible by 3 ($5 + 1 = 6$), divide both sides by 3.
$17 = 5x + 7$ Subtract 7 from both sides.
$10 = 5x$ Divide both sides by 5.
$x = 2$

69. $7(2x - 4) + x = 2(11x - 5)$ Multiply out both sides.
$14x - 28 + x = 22x - 10$
$15x - 28 = 22x - 10$ Add 28 to both sides.
$15x = 22x + 18$ Subtract 22x from both sides.
$-7x = 18$ Divide both sides by -7.

$x = -\frac{18}{7}$ or $-2\frac{4}{7}$ (The equation was a conditional equation.)

73. $5.098(y + 4.5) = -3.487(2y - 1.8)$
$5.098y + 5.098(4.5) = -3.487(2y) + 3.487(1.8)$
$5.098y + 22.941 = -6.974y + 6.2766$ Add 6.974y to both sides.
$12.072y + 22.941 = 6.2766$ Subtract 22.941 from both sides.
$12.072y = -16.6644$ Divide both sides by 12.072.
$y \approx -1.380$

77. We will use $PRT = I$, where $T = 1$, $P = 15{,}400$ and $I = 1000$.
$(15{,}400)(R) = 1000$ Divide both sides by 15,400.

$R = \dfrac{1000}{15{,}400}$ Carry out the computation.

$R \approx 0.064935$ A calculator that displays eight digits gives this approximation.

Because interest rates on bank accounts are usually in steps of one-half percent and because our customer must have a rate that is 0.064935 <u>or better</u>, we round up to the next highest one-half percent. Answer: 6.5%

page 41

Problem Set 3.5

1. $A = \ell w$

    $\dfrac{A}{w} = \dfrac{\ell w}{w}$

    $\ell = \dfrac{A}{w}$

5. $V_1 T_2 = V_2 T_1$

    $\dfrac{V_1 T_2}{T_2} = \dfrac{V_2 T_1}{T_2}$

    $V_1 = \dfrac{V_2 T_1}{T_2}$

9. $V - E + F = 2$ Add E to both sides.
    $V + F = 2 + E$ Subtract 2 from both sides.
    $E = V + F - 2$

13. $C = \dfrac{5}{9}(F - 32)$ Multiply both sides by $\dfrac{9}{5}$.

    $\dfrac{9}{5}C = F - 32$ Add 32 to both sides.

    $F = \dfrac{9}{5}C + 32$

17. $\ell = a + (n - 1)d$ Apply the Distributive Law to remove parentheses.
    $\ell = a + nd - d$ Subtract a from both sides and add d to both sides.
    $\ell - a + d = nd$ Divide both sides by d.

    $n = \dfrac{\ell - a + d}{d}$

21. $y = mx + b$ Subtract b from both sides to isolate the term mx.
    $y - b = mx$ Divide both sides by x.

    $m = \dfrac{y - b}{x}$

25. $x + y + z = 3$ Subtract y and z from both sides.
    $x = 3 - y - z$

29. $vxy = 7$ Divide both sides by vy.

    $\dfrac{vxy}{vy} = \dfrac{7}{vy}$

    $x = \dfrac{7}{vy}$

33. $2v - 3x + 4y = 0$ Add 3x to both sides.
    $2v + 4y = 3x$ Divide both sides by 3.

    $x = \dfrac{2v + 4y}{3}$

37. $-7(v - 2x + 3) = -(2v + x - 9)$
   $-7v + 14x - 21 = -2v - x + 9$
   $14x - 21 = 5v - x + 9$
   $15x - 21 = 5v + 9$
   $15x = 5v + 30$

   $x = \dfrac{5v + 30}{15}$

   $x = \dfrac{v + 6}{3}$

Apply the Distributive Law to remove parentheses. Add 7v, x and 21 to both sides. The goal is to have one term with x on one side, and all other terms on the other side.

Factor the numerator and denominator and reduce.

41. $\dfrac{3x}{7} = \dfrac{15y}{14}$

   $6x = 15y$

   $x = \dfrac{15y}{6}$

   $x = \dfrac{5y}{2}$

Multiply both sides by 14 to remove fractions.

Divide both sides by 6.

Reduce.

45. $x_1 + x = y_1 + y$

   $x = y_1 + y - x_1$

Subtract x_1 from both sides.

(The right side may be arranged in another order.)

49. $xy_1z_1 = y_2z_2$

   $x = \dfrac{y_2z_2}{y_1z_1}$

Divide both sides by y_1z_1.

53. $0.45(2x - 5y) = 3.8(0.5x + 2y)$
   $0.45(2x) - 0.45(5y) = 3.8(0.5x) + 3.8(2y)$
   $0.9x - 2.25y = 1.9x + 7.6y$
   $0.9x = 1.9x + 9.85y$
   $(-1)x = 9.85y$
   $x = -9.85y$

Add 2.25y to both sides.
Subtract 1.9x from both sides.
Multiply both sides by -1.

Solutions To All Review Exercises — Chapter Three

1. 73

2. -1

3. $-\dfrac{5}{17}$

4. 1

5. 3

6. 1

7. 2

8. 2

9. $-(5x - 7y) = -5x + 7y$

10. $+(12x - 7) = 12x - 7$

11. $56\left(\frac{3}{7}x - \frac{5}{8}\right) = 56\left(\frac{3}{7}x\right) - 56\left(\frac{5}{8}\right) = 8(3x) - 7(5) = 24x - 35$

Alternate solution: work inside parentheses, and subtract the two fractions.

$56\left(\frac{3x(8) - 5(7)}{56}\right) = 56\left(\frac{24x - 35}{56}\right) = 24x - 35$

12. $-5(2x - 9) = -10x + 45$

13. $2x(4 - 3x) = 8x - 6x^2$

14. $(13x - 7)(-4) = -52x + 28$

15. $3x$

16. $5r - 7t + 3r + 4t = 5r + 3r - 7t + 4t = 8r - 3t$

17. $-12m + 6(3m - 5) = -12m + 18m - 30 = 6m - 30$

18. $7(2v - 4) - (8v + 1) = 14v - 28 - 8v - 1 = 6v - 29$

19. $11(5x - 7) - 13(4 - 2x) = 55x - 77 - 52 + 26x = 81x - 129$

20. $$\begin{aligned}
2(6x - 5) - [8 - 4(5x + 2)] &= 2(6x - 5) - [8 - 20x - 8] \\
&= 2(6x - 5) - (-20x) \\
&= 12x - 10 + 20x \\
&= 32x - 10
\end{aligned}$$

21. $-\frac{23}{5}$

22. 1

23. -1

24. $\frac{w}{v}$

25. left side:

$5(-4) + 3 = -20 + 3 = -17$

right side:

$7(-4) - 5 = -28 - 5 = -33$

-4 is not a solution.

26. left side:

$-(9(1) - 7) = -(2) = -2$

right side:

$2(4 - 5(1)) = 2(-1) = -2$

1 is a solution.

27. left side:

$\frac{3(-1) - 7}{5} = \frac{-3 - 7}{5} = \frac{-10}{5} = -2$

right side:

$\frac{7(-1)}{4} = \frac{-7}{4} = -\frac{7}{4}$

The left and right sides are not equal. -1 is not a solution.

28. It is sensible to simplify each side before substituting:

left side:

$-4[2v - 3(v - 1)]$

$-4[2v - 3v + 3]$

$-4[-v + 3]$

$4v - 12$

right side:

$18 - (4v + 10)$

$18 - 4v - 10$

$8 - 4v$

Now we substitute -2 for v:

$4(-2) - 12 = -20$ $8 - 4(-2) = 16$

$-20 \neq 16$, so -2 is not a solution.

29. $-17m = -34$

$m = \dfrac{-34}{-17}$

$m = 2$

30. $\dfrac{n}{3} = -12$

$n = -12(3)$

$n = -36$

31. $-\dfrac{2y}{7} = 28$

$y = -\dfrac{7}{2}(28)$

$y = -98$

32. $0.045w = -0.18$

$w = -\dfrac{0.18}{0.045}$

$w = -4$

33. $117 = 0.01t$

$t = \dfrac{117}{0.01}$

$t = 11,700$

34. $0 = -147b$

$b = \dfrac{0}{-147}$

$b = 0$

35. $5x - 8 = 27$
$5x = 35$
$x = 7$

36. $6v + 7 = 11v - 8$
$-5v = -15$
$v = 3$

37. $-8z + 7 - 2z = 5z + 4 + z$
$-10z + 7 = 6z + 4$
$-16z = -3$
$z = \frac{3}{16}$

38. $-r - r - r - r - r = r$
$-5r = r$
$-6r = 0$
$r = 0$

39. $3(2x + 7) = x + 1$
$6x + 21 = x + 1$
$5x = -20$
$x = -4$

40. $(6x - 9) - (3x + 8) = 4(x - 11)$
$6x - 9 - 3x - 8 = 4x - 44$
$3x - 17 = 4x - 44$
$-x = -27$
$x = 27$

41. $7y - 5(2 - y) = 3(2y + 1) + 5$
$7y - 10 + 5y = 6y + 3 + 5$
$12y - 10 = 6y + 8$
$6y = 18$
$y = 3$

42. $9(2y - 3) - 13(5y - 1) = 14(3y - 1)$
$18y - 27 - 65y + 13 = 42y - 14$
$-47y - 14 = 42y - 14$
$-89y = 0$
$y = 0$

43. $0.55(w - 10) = 0.80(w + 3)$
$55(w - 10) = 80(w + 3)$
$55w - 550 = 80w + 240$
$-25w = 790$
$w = -31\frac{3}{5}$ or -31.6

Multiply both sides by 100 to remove decimals.

44. $\frac{t}{8} - 5 = \frac{t}{12} - 6$

$24\left(\frac{t}{8} - 5\right) = 24\left(\frac{t}{12} - 6\right)$

$24\left(\frac{t}{8}\right) - 24(5) = 24\left(\frac{t}{12}\right) - 24(6)$

$3t - 120 = 2t - 144$

$t = -24$

The LCD is 24.

45. $\frac{7m + 27}{6} = \frac{4m + 6}{5}$

$30\left(\frac{7m + 27}{6}\right) = 30\left(\frac{4m + 6}{5}\right)$

$5(7m + 27) = 6(4m + 6)$

$35m + 135 = 24m + 36$
$11m = -99$
$m = -9$

The LCD is 30.

46. $$\frac{167 - 3n}{10} = 6 - \frac{5n - 1}{4}$$ The LCD is 20.

$$20\left(\frac{167 - 3n}{10}\right) = 20\left(6 - \frac{5n - 1}{4}\right)$$ Distribute carefully on the right side.

$$2(167 - 3n) = 120 - 5(5n - 1)$$
$$334 - 6n = 120 - 25n + 5$$
$$209 = -19n$$
$$n = -11$$

47. $$1293y = 1294y + 1295$$
$$-y = 1295$$
$$y = -1295$$

48. $$0.28(x - 5) = 3.1 - 0.78(2 - x)$$ Multiply both sides by 100 to remove fractions.
$$28(x - 5) = 310 - 78(2 - x)$$
$$28x - 140 = 310 - 156 + 78x$$
$$28x - 140 = 154 + 78x$$
$$-50x = 294$$

$$x = -\frac{294}{50} = -5.88$$

49. $$\frac{3v}{10} - \frac{5v}{6} = -\frac{8v}{15}$$ The LCD is 30.

$$30\left(\frac{3v}{10} - \frac{5v}{6}\right) = 30\left(-\frac{8v}{15}\right)$$

$$3(3v) - 5(5v) = 2(-8v)$$
$$9v - 25v = -16v$$
$$-16v = -16v$$
$$0 = 0$$ Solution: all real numbers

50. $$1.8a - 7.8a + 1.97a - 8.3a = 0$$
$$-12.33a = 0$$
$$a = 0$$

51. $$7(y - 3) - 4(y + 2) = 3(y - 7)$$
$$7y - 21 - 4y - 8 = 3y - 21$$
$$3y - 29 = 3y - 21$$
$$-29 = -21$$ This is a contradiction. No solution.

52. $$3(2v + 1) = 2(3v + 1)$$
$$6v + 3 = 6v + 2$$
$$3 = 2$$ This is a contradiction. No solution.

53. $4(w - 1) + (7 - 2w) = 2(w + 1) + 1$
$4w - 4 + 7 - 2w = 2w + 2 + 1$
$2w + 3 = 2w + 3$
$3 = 3$ This is an identity. Solution: all real numbers

54. $2v + 2v + 2v = 5v$
$6v = 5v$ Now subtract 5v from both sides.
$v = 0$ This is a conditional equation, with solution $v = 0$.

55. $v + w = x + y$
$v + w - y = x$
$x = v + w - y$

56. $vw = xy$ Turn the equation around so that x is on the left.
$xy = vw$ (This is not required, but is sensible.)
$x = \dfrac{vw}{y}$

57. $3x - 5y = 7z$
$3x = 7z + 5y$
$x = \dfrac{7z + 5y}{3}$

58. $2(x - y) + 3(2x + y) = 38$
$2x - 2y + 6x + 3y = 38$
$8x + y = 38$
$8x = 38 - y$
$x = \dfrac{38 - y}{8}$

59. five times the number $5m$
four more than that quantity $5m + 4$

$5m + 4 = 49$
$5m = 45$
$m = 9$

60. the sum of a number and seven $m + 7$
one-third of that quantity $\frac{1}{3}(m + 7)$

$\frac{1}{3}(m + 7) = 12$ Multiply both sides by 3.
$m + 7 = 36$
$m = 29$

61. two more than the number

   twice that quantity

   three less than the number

   the opposite of that quantity

$m + 2$

$2(m + 2)$

$m - 3$

$-(m - 3)$ or $3 - m$

$$2(m + 2) = 3 - m$$
$$2m + 4 = 3 - m$$
$$3m = -1$$
$$m = -\frac{1}{3}$$

62. the difference of twice a number minus three

   seven times that quantity

   the sum of the number and one

   three times that quantity

   two less than the previous quantity

$2m - 3$

$7(2m - 3)$

$m + 1$

$3(m + 1)$

$3(m + 1) - 2$

$$7(2m - 3) = 3(m + 1) - 2$$
$$14m - 21 = 3m + 3 - 2$$
$$11m = 22$$
$$m = 2$$

63. five times the number plus two

   one-third of that quantity

   five times a number minus one

   two-sevenths of that quantity

$5m + 2$

$\frac{1}{3}(5m + 2)$

$5m - 1$

$\frac{2}{7}(5m - 1)$

$$\frac{1}{3}(5m + 2) = \frac{2}{7}(5m - 1)$$

We will multiply both sides by the LCD, 21.

$$7(5m + 2) = 6(5m - 1)$$

$$35m + 14 = 30m - 6$$

$$5m = -20$$

$$m = -4$$

64. the number

   one more than the number

   two more than the number

m

$m + 1$

$m + 2$

$$m + (m + 1) + (m + 2) = 93$$
$$3m + 3 = 93$$
$$3m = 90$$
$$m = 30$$

65. The perimeter of a figure is the sum of the lengths of the sides. For a rectangle, this means the sum of twice the width and twice the length.

$2w + 2(3w + 4) = 48$
$8w + 8 = 48$
$8w = 40$
$w = 5$ The width is 5 cm and the length is
                  $3(5) + 4 = 19$ cm.

66. $6(4x + 3) = 54$ Since 54 is divisible by 6, divide both sides by 6.
$4x + 3 = 9$
$4x = 6$

$x = \frac{6}{4} = \frac{3}{2}$ or $1\frac{1}{2}$ or 1.5

67. The area of a triangle is one-half the product of the base and height.

$A = \frac{1}{2}(\text{base})(\text{height})$

$52 = \frac{1}{2}(3b + 4)(8)$

$52 = \frac{1}{2}(8)(3b + 4)$

$52 = 4(3b + 4)$ Since 52 is divisible by 4, we will not distribute but
                          instead divide both sides by 4.

$13 = 3b + 4$
$9 = 3b$
$b = 3$

68. $A = \text{height}\left(\dfrac{\text{base 1} + \text{base 2}}{2}\right)$ The area of a trapezoid. By base 1 and base 2, we
                                                                                       mean the two parallel sides.

$48 = 6\left(\dfrac{2v + 1 + 2v + 3}{2}\right)$

$48 = 3(2v + 1 + 2v + 3)$ Divide both sides by 3, since 48 is divisible by 3.
$16 = 2v + 1 + 2v + 3$
$16 = 4v + 4$
$12 = 4v$
$v = 3$

69. We may approximate the solution by solving
$-4x = 44$
$x = -11$ The correct response is A.

70.   Estimating, we solve $10y = -10$, which gives $y = -1$. It is not yet clear whether the best answer is B or C. Notice that $y = -\dfrac{10.197}{10.3}$ and that the numerator is less than the denominator. This tells is that the fraction $\dfrac{10.197}{10.3}$ is *less than one*. **Answer: C**

71.   We will round to the nearest half:
$$-3x - 4.5 = -3$$
$$-3x = 1.5$$
$$x = -\frac{1.5}{3} = -\frac{1}{2}$$ **Answer: B**

Note: You might have started by rounding to the nearest integer, which certainly is easier. We would have $-3x - 5 = -3$ and the solution to this is $-\frac{2}{3} = -0.6666...$ which suggests either A or B, but it is not clear which.

The reason for this difficulty is that when we round 4.5 to 5, we lose 0.5. Of course, we always lose *something* in rounding, but none of the numbers in the problem are very big, so when you lose 0.5, that's a lot in this problem. This causes our estimate to be off by too much to be useful.

The fact that choices A, B and C are relatively close together suggests that you must be conservative with your estimating, and don't round by large amounts.

72.   Estimating, we solve:
$$4y + 7 = 37$$
$$4y = 30$$
$$y = \frac{30}{4} = \frac{15}{2} \text{ or } 7\frac{1}{2}$$ **Answer: C**

73. $-7.11x + 19.3 = 87.41$
$-7.11x = 68.11$
$x \approx -9.579$

74. $47.38 + 13.8z = 97.4z$
$47.38 = 83.6z$
$z \approx 0.567$

75. $4.983(v + 7.2) = -8.29v + 11.4$
$4.983v + 35.8776 = -8.29v + 11.4$
$13.273v = -24.4776$
$v \approx -1.844$

76. $193.8 = \pi w$
$w = \dfrac{193.8}{\pi} \approx 61.688$

77.   This is worked out in the answer section in your textbook.

78. $\dfrac{t-3}{2} = \dfrac{t+1}{5}$ The LCD is 10.

$10\left(\dfrac{t-3}{2}\right) = 10\left(\dfrac{t+1}{5}\right)$

$5(t-3) = 2(t+1)$ Notice the use of parentheses. Here is where student B went wrong.

$5t - 15 = 2t = 2$

$3t = 17$

$t = \dfrac{17}{3}$ or $5\dfrac{2}{3}$

Solutions To All Mastery Test Problems — Chapter Three

1a. $21x - 28$ 1b. $-20y + 45$ 1c. $3a + 2b$ 1d. $-6a + 3b$

2a. $3v$ 2b. $17x + 2y$ 2c. $-7m + 18m - 27 = 11m - 27$

2d. $-2x + 7y - 6y + 33x = 31x + y$ 3a. $x = 20$

3b. $4y + 7 = 5y + 13$ 3c. $-7w + 13 = -8w - 87$

      $7 = y + 13$ $w + 13 = -87$

      $y = -6$ $w = -100$

3d. $27 - 83z = -(84z + 27)$ 4a. $\dfrac{4}{3}y = 12$

      $27 - 83z = -84z - 27$

      $27 + z = -27$ $y = \dfrac{3}{4}(12) = 9$

      $z = -54$

4b. $-2.1w = 105$ 4c. $-x = 8$

      $w = \dfrac{105}{-2.1} = -50$ $x = -8$

Chapter Three Mastery Test

4d.　$17z = 0$

$$z = \frac{0}{17} = 0$$

4e.　$\frac{3}{7} = \frac{x}{35}$

$$7x = 3(35)$$

$$x = \frac{3(35)}{7}$$

$$x = 3(5) = 15$$

4f.　$\frac{5x}{36} = \frac{3}{4}$

$$x = \left(\frac{36}{5}\right)\left(\frac{3}{4}\right)$$

$$x = \frac{27}{5} \text{ or } 5\frac{2}{5}$$

5a.　$5(x - 2) = 6(x + 3)$

$5x - 10 = 6x + 18$

$-10 = x + 18$

$x = -28$

5b.　$-5y - 9(2y + 3) =$
$$7(3y - 2) - (y - 7) + 23$$
$-5y - 18y - 27 = 21y - 14 - y + 7 + 23$

$-23y - 27 = 20y + 16$

$-27 = 43y + 16$

$-43 = 43y$

$y = -1$

5c.　$\frac{v}{2} - 4 = \frac{v}{3}$

$6\left(\frac{v}{2} - 4\right) = 6\left(\frac{v}{3}\right)$

$3v - 24 = 2v$

$v = 24$

5d.　$\frac{z}{20} + \frac{1}{12} = \frac{z}{15} - \frac{1}{10}$

$$60\left(\frac{z}{20} + \frac{1}{12}\right) = 60\left(\frac{z}{15} - \frac{1}{10}\right)$$

$3z + 5 = 4z - 6$

$5 = z - 6$

$z = 11$

5e.　$\frac{4w + 23}{5} = \frac{19 - w}{7}$

$$(4w + 23)(7) = (5)(19 - w)$$

$28w + 161 = 95 - 5w$

$33w + 161 = 95$

$33w = -66$

$w = -2$

Chapter Three Mastery Test

5f. $-7.7(t - 3) = 2.3(t + 4)$

$(10)[-7.7(t - 3)] = (10)[2.3(t + 4)]$

$-77(t - 3) = 23(t + 4)$

$-77t + 231 = 23t + 92$

$231 = 100t + 92$

$139 = 100t$

$t = 1.39$

6a. $v = wxy$

$\dfrac{v}{wy} = \dfrac{wxy}{wy}$

$x = \dfrac{v}{wy}$

6b. $vw = \dfrac{2}{3}xy$

$\left(\dfrac{3}{2y}\right)vw = \left(\dfrac{3}{2y}\right)\dfrac{2}{3}xy$

$x = \dfrac{3vw}{2y}$

6c. $v + w = x - y$ Add y to both sides.

$v + w + y = x$

$x = v + w + y$

6d. $w = 2x - y$

$w + y = 2x$

$x = \dfrac{w + y}{2}$

6e. $2v = \dfrac{1}{3}(x - y)$

$6v = x - y$

$x = 6v + y$

6f. $(3x + y) - (y - x) = 5(2x + y)$

$3x + y - y + x = 10x + 5y$

$4x = 10x + 5y$

$-6x = 5y$

$x = -\dfrac{5y}{6}$

1. distributive property

2. commutative property of multiplication

3. commutative property of addition

4. associative property of addition

5. associative property of multiplication

6. Look at the digit appearing after the 8; since it is 7, which is greater than five, we round up to 900.

7. The hundredth place is the second place to the right of the decimal. The digit in the third place is 3, which is less than five, so we drop the 3. Answer: 876.54
Notice the difference between this problem and the previous one, where we rounded to the nearest *hundred*.

8. The additive inverse is found by changing the sign: $\frac{2}{3}$

9. The multiplicative inverse is found by inverting, or turning over, the fraction: $-\frac{3}{2}$

10. The absolute value of a negative number is its positive counterpart: $\frac{2}{3}$

    If a number is negative, change the sign to get the absolute value.
    If a number is positive, do nothing. The absolute value is the same as the number.

11. You may find it convenient to write 1.2 as 1.20 before adding. Answer: 1.44

12. You will want to write 1.2 as 1.20 before subtracting. Answer: 0.96

13. 0.24 has two places to the right of the decimal, 1.2 has one, so the product will have three places to the right of the decimal. Add the number of places in the numbers to be multiplied to get the number of places to the right of the decimal in the product. Answer: 0.288

14.
$$
\begin{array}{r}
0\,.\,2 \\
1\,.\,2\,{}_{\wedge}\overline{)\,0\,.\,2\,{}_{\wedge}4\,} \\
\underline{0\quad 0} \\
2\quad 4 \\
\underline{2\quad 4} \\
0
\end{array}
$$
Remember to adjust the decimal points.

15. $\frac{5}{6} - \frac{2}{3}$

    $= \frac{5}{6} - \frac{4}{6}$

    $= \frac{1}{6}$

16. $\frac{5}{6} \div \frac{2}{3}$

    $= \frac{5}{6} \times \frac{3}{2}$

    $= \frac{5(3)}{6(2)}$

    $= \frac{5}{2(2)}$

    $= \frac{5}{4}$

Cumulative Review

17. This is -81 because it is only the 9 that is squared. *After* the squaring is done, the sign is applied. Contrast this problem with the different problem $(-9)^2$, which is $(-9)(-9) = 81$.

18. Answer: 81. See #17 for explanation.

19. Any number times zero is zero. Answer: 0

20. Zero divided by any nonzero number is zero. Answer: 0

21. Division by zero is undefined.

22. You may rewrite this as: $-2 + (-3) + (-4) = -9$

23. This is a product of three numbers, all of which are negative. Taking them two at a time:
$-2(-3)(-4)$
$= 6(-4)$
$= -24$

24. Be careful. We must perform multiplication before addition or subtraction:
$-2 - 3(-4)$
$= -2 - (-12)$
$= -2 + 12$
$= 10$

25. The parentheses indicate what to do first:
$(-2 - 3)(-4)$
$= (-5)(-4)$
$= 20$

26. $28 - 8(11 - 9)$
$= 28 - 8(2)$
$= 28 - 16$
$= 12$

27. $(28 - 8)(11 - 9)$
$= (20)(2)$
$= 40$

28. $12^2 - 8^2$
$= 144 - 64$
$= 80$

29. $(12 - 8)^2$
$= 4^2$
$= 16$

30. $0^{11} = 0$
Notice that $0^{11} = 0 \cdot 0 \cdot 0 \cdot 0 \cdot 0 \cdot 0 \cdot 0 \cdot 0 \cdot 0 \cdot 0 \cdot 0$ (eleven times)
It doesn't matter how many factors of zero we have, the product is still zero.
Zero raised to any positive power is zero.

31. $-1^{10} = -(1^{10}) = -(1) = -1$

32. $(-1)^{10} = (-1)(-1)(-1)(-1)(-1)(-1)(-1)(-1)(-1)(-1) = 1$
Negative one raised to any even power will give one, since each pair is $(-1)(-1) = 1$.
If the exponent were odd, there would be an unpaired factor of -1, so the result would be -1.

33. $(-1)^{45} = -1$ See comments on #32.

page 56
Cumulative Review

34.

$-8 - 2[7 + 4(8 - 13)]$

$= -8 - 2[7 + 4(-5)]$

$= -8 - 2[7 + (-20)]$

$= -8 - 2(-13)$

$= -8 - (-26)$

$= -8 + 26$

$= 18$

35. $\quad 5^2 + 2^5 = 25 + 32 = 57$

36. $\quad$ 8, since $8^2 = 64$.

37. $\quad \sqrt{9 + 16} = \sqrt{25} = 5$

It is incorrect to split up a square root of a sum, so do not write $\sqrt{9} + \sqrt{16}$.

38. $\quad \sqrt{9} + \sqrt{16}$

$= 3 + 4$

$= 7$

39. $\quad$ This is the absolute value of -17, which is 17.

40.

$-|-3 - 4|$

$= -|-3 + (-4)|$

$= -|-7|$

$= -7$

The absolute value of -7 is 7; but after this is done we still must apply the remaining sign. The answer is -7.

41. $\quad \sqrt{4} = 2$, and 2 can be written as ratio of integers, $\frac{2}{1}$, so $\sqrt{4}$ is rational.

42. $\quad \sqrt{4} = 2$ and $\pi \approx 3.14$. Answer: $<$

43.

$(3x - 4y - 7) - (5x - 8y - 11)$

$= 3x - 4y - 7 - 5x + 8y + 11$

$= 3x - 5x - 4y + 8y - 7 + 11$

$= -2x + 4y + 4$

44.

$-2(x + 3y - 2) - 3(x - 2y - 5)$

$= -2x - 6y + 4 - 3x + 6y + 15$

$= -5x + 19$

45.

$2^2 - 5(-2)(-5) + 2(-5)^2$

$= 4 - 5(-2)(-5) + 2(25)$

$= 4 - 50 + 50$

$= 4$

46.

$[3(-10) - 4(-2)] \, [-2(-10) + 5(-2)]$

$= (-30 + 8)(20 - 10)$

$= (-22)(10)$

$= -220$

47. $\quad x - 3 = 15 \qquad$ Add 3 to both sides.

$x = 18$

48. $\quad 3x = 15 \qquad$ Divide both sides by 3.

$x = 5$

page 57

Cumulative Review

49. $\frac{x}{3} = 15$ Multiply both sides by 3.

       $x = 45$

50. $\frac{3x}{4} = -12$ Multiply both sides by $\frac{4}{3}$.

       $x = -\frac{4}{3}(12)$

       $x = -4(4)$
       $x = -16$

51. $2x + 7 = 7x + 22$ Subtract 7 from both sides.
       $2x = 7x + 15$ Subtract 7x from both sides.
      $-5x = 15$ Divide both sides by -5.

       $x = -3$

52. $2(x - 3) = 3(4x - 8)$ Distribute (multiply out both sides).
       $2x - 6 = 12x - 24$ Add 24 to both sides.
       $2x + 18 = 12x$ Subtract 2x from both sides.
       $18 = 10x$ Divide both sides by 10.

       $x = \frac{18}{10}$ Reduce.

       $x = \frac{9}{5}$

53. $2[x - (3x + 4)] = x - 6$ First simply the left side to remove grouping symbols.
       $2[x - 3x - 4] = x - 6$
       $2(-2x - 4) = x - 6$
      $-4x - 8 = x - 6$ Subtract x from both sides.
      $-5x - 8 = -6$ Add 8 to both sides.
      $-5x = 2$ Divide both sides by -5.

       $x = -\frac{2}{5}$

54. $ax + b = c$ Subtract b from both sides.
       $ax = c - b$ Divide both sides by a.

       $x = \frac{c - b}{a}$

55. $2v = 3(5x - y)$ Distribute on the right side.
       $2v = 15x - 3y$ Add 3y to both sides.
       $2v + 3y = 15x$ Divide both sides by 15.

       $x = \frac{2v + 3y}{15}$

Cumulative Review

Problem Set 4.1

1.   Let n represent the number.
        $3n + 7 = 31$ Subtract 7 from both sides.
        $3n = 24$ Divide both sides by 3.
        $n = 8$ The number is 8.

5.   Let n be the number.
        $n - 5$ the number decreased by five
        $2n$ the number doubled

        The statement says that when we decrease the number by 5, we get the same thing as if we double it. This gives the equation:
        $2n = n - 5$ Subtract n from both sides.
        $n = -5$ The number is -5.

9.   Let s be the low score.   13.   Let s be the score on the fourth test.
        $98 - s = 58$ Add s to both sides.
        $98 = 58 + s$ Subtract 58. $$\frac{92 + 83 + 85 + s}{4} = 90$$
        $s = 40$

        The low score was 40. $92 + 83 + 85 + s = 360$
                                                         $260 + s = 360$
                                                         $s = 100$

                                                         The student must earn a score of 100
                                                         on the fourth test to have an average of 90.

17.   n   the first odd integer   21.   n   the smaller integer
        $n + 2$ the next consecutive odd integer $n + 1$ the next consecutive integer
        $n + (n + 2) = 112$ Combine like terms. $5n = (n + 1) + 67$
        $2n + 2 = 112$ $5n = n + 68$
        $2n = 110$ $4n = 68$
        $n = 55$, $n + 2 = 57$ $n = 17$

25.   Let a be the smaller angle   29.   Let c be the cost.
        and $a + 8$ be the larger angle. $\frac{1}{2} c - 100 = 450$ Add 100.
        $a + (a + 8) = 90$
        $2a + 8 = 90$ $\frac{1}{2} c = 550$
        $2a = 82$ $c = 1100$
        $a = 41$, $a + 8 = 49$ The cost is $1100.
        The angles measure 41° and 49°.

33.   Let m be the number of months.
        $25m$ the cost for m months if rented
        850 the cost if purchased

$$850 = 25m \qquad \text{Divide both sides by 25.}$$
$$m = 34 \qquad \text{It would take 34 months.}$$

37. Let m be the number of miles.
     total charge = daily charge for two days + mileage charge
     $$113 = 2(25) + 0.15m$$
     $$113 = 50 + 0.15m$$
     $$63 = 0.15m$$
     $$m = 420 \qquad \text{The renter drove the car 420 miles.}$$

41. Let n be the first number. Then $\frac{1}{3}$ n is the second number, and $5 + \frac{1}{3}$ n is the third number.
     $$40 = n + \frac{1}{3}n + 5 + \frac{1}{3}n$$

     $$35 = \frac{5}{3}n$$
     $$n = 21$$

     The first number is 21, the second number is $\frac{1}{3}(21) = 7$, and the third number is $5 + 7 = 12$.

45. Let w be the width and w + 5 be the length.
     $$78 = 2(w) + 2(w + 5) \qquad \boxed{\text{perimeter = sum of lengths of sides}}$$
     $$78 = 2w + 2w + 10$$
     $$68 = 4w$$
     $$w = 17 \qquad \text{The width is 17 cm.}$$

49. Let w be the width and w + 18 be the length.
     total cost = (length of the trench in meters)($1.35 per meter)
     length of the trench = perimeter of the rectangle = 2w + 2(w+18)
     total cost = [2w + 2(w+18)][1.35]
     $$156.60 = (4w + 36)(1.35) \qquad \text{(Divide both sides by 1.35.)}$$
     $$116 = 4w + 36$$
     $$80 = 4w \; ; \; w = 20 \qquad \text{The width is 20 meters.}$$

Problem Set 4.2

1. See answer section in the text.

5. See answer section in the text.

9. $$\frac{6}{3y} = \frac{5}{6}$$

    $$36 = 15y$$

    $$\frac{36}{15} = y$$

    $$y = \frac{12}{5}$$

13. $$\frac{7z}{10} = \frac{7}{2}$$

     $$7z(2) = 10(7)$$

     $$14z = 70$$

     $$z = \frac{70}{14} = 5$$

17. $\dfrac{5}{2x-3} = \dfrac{3}{x}$

$5x = (2x - 3)(3)$

$5x = 6x - 9$

$x = 9$

21. $\dfrac{\frac{1}{2}}{a} = \dfrac{\frac{1}{3}}{\frac{6}{5}}$

$\left(\dfrac{1}{2}\right)\left(\dfrac{6}{5}\right) = a\left(\dfrac{1}{3}\right)$

$\dfrac{3}{5} = \dfrac{a}{3}$

$3(3) = 5a$

$a = \dfrac{9}{5}$

---

25.   We will show two different ways to work this problem.

Solution #1:

$v = kw$	Find k.

$22 = k(77)$

$\dfrac{22}{77} = k \; ; \quad k = \dfrac{2}{7}$

$v = \dfrac{2}{7}w$	Set up the equation with k as found.

$v = \dfrac{2}{7}(35) = 10$

Solution #2 (setting up a proportion):

$\dfrac{22}{77} = \dfrac{v}{35}$	Reduce $\dfrac{22}{77}$.

$\dfrac{2}{7} = \dfrac{v}{35}$

$2(35) = 7v$

$v = \dfrac{2(35)}{7}$

$v = 10$

---

29.   Let the constant of proportionality be k.

$v = kw$	Substitute in 12 for v and 15 for w to find k.
$12 = k(15)$	Divide both sides by 15 and reduce.

$k = \dfrac{4}{5}$

For *any pair of values v and w, the constant of proportionality is* $\dfrac{4}{5}$. This means that the general equation for this problem is $v = \dfrac{4}{5}$ w.   Although not asked for, you could now find any value of v given a value for w, or find any value of w given a value for v.

---

33.   The ratio is $\dfrac{\text{length of a side on first triangle}}{\text{length of corresponding side on second triangle}}$.

$\dfrac{6}{a} = \dfrac{4}{12}$	Reduce $\dfrac{4}{12}$.
$\dfrac{6}{a} = \dfrac{1}{3}$	Cross-multiply.

$a = 18$

37. The ratio is $\dfrac{\text{defective bulbs}}{\text{number produced}}$.

Let d be the number of defective bulbs in the batch of 10,000 produced.

$\dfrac{2.5 \text{ defective bulbs}}{500 \text{ produced}} = \dfrac{d \text{ defective bulbs}}{10{,}000 \text{ produced}}$	Cross-multiply.
$2.5(10{,}000) = 500d$	Divide both sides by 500.
$\dfrac{2.5(10{,}000)}{500} = d$	Hint: First divide 10,000 by 500.
$d = 50$	We expect 50 defective bulbs out of 10,000.

41. The ratio is $\dfrac{\text{cubic feet of sand}}{\text{cubic feet of concrete}}$.

Let s be the amount of sand needed to make 224 cubic feet of concrete.

$\dfrac{11}{32} = \dfrac{s}{224}$	You may cross-multiply a proportion.
$11(224) = 32s$	Divide by 32 and reduce.
$\dfrac{11(224)}{32} = s$	
$s = 77$	It would take 77 cubic feet of sand.

45. The ratio is $\dfrac{\text{lbs of grass seed}}{\text{square feet of area}}$.

Let x = number of pounds of seed needed for 450 square feet.

$$\dfrac{3}{100} = \dfrac{x}{450}$$

$$3(450) = 100x$$

$$1350 = 100x$$

$$x = 13.5$$

13.5 pounds of seed are needed.

49. The ratio is $\dfrac{\text{defective chips}}{\text{total chips}}$.

Let x = the number of defective chips in a shipment of 10,000.

$$\dfrac{4}{250} = \dfrac{x}{10000}$$

$$4(10{,}000) = 250x$$

$$40000 = 250x$$

$$x = 160$$

160 defective chips would be expected.

53. The ratio is $\dfrac{\text{dosage}}{\text{weight of patient}}$.

We wish to find the dosage, that is, the number of mg to be given *per kg of weight of the patient*. Let n be the number of mg needed per kg of weight.

$\dfrac{293 \text{ mg}}{45 \text{ kg}} = \dfrac{n \text{ mg}}{1 \text{ kg}}$ Solve for n. To the nearest hundredth, we have 6.51.

Repeat this procedure for the remaining two sets of data:

$$\frac{390 \text{ mg}}{60 \text{ kg}} = \frac{n \text{ mg}}{1 \text{ kg}}$$ Solve for n. To the nearest hundredth, we have 6.5.

$$\frac{488 \text{ mg}}{75 \text{ kg}} = \frac{n \text{ mg}}{1 \text{ kg}}$$ Solve for n. To the nearest hundredth, we have 6.51.

Since these figures agree to the nearest tenth, we conclude that the dosage instructions would be to administer 6.5 mg per kg of patient body weight.

57.   The ratio is $\dfrac{\text{length of shadow}}{\text{length of object casting shadow}}$.

Let h be the height of the flagpole that casts a 21-ft shadow.

$$\frac{6}{5} = \frac{21}{h}$$ Solve by cross-multiplying.

$$6h = 5(21)$$

$$h = 17.5$$ The flagpole is 17.5 feet tall.

61.   The ratio is $\dfrac{\text{earned runs}}{\text{outs}}$.

Let n be the number of earned runs that the pitcher would have allowed if he had pitched a whole game (27 outs).

$$\frac{2 \text{ earned runs}}{6 \text{ outs}} = \frac{n \text{ earned runs}}{27 \text{ outs}}$$ Cross-multiply and solve for n.

$$2(27) = 6n$$

$$n = 9.00 \text{ rounded to the nearest hundredth}$$

Problem Set 4.3

To the student: Problem #1 will be discussed at length, as an example of a general approach to these problems.

1.   Since the difference of the two numbers is 5, the smaller number must be 5 less than the larger. The second row is filled in by subtracting 5 from the numbers in the first row.

larger number	20	15	10	5	n
smaller number	15	10	5	0	n−5
their difference	5	5	5	5	5
their sum	35				

In the last row, we fill in the *sum of the larger and smaller numbers.* You will notice that the problem says that the sum of the numbers is 12, *but we are not going to use that fact yet.* Look again at the first column of numbers for direction... simply add the two numbers, and put that sum at the bottom. Notice in the last column, we add n and n−5 to obtain n + (n−5) = 2n − 5.

larger number	20	15	10	5	n
smaller number	15	10	5	0	n−5
their difference	5	5	5	5	5
their sum	35	25	15	5	2n−5

One thing that you are probably noticing is that while we have four pairs of numbers whose difference is 5, the bottom row points out that *none of the pairs add up to 12.* This means that none of these four pairs is a solution, and in fact, you are not expected to stumble upon a solution as you make such a table. The purpose of making the table is to obtain general expressions for the quantities in the problem. Notice the general expressions in the last column.

The second condition in the problem is that the numbers add to 12. *We will use this to write the equation.* You know from your table that the sum of the numbers is also 2n − 5, so

$$12 = 2n - 5$$
$$17 = 2n$$
$$n = \frac{17}{2} \text{ (the larger number)}$$

The smaller number is five less than the larger number, or n − 5 (look at the last column), so

$$n - 5 = \frac{17}{2} - 5 = \frac{17-10}{2} = \frac{7}{2}.$$

Check your answers! See if the numbers satisfy the stated conditions: do they add up to 12?

$$\frac{17}{2} + \frac{7}{2} = \frac{24}{2} = 12 \qquad \checkmark$$

5.

larger number	50	30	26	n
smaller number	0	20	24	50−n
their sum	50	50	50	50
their difference	50	10	2	n−(50−n)

For the second row in this table we concentrate on the condition that the sum of the numbers is always 50, which means that 50 − (larger number) = smaller number. The last row is obtained by calculating (larger number) − (smaller number).

The second condition is that the difference of the numbers is 5.8, and this fact, together with the information in the last column of the table is used to write the equation:

$$5.8 = n - (50-n)$$
$$5.8 = 2n - 50$$

$$55.8 = 2n$$
$$n = 27.9 \quad \text{(the larger number)}$$
$$50-n = 50-27.9 = 22.1 \quad \text{(the smaller number)}$$

9.

multiply column one by 5
↓   multiply column one by 10
↓

bricklayers	2	10	20	1	b
laborers	5	25	50	$\frac{5}{2}$	$\frac{5}{2}b$
total	7	35	70	$\frac{7}{2}$	$\frac{7}{2}b$

↑

divide column one by 2   ↑

multiply previous column by b

There are a total of 98 workers, so $98 = \frac{7}{2}b$; $196 = 7b$; $b = 28$.
The number of bricklayers is 28.
Although not asked for, the number of laborers is $\frac{5}{2}(28) = 5(14) = 70$.

13.

x km

48 km

Let x = the distance traveled by the slower bus and
x + 48 = the distance traveled by the faster bus.

$$688 = \text{sum of their distances}$$
$$688 = x + (x + 48)$$
$$688 = 2x + 48 \qquad \text{The slower bus travels 320 km.}$$
$$640 = 2x$$
$$320 = x$$

17.

		multiply column one by 2 ↓	multiply column one by 20 ↓		
females	5	10	100	1	f
males	6	12	120	$\frac{6}{5}$	$\frac{6}{5}f$
total	11	22	220	$\frac{11}{5}$	$\frac{11}{5}f$
			↑ divide column one by 5	↑ multiply previous column by f	

The total of students is 4686.

$4686 = \frac{11}{5}f$

$23430 = 11f$

$2130 = f$ The number of female students is 2130.

21.

		multiply first column by 10 ↓	divide first column by 2 ↓	
cement	2	20	1	c
gravel	5	50	$\frac{5}{2}$	$\frac{5}{2}c$
sand	4	40	2	2c
mixture	11	110	$\frac{11}{2}$	$\frac{11}{2}c$
				↑ multiply previous column by c

The total number of shovels is 440.

$440 = \frac{11}{2}c$

$880 = 11c$

$c = 80$ 80 shovels of cement are required.

25.   See answer section in the text.   **29.**   See answer section in the text.

1.   slower hiker   faster hiker

$$\xleftarrow{\qquad \underline{x} \qquad \bullet \underline{\qquad\qquad x + 3 \qquad\qquad}} \rightarrow$$

Let x = distance traveled by the slower hiker and
x + 3 = distance traveled by the faster hiker.
The sum of their distances is 18:
x + (x+3) = 18
2x + 3 = 18
2x = 15
x = 7.5   Answer: 7.5 km

5.   Let x = number of the more valuable coins and
35 − x = number of the less valuable coins.

	value per coin ·	number of coins =	total value
worth more	20	x	20x
worth less	8	35 − x	8(35 − x)

The total value of all the coins is $352, so:
8(35 − x) + 20x = 352
280 − 8x + 20x = 352   Answer:
12x = 72   6 of the more valuable coins;
x = 6   29 of the less valuable coins
35 − x = 35 − 6 = 29

9.   Let x = the number of adult tickets
160 − x = the number of student tickets

	value per ticket ·	number of tickets =	income
adult tickets	5	x	5x
student tickets	2	160 − x	2(160 − x)

(income from adult tickets) + (income from student tickets) = total income
5x + 2(160−x) = 680
5x + 320 − 2x = 680
3x = 360   Answer: 120 adult tickets;
x = 120   40 student tickets

See comment that follows:

Notice that we are given that there are three times as many adult tickets as student tickets, but did not use it. Another approach: Let x = the number of student tickets and 3x = the number of adult tickets. You will arrive at the equation 2(x) + 5(3x) = 680, which will yield the same answers as above. Working the problem this way, you do not use the fact that there are 160 tickets. The important thing is to be sure that all the information given in the problem, whether you use it or not, is consistent with the answer.

13. Let p be the amount of money invested at 6%.

	principal	rate	time	dollars of interest
6% investment	p	0.06	1	0.06p
8% investment	5000 − p	0.08	1	0.08(5000 − p)

Notice that p dollars are invested at 6%, so there must be 5000 − p dollars left to invest at 8%. The basic formula used here to complete the right-most column of the table is:

$$(\text{principal})(\text{rate})(\text{time}) = \text{dollars of interest}$$

The principals are the amounts invested, which in this problem are dollars. The rate is expressed as a decimal value in calculations; that is, write 0.06 and not 6 when using a rate of 6%. The time is the amount of time in years (since the interest rate is an *annual* one) that the principal stays in the account. The last column in the table contains the *dollar amount of interest earned*, which should not be confused with the interest rate.

Generally speaking when making such a table, you will not use all the facts given in the problem. Such remaining facts are often used after making the table, to *write the equation*. Rereading the problem, we see that the combined interest earned at the end of the year is $318.00. Adding the two dollar amounts of interest earned will yield 318:

$$0.06p + 0.08(5000 - p) = 318$$

You can save some work if you multiply both sides by 100, but be sure you do this correctly:

$$6p + 8(5000 - p) = 31800$$
$$6p + 40000 - 8p = 31800$$
$$-2p = -8200$$
$$p = 4100$$
$$5000 - p = 5000 - 4100 = 900$$

$4100 was invested at 6%; $900 was invested at 8%.

section   4.4

17.   Let n be the amount of nuts costing $1.90 per pound.   That leaves 25 − n pounds of the less expensive nuts to use.

	pounds of nuts	cost per pound	cost in dollars
expensive nuts	n	1.90	1.90n
cheap nuts	25 − n	1.65	1.65(25 − n)

The total cost of the two amounts of nuts is $44.50.

$1.90n + 1.65(25 - n) = 44.50$ Multiply both sides by 100.
$190n + 165(25 - n) = 4450$ (note 25 − n was not multiplied by 100)
$190n + 4125 - 165n = 4450$
$25n = 325$
$n = 13$ We use 13 pounds of nuts costing $1.90/lb,
$25 - n = 25 - 13 = 12$ and 12 pounds of nuts costing $1.65/lb.

21a.   Note diagrams provided in your book.
        Let t be the time needed for the boats to be 30 km apart.

	rate	time	distance
slower boat	8 km/hr	t	
faster boat	12 km/hr	t	

If you were tempted to use second variable here for distance, STOP! We don't have the machinery to solve a problem with two variables at this stage in the course, so let's try something else.

With most of these problems you filled out the columns this way: one of the columns was filled out using your chosen variable, one was filled out using some given constants (like rates, in this problem; in another problem it might be something else) and the third was filled out using some basic principle or formula.   A basic principle relating rate and time is:
rate × time = distance.   **Use this formula to fill in the last column in the table:**

	rate	time	distance
slower boat	8 km/hr	t	8t
faster boat	12 km/hr	t	12t

The sum of the two distances is 30 km, so

$8t + 12t = 30$

$20t = 30$

$t = 1.5$ It will take 1.5 hours.

21b. This problem has the same chart as 21a; the difference will be in writing the equation. Since the difference of their distances is 30 km, we must subtract the two distances. Be sure you subtract in the correct order!

larger distance − smaller distance = 30

$12t − 8t = 30$ The larger distance is the one for the *faster* boat.

$4t = 30$

$t = 7.5$ It will take 7.5 hours.

25.

	principal	rate	time	dollars of interest
6% investment	p	0.06	1	0.06p
8% investment	p	0.08	1	0.08p
9% investment	120000-2p	0.09	1	0.09(120,000 − 2p)

Let p be the amount of dollars invested at 6% and the amount invested at 8%. Notice that this leaves $120,000 − 2p$ dollars to be invested at 9%.

$0.06p + 0.08p + 0.09(120,000 − 2p) = 8800$ Multiply both sides by 100.

$6p + 8p + 9(120,000 − 2p) = 880,000$

$14p + 1,080,000 − 18p = 880,000$

$− 4p = − 200,000$

$p = 50,000$ $120,000 − 2p = 120,000 − 2(50,000) = 20,000$

$50,000 invested at 6%, $50,000 invested at 8%, and $20,000 invested at 9%.

29. Let x be the amount of liters of pure water to be added.

	liters of soln	percent* water	liters of water
concentrate	12	0.15	0.15(12)
added water	x	1.00	x
mixture	12 + x	0.5	0.5(12 + x)

*We have expressed the percentages as their equivalent decimal values, which is what we must use in the calculations.

Since the mixture is made up of concentrate and added water, the sum of the liters of water in the concentrate and the liters of added water is the amount of water that will be in the final mixture:

$0.15(12) + x = 0.50(12+x)$ Multiply both sides by 100 to remove decimals.

$15(12) + 100x = 50(12 + x)$

$180 + 100x = 600 + 50x$

$50x = 420$

$x = 8.4$ We would need to add 8.4 L of pure water.

section 4.4

1. See answer section in the text.
2. means: m + 2, 8   extremes: m + 1, 8
3. See answer section in the text.
4. means: a, 2   extremes: 5a + 2, 1

If you had trouble with #1 − 4, review the definitions of means and extremes of proportions.

5. $\dfrac{3}{5} = \dfrac{18}{v}$ You may cross-multiply a proportion.

  $3v = 5(18)$

  $v = \dfrac{5(18)}{3} = 30$

6. $\dfrac{4}{x+1} = \dfrac{10}{2x+5}$ You may cross-multiply a proportion.

  $4(2x + 5) = 10(x + 1)$
  $8x + 20 = 10x + 10$
  $-2x = -10$
  $x = 5$

7. $\dfrac{3x+1}{14} = \dfrac{9x-2}{35}$ You may cross-multiply a proportion.

  $35(3x + 1) = 14(9x - 2)$
  $105x + 35 = 126x - 28$
  $-21x = -63$

  $x = \dfrac{-63}{-21} = 3$

8. $\dfrac{5}{0.5x+6} = \dfrac{7}{1.5x+2}$ You may cross-multiply a proportion.

  $5(1.5x + 2) = 7(0.5x + 6)$
  $7.5x + 10 = 3.5x + 42$
  $4x = 32$
  $x = 8$

9. $y = kx$ First find k.

  $28 = k(8)$

  $k = \dfrac{28}{8} = \dfrac{7}{2}$

  $y = \dfrac{7}{2}x$ This is the complete equation, with the value of k included.

  $y = \dfrac{7}{2}(10) = 35$

Chapter Four Review

10. $y = kx$ First find k.

        $4 = k(14)$

        $k = \frac{2}{7}$

        $y = \frac{2}{7}x$ This is the complete equation, with the value of k included.

        $21 = \frac{2}{7}x$

        $x = 21(\frac{7}{2})$

        $x = \frac{147}{2}$ or $73\frac{1}{2}$ or 73.5

11. n the number
        $7n$ seven times the number
        $9n$ nine times the number

        $7n + 8 = 9n + 2$
        $-2n = -6$
        $n = 3$

12. n the number
        $\frac{2}{3}n$ two-thirds of the number
        $n - 6$ six less than the number (reduce the number by 6)

        $\frac{2}{3}n = n - 6$

        $3(\frac{2}{3}n) = 3(n - 6)$

        $2n = 3n - 18$

        $n = 18$

13.   Let x be the grade on the fourth test.

        $\frac{83 + 91 + 78 + x}{4} = 85$

        $\frac{252 + x}{4} = 85$

        $252 + x = 340$

        $x = 88$ The student must earn a grade of 88 on the fourth test.

Chapter Four Review

14.   n   the first integer
       n + 1   the next consecutive integer

$$n + (n+1) = 39$$
$$2n + 1 = 39$$
$$2n = 38$$
$$n = 19 \qquad \text{The two numbers are 19 and 20.}$$
$$n + 1 = 20$$

15.   n   the first integer
       n + 2   the next consecutive even integer

$$5n - 3(n + 2) = n + 10$$
$$5n - 3n - 6 = n + 10$$
$$2n - 6 = n + 10$$
$$n = 16 \qquad \text{The numbers are 16 and 18.}$$
$$n + 2 = 18$$

16.   n   the first integer
       n + 2   the second consecutive odd integer
       (n + 2) + 2   the third consecutive odd integer; this simplifies to n + 4

$$n + (n + 2) + (n + 4) = 99$$
$$3n + 6 = 99$$
$$3n = 93$$
$$n = 31 \qquad \text{The consecutive odd integers are 31, 33 and 35.}$$

17.   x   the larger angle
       x − 11   the smaller angle

$$x + (x - 11) = 180 \qquad \text{The measures of supplementary angles add to } 180°.$$
$$2x - 11 = 180$$
$$2x = 191$$
$$x = 95.5$$
$$95.5 - 11 = 84.5 \qquad \text{The angles are } 95.5° \text{ and } 84.5°.$$

As a check, be sure that $95.5° + 84.5° = 180°$. √

18.   x   the larger angle
       x − 6   the smaller angle

$$x + (x - 6) = 90 \qquad \text{The measures of complementary angles add to } 90°.$$
$$2x - 6 = 90$$
$$2x = 96$$
$$x = 48$$
$$48 - 6 = 42 \qquad \text{The angles are } 48° \text{ and } 42°.$$

As a check, be sure that $48° + 42° = 90°$. √

Chapter Four Review

19.   w   width of the rectangle
          2w + 4   length of the rectangle

The perimeter of a rectangle is the
sum of the lengths of the sides.

$2w + 2(2w + 4) = 50$ You can save a bit of work by dividing both sides by 2.
$w + 2w + 4 = 25$
$3w = 21$
$w = 7 \; ; \; 2(7) + 4 = 18$ The width is 7 cm and the length is 18 cm.

20.   x   length of the shorter piece
          x + 52   length of the longer piece

$x + (x + 52) = 408$ The sum of the two pieces is the total length of the
                                  cable.

$2x + 52 = 408$
$2x = 356$

$x = 178$ The shorter piece is 178 meters in length.
$178 + 52 = 230$ The longer piece is 230 meters in length.

21.   d   the shorter distance
          d + 3   the longer distance

$d + (d + 3) = 17$ The sum of the distances is 17 km.
$2d + 3 = 17$
$2d = 14$

$d = 7$ The shorter distance is 7 km.
$d + 3 = 10$ The longer distance is 10 km.

22.   n   the number of nickels
          2n   the number of dimes

5n   the value of the nickels in cents (we use cents to
                                     avoid decimals; but you don't have to do it
                                     this way)
10(2n)   the value of the dimes in cents

$5n + 10(2n) = 11200$ The total amount of money is 11200 *cents.*
$25n = 11200$

$n = 448$ There are 448 nickels and 896 dimes.
$2n = 896$

23. a   the number of adult tickets
    400 − a   the number of child tickets

$5a + 3(400 - a) = 1750$ The total number of tickets is 1750.
$5a + 1200 - 3a = 1750$
$2a = 550$
$a = 275 \ ; \ 400 - 275 = 125$ They sold 275 adult tickets and 125 child tickets.

24.   We will be using the formula $D = R \cdot T$ in each case.

(a) The boats travel in opposite directions.

      slower boat   faster boat
       7 km/h   11 km/h
<------------ • ------------------------------------->

Let t be the number of hours before the boats are 27 km apart if they are traveling in opposite directions. This means that the sum of their distances is 27 km.

distance for the slower boat: 7t } $7t + 11t = 27$
distance for the faster boat: 11t

$18t = 27$

$t = \dfrac{27}{18} = \dfrac{3}{2}$ or $1\dfrac{1}{2}$ It will take $1\dfrac{1}{2}$ hours.

(b) The boats travel in the same direction. The faster boat goes farther.

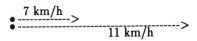
         •------------->
     •------------------------------------->
            11 km/h

Let t be the number of hours before the boats are 27 km apart if they are traveling in the same direction. This means that the faster boat will be ahead by 27 km, and we will set the *difference* of their distances to be 27.

distance for the slower boat: 7t } $11t - 7t = 27$
distance for the faster boat: 11t

If you are not sure why we subtracted in this order, keep in mind that the faster boat will go a longer distance than the slower boat, so 11t is larger than 7t. We must subtract smaller from larger to get the *positive* difference of 27.

$4t = 27$

$t = \dfrac{27}{4}$ or $6\dfrac{3}{4}$ It will take $6\dfrac{3}{4}$ hours.

Chapter Four Review

25.   Let x = the number of liters of a solution of 8% sulfuric acid and
       40 − x = the number of liters of a solution of 12% sulfuric acid.

	strength of solution	amount of solution	amount of sulfuric acid
8% solution	0.08	x	0.08x
12% solution	0.12	40 − x	0.12(40 − x)
9.5% solution	0.095	40	0.095(40)

$0.08x + 0.12(40 − x) = 0.095(40)$
$0.08x + 4.8 − 0.12x = 3.8$
$−0.04x = −1$
$x = 25$ liters of 8% solution

26.   Let p be the amount put into the savings account.

	principal	rate	time	dollars of interest
6% investment	p	0.06	1	0.06p
9% investment	6000 − p	0.09	1	0.09(6000 − p)

$$0.06p + 0.09(6000 − p) = 510$$
You can save some work if you multiply both sides by 100, but be sure you do this
correctly:
$6p + 9(6000 − p) = 51000$
$6p + 54000 − 9p = 51000$

$−3p = −3000$
$p = 1000$ **$1000 was invested in the savings account.**

27.   Let d be the distance the spring is stretched and F is the force on the spring.
       $F = kd$ is the general formula, with k being the constant of variation.
       Using $F = 8$ and $d = 6$, we will find k: $8 = k(6); k = \frac{4}{3}$

       The formula is $F = \frac{4}{3} d$.

       (a) Find d given $F = 12$.   (b)   Find F, given $d = 7.5$.
       $12 = \frac{4}{3}(d)$ $F = \frac{4}{3}(7.5) = 10$ Newtons
       $d = 9$ cm

       Notice that you only needed to find k once. The general formula $F = \frac{4}{3} d$ holds for this spring
       for any questions of the type (a) or (b).

28. The ratio is $\dfrac{1 \text{ inch}}{20 \text{ miles}}$. Let d be the distance, in miles, between Louisville and Columbus.

$\dfrac{1}{20} = \dfrac{10.5}{d}$ Cross multiply.

$d = 20(10.5) = 210$ The distance is 210 miles.

29. Let x = the number of defective tubes in the shipment of 500.

The ratio used will be $\dfrac{\text{defective tubes}}{\text{all tubes}}$.

$\dfrac{3}{75} = \dfrac{x}{500}$ Cross multiply.

$3(500) = 75x$

$1500 = 75x$

$x = 20$ We would expect 20 defective tubes.

30. We will start with $\dfrac{a}{3} = \dfrac{7.5}{5}$. Solve for a and you will get a = 4.5.

Before we go on, some comments:

(1) Notice that you could also have written $\dfrac{3}{a} = \dfrac{5}{7.5}$, but $\dfrac{3}{a} = \dfrac{7.5}{5}$ would be incorrect.

(2) It is also true that $\dfrac{a}{3} = \dfrac{b}{9}$, but this equation involves two unknowns, a and b, and we would not be able to determine either using that equation.

Continuing, we can use the same ratio $\dfrac{7.5}{5}$ that we used to find a, to help us find b and c:

$\dfrac{b}{9} = \dfrac{7.5}{5}$ gives b = 13.5

$\dfrac{c}{5} = \dfrac{7.5}{5}$ gives c = 7.5

31. The ratio is $\dfrac{\text{height of the plant}}{\text{length of the shadow}}$. Let h be the height of the tree.

$\dfrac{0.64}{6.72} = \dfrac{h}{46.2}$ Solving gives h = 4.4, so the tree is 4.4 meters tall.

(This problem can also be done by drawing the tree as a vertical line segment to the right of the shrub, and then using similar triangles.)

32.   Let a be the number of mL of chemical A.

mL of chemical A	3	6	15	1	a
mL of chemical B	7	14	35	$\frac{7}{3}$	$\frac{7}{3}$ a
mL of mixture	10	20	50	$\frac{10}{3}$	$\frac{10}{3}$ a

$\frac{10}{3}$ a $= 45$

a $= 13.5$ Use 13.5 mL of chemical A.

$\frac{7}{3}$ $(13.5) = 31.5$ Use 31.5 mL of chemical B.

33.

larger number	58	55	50	40	n
smaller number	0	3	8	18	58−n
their sum	58	58	58	58	58
their difference	58	52	42	22	2n−58

The difference of the numbers is 14:

$2n - 58 = 14$

$2n = 72$

$n = 36$ (larger number) $58 - n = 58 - 36 = 22$ (smaller number)

34.   n   the first even integer
       n + 2   the second consecutive even integer
       n + 4   the third consecutive even integer

       n + (n + 4) = n + 2
       2n + 4 = n + 2
       n = −2   The three consecutive even integers
                                  are −2, 0, and 2.

35.   a   the number of degrees in angle A
       a + 1   the number of degrees in angle B
       a + 2   the number of degrees in angle C

       a + (a + 1) + (a + 2) = 180   The sum of the measures of the angles of a
       3a + 3 = 180   triangle is 180°.
       3a = 177
       a = 59   The angles are 59°, 60°, and 61°.

1a.   Let the number be represented by n.

three times the number: 3n   five less than three times the number: 3n − 5
seven times the number: 7n   fifteen less than seven times the number: 7n − 15

$3n - 5 = 7n - 15$
$3n = 7n - 10$
$-4n = -10$

$n = \frac{5}{2}$ or $2\frac{1}{2}$

1b.   Let the two consecutive odd integers be x and x + 2. Their sum will be x + (x + 2).

$3(x + x + 2) = 84$
$3(2x + 2) = 84$ Notice that 84 is divisible by 3.
$2x + 2 = 28$
$2x = 26$
$x = 13$ $x + 2 = 15$ The numbers are 13 and 15.

2a.   Let n be the number.

one more than the number: n + 1   the ratio of that quantity to three: $\frac{n + 1}{3}$

one less than the number: n − 1   the ratio of that quantity to two: $\frac{n - 1}{2}$

$\frac{n + 1}{3} = \frac{n - 1}{2}$

$(n + 1)(2) = (3)(n - 1)$
$2n + 2 = 3n - 3$
$2n + 5 = 3n$
$n = 5$

2b.   The ratio is $\frac{\text{ounces of gold}}{\text{tons of ore}}$.

Let x = the number of ounces of gold
          expected in 189 tons of ore.

$\frac{4}{14} = \frac{x}{189}$

$4(189) = 14x$

$x = \frac{4(189)}{14} = \frac{2(2)(7)(27)}{2(7)} = 54$

54 ounces of gold can be expected per week.

3a. $y = kx$ Let x = 20, y = 12.
$12 = k(20)$

$k = \frac{12}{20} = \frac{3}{5}$; $y = \frac{3}{5}x$

(i) $y = \frac{3}{5}(15) = 3(3) = 9$

(ii) $15 = \frac{3}{5}x$; $15(5) = 3x; 25 = x$

page 79

Chapter Four Mastery Test

3b.   Let n = the number of bass in the lake, and s = the surface area of the water.

$$n = ks$$ Let $s = \frac{1}{2}$ and n = 250 to find k.

$$250 = k\left(\frac{1}{2}\right)$$
$$500 = k$$

The complete equation is n = 500s.

$$n = 500(25) = 12,500$$

There are 12,500 bass.

4a.

larger number	3	10	20	30	a
smaller number	0	7	17	27	a − 3
their difference	3	3	3	3	3
their sum	3	17	37	57	2a − 3

$$2a - 3 = 73$$
$$2a = 76$$
$$a = 38 \; ; \quad a - 3 = 38 - 3 = 35$$

The numbers are 38 and 35.

4b.

Flight Attendants	5	10	20	30	a
Pilots	2	4	8	12	$\frac{2}{5}a$
Total	7	14	28	42	$\frac{7}{5}a$

$$\frac{7}{5}\,a = 154$$

$$a = \frac{5}{7}(154) = 5(22) = 110$$

There were 110 flight attendants on duty that day.

Chapter Four Mastery Test

5a. Let n = the number of one-dollar bills and 40 − n = the number of five-dollar bills.

total value of the one-dollar bills = (n)(1) = n
total value of the five-dollar bills = (40 − n)(5) = 200 − 5n

$100 = n + (200 − 5n)$
$100 = 200 − 4n$
$−100 = −4n$
$25 = n;\ 40 − n = 15$ There are 25 one-dollar bills and 15 five-dollar bills.

5b. Let ℓ = the number of liters of the 10% solution needed.

	Strength •	Solution =	Amt of Nitric Acid
10% solution	0.10	ℓ	0.10ℓ
5% solution	0.05	$25 − \ell$	$0.05(25−\ell)$
mixture	0.09	25	$0.09(25)$

$0.10\ell + 0.05(25 − \ell) = 0.09(25)$ Multiply both sides by 100 to remove decimals.
$10\ell + 5(25 − \ell) = 9(25)$
$10\ell + 125 − 5\ell = 225$
$5\ell = 100$
$\ell = 20$ Use 20 liters of the 10% solution and 5 liters of the 5% solution.
$25 − \ell = 5$

Problem Set 5. 1 Note: See answer section in the text for graphs.

1. See answer section in the text.
9. See answer section in the text.
17. See answer section in the text.

5. See answer section in the text.
13. See answer section in the text.

21. $y - 7 \leq -3$
$y - 7 + 7 \leq -3 + 7$
$y \leq 4$

25. $5y - 8 < 4y - 8$
$5y < 4y$ subtract 4y
$y < 0$

29. $-3 > y$ means $y < -3$

$\boxed{\text{If } a > b \text{ then } b < a.}$

33. $5m - 7 \leq 6m - 9$
$5m + 2 \leq 6m$ subtract 5m
$2 \leq m$ or $m \geq 2$

37. $5 - 9v < 6 - 10v$
$5 + v < 6$
$v < 1$

41. $5(x - 2) \leq 4(x - 2)$
$5x - 10 \leq 4x - 8$
$5x \leq 4x + 2$
$x \leq 2$

45. We can estimate the solution by writing

Answer: A

$x - 0.7 > 0.4$
$x > 0.4 + 0.7$
$x > 1.1$

49. $5(w - 2) + 14 > 4(w + 3)$
$5w - 10 + 14 > 4w + 12$
$5w + 4 > 4w + 12$
$w + 4 > 12$
$w > 8$

53. $4(6m + 7) - 2m \geq 7(3m + 1) + 1$
$24m + 28 - 2m \geq 21m + 7 + 1$
$22m + 28 \geq 21m + 8$
$m + 28 \geq 8$
$m \geq -20$

57. $3y - 4(5 + 2y) > y - 6(3 + y)$
$3y - 20 - 8y > y - 18 - 6y$
$-5y - 20 > -5y - 18$
$-20 > -18$
False; no solution.

61. five minus two times x: $5 - 2x$
seven minus three times x: $7 - 3x$
$5 - 2x \leq 7 - 3x$
$5 + x \leq 7$
$x \leq 2$

65. the quantity m plus five $m + 5$
three times that quantity $3(m + 5)$
twelve more than twice m $12 + 2m$

$3(m + 5)$ is at most $12 + 2m$; this means that $3(m + 5)$ is less than or equal to $12 + 2m$.

$3(m + 5) \leq 12 + 2m$
$3m + 15 \leq 12 + 2m$
$m + 15 \leq 12$
$m \leq -3$

69. $x + y < z$ Subtract y from both sides.

   $x < z - y$

Problem Set 5.2   Note: See answer section in the text for graphs.

1. $2x \geq 4$

   $\frac{2x}{2} \geq \frac{4}{2}$

   $x \geq 2$

5. $\frac{1}{2}v < -4$

   $2(\frac{1}{2}v) < 2(-4)$

   $v < -8$

9. $-\frac{3}{4}w < -12$

   $(-\frac{4}{3})(-\frac{3}{4}w) > (-\frac{4}{3})(-12)$

   $w > \frac{4(12)}{3}$

   $w > 16$

13. $-\frac{n}{4} > \frac{1}{2}$

   $(-4)(-\frac{n}{4}) < (-4)\frac{1}{2}$

   $n < -2$

17. $-\frac{t}{3} \geq 0$

   $(-3)(-\frac{t}{3}) \leq (-3)(0)$

   $t \leq 0$

21. $-3m + 4 < 7$

   $-3m < 3$

   $m > -1$

25. $9x + 7 > 5x - 13$

   $4x + 7 > -13$

   $4x > -20$

   $x > -5$

29. $5 - 9y \leq 19 - 2y$

   $5 - 7y \leq 19$

   $-7y \leq 14$

   $y \geq -2$

33. $2(3t - 4) > 2(t - 2)$

   $6t - 8 > 2t - 4$

   $4t - 8 > -4$

   $4t > 4$

   $t > 1$

37. $3(2y - 1) \leq 5(3y + 4) + 4$

   $6y - 3 \leq 15y + 20 + 4$

   $6y - 3 \leq 15y + 24$

   $6y \leq 15y + 27$

   $-9y \leq 27$

   $y \geq -3$

41.   See answer section in the text. You may add or subtract the same quantity to both sides of an inequality without altering the inequality sign.

45. See answer section in the text. You may add or subtract the same quantity to both sides of an inequality without altering the inequality sign.

49. We may estimate this problem by rounding *down* to the nearest whole number:
$5x \leq 35$
$x \leq 7$ Answer: C

53.
$-3(5 - 2x) > -2(5x + 1) + 3$
$-15 + 6x > -10x - 2 + 3$
$-15 + 6x > -10x + 1$
$6x > -10x + 16$
$16x > 16$
$x > 1$

57. the quantity x plus 5: $x + 5$
twice that quantity: $2(x + 5)$
$2(x + 5) < 6$
$2x + 10 < 6$
$2x < -4$
$x < -2$

61.
$F \geq 41$ Start out with a simple statement that you know is true.

$\frac{9}{5}C + 32 \geq 41$ Since $F = \frac{9}{5}C + 32$, replace F with $\frac{9}{5}C + 32$.

$\frac{9}{5}C \geq 9$

$C \geq 9(\frac{5}{9})$

$C \geq 5$ The temperature must be at least 5° Celsius.

65.
$2x + 3y < 5y - 2x$ Subtract 3y from both sides.

$2x < 2y - 2x$ Add 2x to both sides.

$4x < 2y$ Divide both sides by 4.

$x < \frac{2y}{4}$ Reduce the fraction on the right.

$x < \frac{y}{2}$

69. Divide -103.73 by 4.715. Answer: $x \geq -22$

Problem Set 5.3   Note: See answer section in the text for graphs.

1. See answer section in the text.   5. See answer section in the text.

9. See answer section in the text.   13. See answer section in the text.

17. These are the numbers that are between -2 and approximately 1.4, including -2 but not including $\sqrt{2}$. (The estimates $\sqrt{2} \approx 1.4$ and $\sqrt{3} \approx 1.7$ are useful and can be found with your calculator.) See answer section in the text.

21. $-7 \le n - 8 \le -6$

$-7 + 8 \le n - 8 + 8 \le -6 + 8$

$1 \le n \le 2$

25. $-4 < -4x \le 8$

$\dfrac{-4}{-4} > \dfrac{-4x}{-4} \ge \dfrac{8}{-4}$

$1 > x \ge -2$
$-2 \le x < 1$

29. $3 \le 2w + 3 < 11$

$3 \le 2w + 3 - 3 < 11 - 3$

$0 \le 2w < 8$

$\dfrac{0}{2} \le \dfrac{2w}{2} < \dfrac{8}{2}$

$0 \le w < 4$

33. $-4 < 6 - 5t < 11$

$-4 - 6 < -5t < 11 - 6$

$-10 < -5t < 5$

$2 > t > -1$

preferred: $-1 < t < 2$

37. $-2 < \dfrac{m - 3}{2} \le -1$

$-4 < m - 3 \le -2$

$-1 < m \le 1$

41. $-2 \le -3x + 1 < 0$

$-3 \le -3x < -1$

$1 \ge x > \dfrac{1}{3}$

preferred: $\dfrac{1}{3} < x \le 1$

45. The average is the sum of the three prices divided by 3. Let p be the price of the stock on the third day.

$\dfrac{75 + 78 + p}{3} \ge 80$

$75 + 78 + p \ge 240$
$153 + p \ge 240$
$p \ge 87$ 　　　　　The price on the third day must be at least \$87.

49. $v + 1 < v$ 　　　　Subtract v from both sides.
$1 < 0$ 　　　　This is a contradiction. There is no solution.

53. $\dfrac{2}{3}(t + 2) \le \dfrac{5}{6}(t + 3)$

$6\left(\dfrac{2}{3}(t + 2)\right) \le 6\left(\dfrac{5}{6}(t + 3)\right)$

$4(t + 2) \le 5(t + 3)$
$4t + 8 \le 5t + 15$
$-t \le 7$
$t \ge -7$

57. $10 < 5(w + 3) \le 35$ 　　Divide by 5.
$2 < w + 3 \le 7$
$-1 < w \le 4$

61. You can solve the given inequality by the usual method, but as a mental exercise, let us proceed as follows:
We are given that a quantity (negative one-half of x) is between -4 and 6. Twice that quantity (negative x) would then fall between -8 and 12. If we change the sign of the quantity, we are now speaking of x, which would then fall between 8 and -12.
Answer: C

65. Let ℓ be the length of the rectangle.
$2\ell + 2w$ is the perimeter, which must be at most 44 meters.

$2\ell + 2(9) \leq 44$ Use 9 for w.
$2\ell + 18 \leq 44$
$2\ell \leq 26$
$\ell \leq 13$ The length must be at most 13 meters.

69. The difference between the average annual claim and the yearly premium is $19, is a loss to the company. The company can only make a profit if the investment yields more than $19 per policy. We must take a closer look at what the investment can yield.
The insurance company collects the yearly premium, $390, and invests it. The return is expected to be between 4.3% and 4.8% of $390.
Let d be the dollar amount of return on the investment of $390.

$$0.043(390) \leq d \leq 0.048(390)$$
$$16.77 \leq d \leq 18.72$$

Since the investment can yield at most $18.92, which is less than $19.00, the company will lose money on this policy.

73. $-6 \leq -3(x - y) < 9$ Divide each member by -3. Reverse inequalities.
$2 \geq x - y > -3$
$2 + y \geq x > -3 + y$

$\left.\begin{array}{l} -3 + y < x \leq 2 + y \\ \text{or: } y - 3 < x \leq y + 2 \end{array}\right\}$ These are equivalent.

Solutions To All Review Exercises $-$ Chapter Five

1. $x > -2$ 2. $x \leq -2$

3. $x \geq 0$ 4. $x < 0$

5. $-1 < x \leq 2$ 6. $1 < x < 3$

7. $-2 \leq x \leq 2$ 8. $3 \leq x < 5$

9. $x \quad > -1.5$
      $-6 \quad > -1.5$ False. -6 is not a solution.

$-3 > -1.5$	False.	-3	is not a solution.
$-1.5 > -1.5$	False.	-1.5	is not a solution.
$-\frac{1}{3} > -1.5$	True.	$-\frac{1}{3}$	is a solution.
$0 > -1.5$	True.	0	is a solution.
$1 > -1.5$	True.	1	is a solution.
$\sqrt{5} > -1.5$	True.	$\sqrt{5}$	is a solution. (Note: $\sqrt{5} \approx 2.2$)
$4 > -1.5$	True.	4	is a solution.

10. $x < 0$

$-6 < 0$	True.	-6	is a solution.
$-3 < 0$	True.	-3	is a solution.
$-1.5 < 0$	True.	-1.5	is a solution.
$-\frac{1}{3} < 0$	True.	$-\frac{1}{3}$	is a solution.
$0 < 0$	False.	0	is not a solution.
$1 < 0$	False.	1	is not a solution.
$\sqrt{5} < 0$	False.	$\sqrt{5}$	is not a solution.
$4 < 0$	False.	4	is not a solution.

11. $x \leq 0$
Please see the solution to #10. The only difference will occur when we test 0:
$0 \leq 0$ True   0 is a solution
The numbers that are solutions are -6, -3, -1.5, $-\frac{1}{3}$, <u>and</u> 0.

12. $x \geq -\frac{1}{7}$

$-6 \geq -\frac{1}{7}$	False.	-6	is not a solution.
$-3 \geq -\frac{1}{7}$	False.	-3	is not a solution.
$-1.5 \geq -\frac{1}{7}$	False.	-1.5	is not a solution.
$-\frac{1}{3} \geq -\frac{1}{7}$	False.	$-\frac{1}{3}$	is not a solution.
$0 \geq -\frac{1}{7}$	True.	0	is a solution.
$1 \geq -\frac{1}{7}$	True.	1	is a solution.
$\sqrt{5} \geq -\frac{1}{7}$	True.	$\sqrt{5}$	is a solution.
$4 \geq -\frac{1}{7}$	True.	4	is a solution.

13. $-1 < x < 1$ This represents the set of numbers between -1 and 1. The only numbers in the list between -1 and 1 are $-\frac{1}{3}$ and 0. Notice that 1 is not included.

14. $-3 \leq x \leq 3$ This represents the set of numbers between -3 and 3, including -3 and 3. The numbers from the list satisfying these requirements are -3, -1.5, $-\frac{1}{3}$, 0, 1, $\sqrt{5}$. Note: $\sqrt{5} \approx 2.2$.

15. $v + 7 < 5$ See answer section in the text.
$v < -2$

16. $v - 11 > -10$
$v > 1$

#16

17. $-10 \geq 2x$ See answer section in the text.
$2x \leq -10$
$x \leq -5$

18. $-3x \leq 12$
$x \geq -4$

#18

19. $3y + 7 > 5y + 13$ See answer section in the text.
$-2y > 6$
$y < -3$

20. $7 \geq 7 - 9y$
$9y + 7 \geq 7$
$9y \geq 0$
$y \geq 0$

#20

21. $-\frac{x}{2} < \frac{1}{4}$

$-2\left(-\frac{x}{2}\right) > -2\left(\frac{1}{4}\right)$

$x > -\frac{1}{2}$

22. $\frac{3x}{7} + \frac{4}{5} > \frac{3x}{5} + \frac{2}{7}$ The LCD is 35.

$35\left(\frac{3x}{7} + \frac{4}{5}\right) > 35\left(\frac{3x}{5} + \frac{2}{7}\right)$

$5(3x) + 7(4) > 7(3x) + 5(2)$
$15x + 28 > 21x + 10$
$-6x > -18$
$x < 3$

Chapter Five Review

23.	$7y + 4y \geq 2(3y + 8)$ Distribute to remove parentheses.
$7y + 14 \geq 6y + 16$ Collect like terms.
$y \geq 2$

24.	$-5(2y - 2) \leq 7 - 11y$ Distribute to remove parentheses.
$-10y + 10 \leq 7 - 11y$ Collect like terms.
$y + 10 \leq 7$
$y \leq -3$

25.	$4(2w - 3) \leq 2(3w - 8)$ Distribute to remove parentheses.
$8w - 12 \leq 6w - 16$ Collect like terms.
$8w \leq 6w - 4$
$2w \leq -4$
$w \leq -2$

26.	$3(6w - 4) \geq 5(3w - 6)$ Distribute to remove parentheses.
$18w - 12 \geq 15w - 30$ Collect like terms.
$3w - 12 \geq -30$
$3w \geq -18$
$w \geq -6$

27.	$2(3t - 4) > 3(t - 6) + 1$ Distribute to remove parentheses.
$6t - 8 > 3t - 18 + 1$ Collect like terms.
$6t - 8 > 3t - 17$
$3t - 8 > -17$
$3t > -9$
$t > -3$

28.	$2(11t - 3) < 5(3t + 2) - 20$ Distribute to remove parentheses.
$22t - 6 < 15t + 10 - 20$ Collect like terms.
$22t - 6 < 15t - 10$
$7t - 6 < -10$
$7t < -4$
$t < -\frac{4}{7}$

29.	$5(x - 4) + 6 < x + 4 - 30$ Distribute to remove parentheses.
$5x - 20 + 6 < x - 26$ Collect like terms.
$5x - 14 < x - 26$
$5x < x - 12$
$4x < -12$
$x < -3$

Chapter Five Review

30. $7(x - 3) + 6 \geq 12(x + 4) + 2$ Distribute to remove parentheses.
$7x - 21 + 6 \geq 12x + 48 + 2$ Collect like terms.
$7x - 15 \geq 12x + 50$
$7x \geq 12x + 65$
$-5x \geq 65$ BE CAREFUL.   Reverse the inequality.
$x \leq -13$

31. $4(y - 1) - 7(y + 1) \leq -3(y + 2) - 5$ Distribute to remove parentheses.
$4y - 4 - 7y - 7 \leq -3y - 6 - 5$
$-3y - 11 \leq -3y - 11$ This is an identity.
$-11 \leq -11$ The solution set is all real numbers.

32. $1 - 4(y - 2) \geq 3(y - 7) - 5(4 - y)$ Distribute to remove parentheses.
$1 - 4y + 8 \geq 3y - 21 - 20 + 5y$ Collect like terms.
$9 - 4y \geq 8y - 41$
$-4y \geq 8y - 50$
$-12y \geq -50$ BE CAREFUL. Reverse the inequality.
$y \leq \frac{25}{6}$ or $4\frac{1}{6}$

33. $\frac{v}{2} - \frac{3v + 4}{4} > \frac{v + 40}{4}$ The LCD is 4. Multiply both sides by 4.

$4\left(\frac{v}{2} - \frac{3v + 4}{4}\right) > 4\left(\frac{v + 40}{4}\right)$

$2v - (3v + 4) > v + 40$
$2v - 3v - 4 > v + 40$
$-v - 4 > v + 40$
$-4 > 2v + 40$
$-44 > 2v$
$-22 > v$ or $v < -22$

34. $\frac{v}{8} + 6 > \frac{v}{12} + 5$ The LCD is 24. Multiply both sides by 24.

$24\left(\frac{v}{8} + 6\right) > 24\left(\frac{v}{12} + 5\right)$

$3v + 144 > 2v + 120$
$v + 144 > 120$
$v > -24$

35. $9 < x + 7 \leq 13$ Subtract 7 from each member.
$2 < x \leq 6$.

36. $0 \leq \frac{4}{5}x < 20$ Multiply by $\frac{5}{4}$.

$0 \leq x < 25$

37. $42 \leq \frac{-3m}{7} < 60$ Multiply each member by 7.

$7(42) \leq 7\left(\frac{-3m}{7}\right) < 7(60)$

$294 \leq -3m < 420$

$\frac{294}{-3} \geq \frac{-3m}{-3} > \frac{420}{-3}$ BE CAREFUL. Reverse the inequalities.

$-98 \geq m > -140$
$-140 < m \leq -98$ This is preferred.

38. $-5 \leq 9 - x \leq 5$ Subtract 9 from each member.
$-14 \leq -x \leq -4$ Now divide by -1.   Reverse inequalities!
$14 \geq x \geq 4$ We need to rewrite this.
$4 \leq x \leq 14$ This is preferred.

39. $-7 \leq 5x + 3 \leq 13$ Subtract 3 from each member.
$-10 \leq 5x \leq 10$ Divide by 5.
$-2 \leq x \leq 2$

40. $-10 < 5 - 3x < 8$ Subtract 5.
$-15 < -3x < 3$ Divide by -3. Reverse the inequalities!
$5 > x > -1$ This is technically correct, but a poor way to leave the answer.
$-1 < x < 5$ This is better. Notice the numbers are in order from left to right.

41. $11 < 9 - 4(n - 3) \leq 29$ Distribute to remove parentheses.
$11 < 9 - 4n + 12 \leq 29$ Collect like terms.
$11 < 21 - 4n \leq 29$
$-10 < -4n \leq 8$ Divide and reverse the inequalities!

$\frac{-10}{-4} > n \geq \frac{8}{-4}$

$\frac{5}{2} > n \geq -2$ or $-2 \leq n < \frac{5}{2}$ The second form is preferred.

42. $-4 < \frac{2 - 5w}{7} < 4$ Multiply by 7.

$-28 < 2 - 5w < 28$
$-30 < -5w < 26$

$6 > w > -\frac{26}{5}$ or $-\frac{26}{5} < w < 6$ The second form is preferred.

43. $3v + 3v > 5v$ Combine terms on the left.
$6v > 5v$ Subtract 5v from both sides.
$v > 0$ This is a conditional inequality.

44. $3v + 3v \le 5v$ Combine terms on the left.
$6v \le 5v$ Subtract 5v from both sides.
$v \le 0$ This is a conditional inequality.

45. $3(v + 1) < 3v + 4$ Distribute to remove parentheses.
$3v + 3 < 3v + 4$
$3 < 4$ This is an absolute inequality.   Solution: all real numbers

46. $5(3v - 1) > 3(5v - 1)$
$15v - 5 > 15v - 3$
$-5 > -3$ This is a contraction. No solution.

47. $4(x + 3) > 8$
$4x + 12 > 8$
$4x > -4$
$x > -1$

48. $3(x - 17) \le 2(x + 11)$ Notice "the quantity" indicates the need for
$3x - 51 \le 2x + 22$ parentheses.
$x - 51 \le 22$
$x \le 73$

49. three x minus two: $3x - 2$
$7 \le 3x - 2 < 10$
$9 \le 3x < 21$
$3 \le x < 7$

50. Let p be the points needed in the fourth game.

$$\frac{17 + 27 + 18 + p}{4} \ge 20$$

$$\frac{62 + p}{4} = 20$$

$$62 + p = 80$$

$$p \ge 18$$ The player will need to score at least 18 points.

51. Let x be the fourth week's earnings.

$$\frac{560 + 450 + 480 + x}{4} \ge 500$$

$$\frac{1490 + x}{4} \geq 500$$

$$1490 + x \geq 2000$$

$$x \geq 510 \qquad\qquad \text{She must earn at least \$510.}$$

52. Let ℓ be the length of the rectangle.
The perimeter is $\quad 2\ell + 2(12)$
$$= 2\ell + 24$$

$44 < 2\ell + 24 < 64$
$20 < 2\ell < 40$
$10 < \ell < 20$ The length must be between 10 cm and 20 cm.

53. We may approximate the answer by using:
$-5v > 35$
$v < -7$ Answer: B

54. $2v + 19 \leq 47$
$2v \leq 28$
$\quad v \leq 14$ Answer: E

55. $-4x - 5 < 11$
$-4x < 16$
$\quad x > -4$ Answer: D

56. $\quad -9 \leq 10x < 30$
$-0.9 \leq \quad x < 3$ Answer: A

57. $v - w < x - y$
$v - w + y < x$
or $\quad x > v - w + y$ This answer is preferred.

58. $\quad 5x + 3y \geq 7x - 8y$

$-2x + 3y \geq -8y$

$-2x \geq -11y$

$\quad x \leq \frac{11}{2}y$

Chapter Five Review

59.
$$2(3x - 4y) \leq 4(2x + 5y)$$
$$3x - 4y \leq 2(2x + 5y)$$
$$3x - 4y \leq 4x + 10y$$
$$-x - 4y \leq 10y$$
$$-x \leq 14y$$
$$x \geq -14y$$

Why not divide both sides by 2?
Now distribute to remove parentheses.

Don't forget to reverse the inequalities.

60.
$$4y \leq y - 3x < 10y$$
$$3y \leq -3x < 9y$$
$$-y \geq x > -3y$$
$$\text{or} \quad -3y < x \leq -y$$

Subtract y from each member.

This answer may be preferred.

61.
$$5.4x + 2.004 \leq 21.12$$
$$5.4x \leq 19.116$$
$$x \leq 3.54$$

62.
$$-6.25 < 3.9 - 3.5w < -2.75$$
$$-10.15 < -3.5w < -6.65$$
$$2.9 > w > 1.9$$
$$1.9 < w < 2.9$$

This is preferred.

63. See answer section in the text for the solution.

64.
$$-2y + 3 > 5$$
$$-2y > 2$$
$$y < -1$$

We do not reverse inequalities due to *subtraction*.

Solutions To All Mastery Test Problems — Chapter Five

1. See answer section in the text for the graphs.

2a. $x \geq 7$ 　　2b. $y < 19$ 　　2c. $-5 \leq v < 15$ 　　2d. $0 \leq w \leq \pi$

3a.
$$m + 9 \leq 21$$
$$m \leq 12$$

3b.
$$21n + 11 < 4(5n + 7)$$
$$21n + 11 < 20n + 28$$
$$n + 11 < 28$$
$$n < 17$$

3c.
$$3(5n - 9) > 2(7n + 6)$$
$$15n - 27 > 14n + 12$$

3d.
$$8y + 3(4 - 5y) \geq y - 3(3y + 5)$$
$$8y + 12 - 15y \geq y - 9y - 15$$

$$n - 27 > 12$$
$$n > 39$$

$$-7y + 12 \geq -8y - 15$$
$$y + 12 \geq -15$$
$$y \geq -27$$

4a. $-11y \leq 165$

$$y \geq \frac{165}{-11}$$

$$y \geq -15$$

4b. $\frac{3v}{7} > 105$

$$v > \left(\frac{7}{3}\right)105$$

$$v > 7(35)$$

$$v > 245$$

4c. $8(w + 3) < 4(w + 4) - (10 - 2w)$

$$8w + 24 < 4w + 16 - 10 + 2w$$
$$8w + 24 < 6w + 6$$
$$2w < -18$$
$$w < -9$$

4d. $-5 \leq \frac{4x - 9}{3}$

$$-15 \leq 4x - 9$$
$$-6 \leq 4x$$

$$-\frac{3}{2} \leq x \text{ or } x \geq -\frac{3}{2}$$

5a. $-30 < -6x \leq 48$

$$\frac{-30}{-6} > x \geq \frac{48}{-6}$$

$$5 > x \geq -8$$

$$-8 \leq x < 5$$

5b. $-30 \leq x - 6 < 48$

$$-24 \leq x < 54$$

5c. $-1 \leq 4x + 3 \leq 19$

$$-4 \leq 4x \leq 16$$
$$-1 \leq x \leq 4$$

5d. $-6 < \frac{3(x - 4)}{2} < -\frac{9}{2}$

$$-12 < 3(x - 4) < -9$$
$$-4 < x - 4 < -3$$
$$0 < x < 1$$

6a. $y + y < y$

$y < 0$ (Subtract y from both sides.)

conditional inequality

6b. $2(3x + 5) \geq 3(2x - 6)$

$$6x + 10 \geq 6x - 18$$
$10 \geq -18$ absolute inequality

solution: all real numbers

page 95

Chapter Five Mastery Test

6c. $3(8v + 1) > 4(6v + 5)$

$24v + 3 > 24v + 20$

$3 > 20$ contradiction; no solution

6d. $2v + 3 \leq 3v + 3$

$3 \leq v + 3$

$0 \leq v; \quad v \geq 0$ conditional inequality

7a. Let s = the score on the third game. The average of the scores equals $\frac{\text{sum of the scores}}{\text{number of scores}}$.

$$\frac{168 + 184 + s}{3} \geq 180$$

$168 + 184 + s \geq 540$

$352 + s \geq 540$

$s \geq 188$ She must achieve at least 188 on the third game.

7b. Let s = the dollar amount of sales for the fourth week of March.

$$\frac{7500 + 9000 + 8100 + s}{4} \geq 8500$$

$$\frac{24600 + s}{4} \geq 8500$$

$24600 + s \geq 34000$

$s \geq 9400$ The sales for the fourth week must be at least $9400.

7c. Let s = the length of the third side. The perimeter is the sum of the lengths of the sides.

$52 \leq 17 + 31 + s \leq 90$

$52 \leq 48 + s \leq 90$

$4 \leq s \leq 42$ The third side must measure at least 4 cm and at most 42 cm.

7d. Let d be the distance the trucker travels on Tuesday.

We start out with:

$350 \leq d \leq 500$ Remember that $d = r \cdot t$, so substitute $r \cdot t$ for d.

$350 \leq rt \leq 500$ The rate is 50 miles per hour, so replace r with 50.

$350 \leq 50t \leq 500$ Now divide all members by 50.

$7 \leq t \leq 10$

The trucker will take from 7 to 10 hours to drive the route.

Chapter Five Mastery Test

Problem Set 6.1

1. See answer section in the text.

5. See answer section in the text.

9. See answer section in the text.

13a. The order of operations requires us to exponentiate before multiplying by -1, so it is only 1 that is raised to the 12th power. *Then* we multiply by -1. Answer: -1

13b. Contrast this with 13a. The parentheses indicate that the number -1 is to be raised to the 12th power, giving 1.

13c. -1 raised to an odd power is -1.

13d. This is equivalent to $(10)(10)(10)(10) = 10,000$.

17. $y\,y^7 = y^{1+7} = y^8$

21. $(-5y^2)(4y^3) = (-5)(4)\,y^2y^3 = -20y^5$

25. $(-3b)(-4b)(6b^2) = (-3)(-4)(6)\,b^{1+1+2} = 72\,b^4$

29. Multiply the exponents. Answer: x^{220}

33. $-(2n)^6 = -2^6 n^6 = -64n^6$

37. $(-3bcd)^2 = (-3)^2 b^2 c^2 d^2 = 9b^2 c^2 d^2$

41. $\left(\frac{2}{v}\right)^3 = \frac{(2)^3}{v^3} = \frac{8}{v^3}$

45. $-\left(\frac{z}{5}\right)^4 = -\frac{z^4}{5^4} = -\frac{z^4}{625}$

49. $(2x)^2(3x)^3 = 2^2 x^2 \cdot 3^3 x^3$
$= 4x^2 \cdot 27x^3 = 108\,x^5$

53a. $[\,(2)(3)]^2 = 6^2 = 36$

It is easiest to work inside the brackets first.

53b. $xy^2 = 2\,(3)^2 = 2\,(9) = 18$

Notice that only y is raised to the second power.

57a. $x^3 - y^3 = 2^3 - 3^3 = 8 - 27 = -19$

57b. $(x - y)^3 = (2 - 3)^3 = (-1)^3 = -1$

61. We may estimate by using 1 for 1.01.

$1^2 + 1^2 + (1 + 1)^2 = 1 + 1 + 2^2 = 6$, so the correct response is B.

65. $(x^2y)^3 = (x^2)^3 y^3 = x^6 y^3$

69. $\left(\frac{xy}{z}\right)^4 = \frac{(xy)^4}{z^4} = \frac{x^4 y^4}{z^4}$

73. $(-3)^2 + (-5)^2 = 9 + 25 = 34$

77. $x^m x^2 = x^{m+2}$

Problem Set 6.2

1a. $\frac{10^8}{10^5} = 10^{8-5} = 10^3 = 1000$

1b. $\frac{9^9}{9^7} = 9^{9-7} = 9^2 = 81$

1c. $\dfrac{7^4}{7^6} = \dfrac{1}{7^{6-4}} = \dfrac{1}{7^2} = \dfrac{1}{49}$

1d. $\dfrac{8^8}{8^9} = \dfrac{1}{8^1} = \dfrac{1}{8}$

1e. $\dfrac{11^6}{11^6} = 11^{6-6} = 11^0 = 1$ Notice also that any number divided by itself is one.

5a. $2^0 + 8^0 = 1 + 1 = 2$

5b. $(2+8)^0 = 10^0 = 1$

5c. $2^0 - 8^0 = 1 - 1 = 0$

5d. $(2-8)^0 = (-6)^0 = 1$

5e. $(8-2)^0 = 6^0 = 1$

9a. Any nonzero number raised to the zero power is 1.

9b. 0^0 does not have meaning, so we say it is undefined.

13. $\dfrac{x^{15}}{x^5} = x^{15-5} = x^{10}$

17. Any nonzero number divided by itself is 1.

21. $\dfrac{t^8}{t^{12}} = \dfrac{1}{t^{12-8}} = \dfrac{1}{t^4}$

25. $-\dfrac{a^6}{a^9} = -\dfrac{1}{a^{9-6}} = -\dfrac{1}{a^3}$

29. $\dfrac{14m^5}{21m^2} = \dfrac{2}{3}m^{5-2} = \dfrac{2}{3}m^3$ or $\dfrac{2m^3}{3}$

33. $\dfrac{15a^2b^3}{20a^2b^2} = \dfrac{3}{4}a^{2-2}b^{3-2} = \dfrac{3}{4}b$ or $\dfrac{3b}{4}$

37. $-\dfrac{27v^4w^9}{45v^6w^6} = -\dfrac{3}{5}\left(\dfrac{1}{v^{6-4}}\right)w^{9-6} = -\dfrac{3w^3}{5v^2}$

41. $\left(\dfrac{v^8}{v^5}\right)^4 = (v^3)^4 = v^{12}$

45. $[(3x^2)(2x^3)]^2 = (6x^5)^2 = 6^2(x^5)^2$
$= 36x^{10}$

49a. $(4x)^0 = 1$

49b. $4x^0 = 4(1) = 4$

49c. $(4x - 3y)^0 = 1$

49d. $(4x)^0 - (3y)^0 = 1 - 1 = 0$

49e. $4x^0 - 3y^0 = 4(1) - 3(1) = 4 - 3 = 1$

53. We may estimate using $\dfrac{8^4}{8^2}$, which is 8^2. The best approximation is 64, answer D.

57. $\dfrac{2^{42}}{2^{45}} = \dfrac{1}{2^3} = \dfrac{1}{8}$

61. $\left(\dfrac{10x^7}{5x^4}\right)^5 = (2x^3)^5 = 2^5x^{15} = 32x^{15}$

65. $(5x^3)^2(2x^5)^3 = 5^2x^6 2^3x^{15} = 25(8)x^{21} = 200x^{21}$

69. $\dfrac{36a^7b^8}{12a^3b^3} = \dfrac{36}{12} \cdot a^{7-3} \cdot b^{8-3} = 3a^4b^5$

73. $(-2)(-5)^2 = (-2)(25) = -50$
Notice only -5 is squared.

77. $(-1)^{m+1-m} = -1$

81. $x^{m+2-m} = x^2$

1a. $3^{-2} = \frac{1}{3^2} = \frac{1}{9}$

1b. $2^{-5} = \frac{1}{2^5} = \frac{1}{32}$

1c. $-2^5 = (-1)2^5 = (-1)32 = -32$
Notice that 2 is raised to the fifth power, not (-2).

1d. $-5^2 = (-1)5^2 = (-1)25 = -25$
Notice that 5 is raised to the second power, not (-5).

With problems like 1c and 1d, remember that the presence of the minus sign indicates multiplication by -1. The order of operations for arithmetic tells us to perform exponentiation before multiplication.

5a. $10^{-2} = \frac{1}{10^2} = \frac{1}{100}$

5b. $10^{-3} = \frac{1}{10^3} = \frac{1}{1000}$

5c. $10^{-4} = \frac{1}{10^4} = \frac{1}{10000}$

5d. $-10^{-2} = (-1)10^{-2} = (-1)\left(\frac{1}{10^2}\right)$
$= (-1)\left(\frac{1}{100}\right) = -\frac{1}{100}$

9a. $\left(\frac{1}{2} + \frac{1}{5}\right)^{-1} = \left(\frac{5+2}{10}\right)^{-1}$
$= \left(\frac{7}{10}\right)^{-1} = \frac{10}{7}$

9b. $\left(\frac{1}{2}\right)^{-1} + \left(\frac{1}{5}\right)^{-1} = \frac{2}{1} + \frac{5}{1} = 7$

9c. $\left(\frac{1}{2} + \frac{1}{5}\right)^{-2} = \left(\frac{7}{10}\right)^{-2} = \left(\frac{10}{7}\right)^2$
$= \frac{10^2}{7^2} = \frac{100}{49}$

13a. $(3x)^{-2} = \frac{1}{(3x)^2} = \frac{1}{3^2 x^2}$
$= \frac{1}{9x^2}$

13b. $3x^{-2} = (3)(x^{-2}) = 3\left(\frac{1}{x^2}\right) = \frac{3}{x^2}$

17a. $4.58 \times 10^4 = 4.58(10,000)$
$= 45,800$

17b. $4.58 \times 10^{-4} = 4.58 \times \frac{1}{10^4}$
$= \frac{4.58}{10,000} = 0.000458$

Notice that when we multiply 4.58 by a positive power of 10, such as 10^4 (which is 10,000), we obtain a larger quantity; hence the decimal point must move to the right. When multiplying by a negative power of ten such as 10^{-4} (which is the same as division by 10,000) we obtain a smaller quantity, and so the decimal point moves to the left.

17c. We are multiplying by a positive power of 10, so we will obtain a larger quantity. Move the decimal point 6 places to the right: 4,580,000

17d. We are multiplying by a negative power of 10 (which means we are dividing by 10^6) and so will obtain a smaller quantity. Move the decimal point 6 places to the left: 0.000 004 58

21a. $5,863,000 = 5.863 \times 10^6$. The exponent must reflect the number of places that the decimal point has moved. (5,863,000 has an understood decimal point: 5,863,000.0, so the decimal point has moved 6 places.) If you are not sure whether the exponent should be 6 or −6, remember that 5.863×10^{power} must equal $5,863,000$, a number larger in quantity than 5.863. Therefore, a positive exponent is needed.

21b. $97,000,000 = 9.7 \times 10^7$. The decimal point moved 7 places. A positive exponent is required because 9.7 would have to be multiplied by 10,000,000 to obtain 97,000,000.

21c. $0.0717 = 7.17 \times 10^{-2}$. The decimal point moved 2 places. A negative exponent is required because 0.0717 is smaller than 7.17; in other words, 7.17 would have to be divided by 100 to obtain 0.0717.

21d. $0.936 = 9.36 \times 10^{-1}$. The decimal point moved 1 place. A negative exponent is required because 0.936 is smaller than 9.36 ; that is, 9.36 would have to be divided by 10 in order to obtain 0.936.

25. See answer section in the text.   29.   See answer section in the text.

33. $(10,013)(0.00007943) \approx (10^4)(8 \times 10^{-5}) = (8)(10^4)(10^{-5})$
$= 8(10^{-1}) = 0.8$

37. $\dfrac{0.000005034}{0.00009893} \approx \dfrac{0.000005}{0.0001} \approx \dfrac{5 \times 10^{-6}}{1 \times 10^{-4}} = 5\left(\dfrac{10^{-6}}{10^{-4}}\right) = 5\left(10^{-6-(-4)}\right)$

$= 5(10^{-2}) = 0.05$

41. $(4.32 \times 10^{15})(8.49 \times 10^{-17}) = 0.366768$; rounding to three significant digits gives 0.367.

Suggestions on how to compute this:
First of all, we can either rewrite it and then compute, or leave it as is and compute.

It can be rewritten as: $(4.32)(8.49) \times 10^{-2} = \dfrac{4.32(8.49)}{100}$. Now use your calculator.

Without rewriting it, we can use the scientific notation feature found on many calculators.
scientific calculator: 4.32 $\boxed{\text{EXP}}$ 15 $\boxed{\times}$ 8.49 $\boxed{\text{EXP}}$ 17 $\boxed{\pm}$ $\boxed{=}$
graphics calculator: 4.32 $\boxed{\text{EE}}$ 15 $\boxed{\times}$ 8.49 $\boxed{\text{EE}}$ $\boxed{(-)}$ 17 $\boxed{\text{Enter}}$

Please note: there are many models of calculators available. These instructions work on commonly used models, but yours may be different. Ask your instructor if you need help.

45. $\dfrac{v^{12}}{v^{-3}} = v^{12-(-3)} = v^{15}$ 49. $\dfrac{w^{-21}}{w^7} = \dfrac{1}{w^{7+21}} = \dfrac{1}{w^{28}}$

53. $y\, y^0\, y^{-9} = y^{1+0-9} = y^{-8} = \dfrac{1}{y^8}$ 57. $(4v^{-2})^{-3} = 4^{-3}v^6 = \dfrac{v^6}{4^3} = \dfrac{v^6}{64}$

61. $(2x^{-3}y^4)^3(3x^4y^{-2})^2$

$= 2^3x^{-9}y^{12}\ 3^2x^8y^{-4}$

$= 8(9)\ x^{-1}y^8$

$= \dfrac{72y^8}{x}$

65. $\left(\dfrac{m^3n^{-7}}{m^7n^{-11}}\right)^{-3}$

$= \left(\dfrac{n^{-7+11}}{m^{7-3}}\right)^{-3}$

$= \left(\dfrac{n^4}{m^4}\right)^{-3}$

$= \left(\dfrac{m^4}{n^4}\right)^3$

$= \dfrac{m^{12}}{n^{12}}$

69. $((-2)(-3))^{-1}$

$= (6)^{-1}$

$= \dfrac{1}{6}$

73. $\left(\dfrac{-2}{-3}\right)^{-1}$

$= \left(\dfrac{2}{3}\right)^{-1}$

$= \dfrac{3}{2}$

77. This is 5.768×10^8 , which in standard decimal notation is 576,800,000.

Problem Set 6.4

1. See answer section in the text.

5. See answer section in the text.

9. It is not a monomial because of the negative exponent (-2) on the variable y.

13. See answer section in the text.

17. See answer section in the text.

21. See answer section in the text.

25. See answer section in the text.

29. $(6x - 7) + (8x + 5)$
$= 6x - 7 + 8x + 5$
$= 6x + 8x - 7 + 5$
$= (6 + 8)x - 2$
$= 14x - 2$

33. See answer section in the text.

37. $\begin{array}{r} 8x^2 - 3x + 4 \\ (+)\ \underline{-2x^2 + 2x + 9} \\ 6x^2 - x + 13 \end{array}$

41. $-w^2 + 4 + 4w^2 - 7w + 6$
$= -w^2 + 4w^2 - 7w + 4 + 6$
$= 3w^2 - 7w + 10$

45. $2x^2 - 6x - 10 - 7x^2 - 14x + 7$
$= -5x^2 - 20x - 3$

49. See answer section in the text.

53. Round in order to estimate:
$(5x^2 + 8x - 8) - (2x^2 - 2x + 2)$
$= 3x^2 + 10x - 10$ Answer: D

57. $x^4 - 2xy^3 + x^2y^2 + xy^3 - 3y^4 - 2x^4 - xy^3 + 5x^2y^2 + 7y^4 - 7x^4 - 3x^2y^2 - 2y^4$

$= x^4 - 2x^4 - 7x^4 \ - 2xy^3 + xy^3 - xy^3 \ + x^2y^2 + 5x^2y^2 - 3x^2y^2 - 3y^4 + 7y^4 - 2y^4$

$= -8x^4 - 2xy^3 + 3x^2y^2 + 2y^4$

$= -8x^4 + 3x^2y^2 - 2xy^3 + 2y^4$

61. Call the polynomial P. The problem states that if we subtract P from something, we obtain a certain difference.

$(3v^2 - 4v + 11) - P = (2v^2 - 5v - 12)$ Now add P to both sides.
$(3v^2 - 4v + 11) = P + (2v^2 - 5v - 12)$ Solve for P.
$P = (3v^2 - 4v + 11) - (2v^2 - 5v - 12)$ Simplify the right side.
$P = 3v^2 - 4v + 11 - 2v^2 + 5v + 12$
$P = 3v^2 - 2v^2 - 4v + 5v + 11 + 12$
$P = v^2 + v + 23$

65. $-(3)^2 + 8(3)(-7) + 3(-7)^2$
$= (-1)3^2 - 168 + 3(49)$
$= -9 - 168 + 147$
$= -30$

69. The shaded region is the area of the larger square minus the area of the smaller square.
larger square's area: $x \cdot x = x^2$
smaller square's area: $2 \cdot 2 = 4$
area of shaded region: $x^2 - 4$
For $x = 5$, $5^2 - 4 = 25 - 4 = 21$.

73. There are many satisfactory answers. Examples: $2x$ and $-2x$, or $\frac{1}{3}x$ and $-\frac{1}{3}x$. The coefficients must be additive inverses (2 and -2 or $\frac{1}{3}$ and $-\frac{1}{3}$, etc.) so that when the monomials are added the first-degree terms (the terms which contain x to the first power) drop out. Notice that the sum $2x + (-2x) = 0$, a constant.

77. See answer section in the text. Be sure to add like terms and watch the signs.

Problem Set 6.5

1. $2x(-5x) = (2)(-5)(x)(x)$
$= -10x^2$

5. $(a^3b^4)(4a^6b) = 4a^3a^6b^4b$
$= 4a^{3+6}b^{4+1} = 4a^9b^5$

9. $-4v^2(8v + 3)$
$= -4v^2(8v) + (-4v^2)(3)$
$= (-4)(8)v^2v + (-4)(3)v^2$

13. $-11x^2(4x^2 - 5x - 8)$
$= (-11x^2)(4x^2) + (-11x^2)(-5x)$
$\qquad\qquad + (-11x^2)(-8)$

$$= -32v^3 - 12v^2 \qquad\qquad = -44x^4 + 55x^3 + 88x^2$$

17.
$$
\begin{array}{r}
6x + 5 \\
(\times) \quad 9x + 3 \\
\hline
54x^2 + 45x \\
18x + 15 \\
\hline
54x^2 + 63x + 15
\end{array}
$$

21.
$$
\begin{array}{r}
x^2 - 3x - 4 \\
(\times) \quad x + 2 \\
\hline
x^3 - 3x^2 - 4x \\
2x^2 - 6x - 8 \\
\hline
x^3 - x^2 - 10x - 8
\end{array}
$$

25.
$$
\begin{array}{r}
x^2 - xy + y^2 \\
(\times) \quad 2x + y \\
\hline
2x^3 - 2x^2y + 2xy^2 \\
x^2y - xy^2 + y^3 \\
\hline
2x^3 - x^2y + xy^2 + y^3
\end{array}
$$

29. $(y + 2)(y + 3)$
$$= (y + 2)(y) + (y + 2)(3)$$
$$= (y)(y) + (2)(y) + (y)(3) + (2)(3)$$
$$= y^2 + 2y + 3y + 6$$
$$= y^2 + 5y + 6$$

33. $(x + y)(2x - y)$
$$= (x + y)(2x) + (x + y)(-y)$$
$$= (x)(2x) + (y)(2x)$$
$$\qquad\qquad + (x)(-y) + (y)(-y)$$
$$= 2x^2 + 2xy - xy - y^2$$
$$= 2x^2 + xy - y^2$$

37. $(2m - 5)(m^2 + 3m + 2)$
$$= (2m - 5)(m^2) + (2m - 5)(3m)$$
$$\qquad\qquad + (2m - 5)(2)$$

$$= 2m^3 - 5m^2 + 6m^2 - 15m$$
$$\qquad\qquad + 4m - 10$$
$$= 2m^3 + m^2 - 11m - 10$$

41. $-2a(a - 7)(a + 7)$
$$= -2a(a^2 - 49)$$
$$= -2a^3 + 98a$$

45. We can estimate by rounding. We will first round to the nearest integer, because that gives numbers that are easiest to work with:
$2x^2(2x - 5)$
$= 4x^3 - 10x^2$
This does not match any of the choices, but is closest to choice B.   To be sure that this will be the correct choice, go back to the original problem. We see that a lot has been lost by rounding 2.49 to 2.   Let us instead round to the nearest tenth:
$2x^2(2.5x - 5)$
$= 5x^3 - 10x^2$ which now matches choice B.

Another method: Notice that the best estimate, whatever it is, must be of the form: _____ x^3 − _____ x^2 , *which eliminates choices A and D.   Rounding to the nearest integer, as we originally did, would eliminate choice C, because we know that the second coefficient must be close to 10 or − 10.   This leaves B as the only choice.*
This process of elimination combined with rounding does not especially save any time, but can help you avoid mistakes by reducing the number of possible choices.

49.

$$n^4 - 2n^3 + n^2 - 7n + 8$$

$(\times)$ $\qquad \dfrac{n + 3}{}$.

$$n^5 - 2n^4 + n^3 - 7n^2 + 8n$$

$$\underline{\qquad 3n^4 - 6n^3 + 3n^2 - 21n + 24 \qquad}$$

$$n^5 + n^4 - 5n^3 - 4n^2 - 13n + 24$$

Notice how we vertically line up like terms.

53. $4(x - 7)(2x + 3)$

$= 4[\,(x - 7)(2x) + (x - 7)(3)\,]$

$= 4[\,x(2x) - 7(2x) + x(3) - 7(3)\,]$

$= 4[\,2x^2 - 14x + 3x - 21\,]$

$= 4(\,2x^2 - 11x - 21)$

$= 8x^2 - 44x - 84$

57. $(v + 2)^3$

$= (v + 2)(v + 2)(v + 2)$

$= (v^2 + 4v + 4)(v + 2)$

$= (v^2 + 4v + 4)v + (v^2 + 4v + 4)2$

$= v^3 + 4v^2 + 4v + 2v^2 + 8v + 8$

$= v^3 + 6v^2 + 12v + 8$

61. The area of a triangle is one-half the product of the base and height.
The base will have length $2x + 4$ and the height is measured perpendicular to the base, and so has length $x - 1$.

$$\text{area} = \tfrac{1}{2}(2x + 4)(x - 1) = (x + 2)(x - 1)$$

$$= (x + 2)(x) + (x + 2)(-1)$$
$$= x^2 + 2x - x - 2$$
$$= x^2 + x - 2$$

65. Notice that the semicircle that is atop the figure is the same size as the one cut out underneath. This means that the total area of the shaded region is the same as the rectangle with width $2x + 1$ and height $2x - 1$, since the semicircles have the effect of subtracting some area that is then just added back. The area of a rectangle is width $\times$ height, so we have:

$$\text{area} = (2x + 1)(2x - 1)$$
$$= 4x^2 - 1$$

69. $(x + y)^2 - (x - y)^2$ We will expand each term, and then combine the results later:

$(x + y)^2$
$= (x + y)(x) + (x + y)(y)$
$= x^2 + yx + xy + y^2$
$= x^2 + 2xy + y^2$

$(x - y)^2$
$= (x - y)(x) + (x - y)(-y)$
$= x^2 - yx - xy + y^2$
$= x^2 - 2xy + y^2$

$(x + y)^2 - (x - y)^2 = x^2 + 2xy + y^2 - (x^2 - 2xy + y^2)$
$\qquad\qquad\qquad = x^2 + 2xy + y^2 - x^2 + 2xy - y^2$
$\qquad\qquad\qquad = 4xy$

73. You can start by multiplying the first two factors together:

$$(-5xy)(2x^2 + 3xy + y^2)(4x - y)$$

$$(-5xy)(2x^2 + 3xy + y^2) = (-5xy)(2x^2) + (-5xy)(3xy) + (-5xy)(y^2)$$
$$= -10x^3y - 15x^2y^2 - 5xy^3$$

Now multiply by $4x - y$:

$$(-10x^3y - 15x^2y^2 - 5xy^3)(4x - y)$$
$$= (-10x^3y - 15x^2y^2 - 5xy^3)(4x) + (-10x^3y - 15x^2y^2 - 5xy^3)(-y)$$
$$= (-10x^3y)(4x) + (-15x^2y^2)(4x) + (-5xy^3)(4x)$$
$$\quad + (-10x^3y)(-y) + (-15x^2y^2)(-y) + (-5xy^3)(-y)$$
$$= -40x^4y - 60x^3y^2 - 20x^2y^3 + 10x^3y^2 + 15x^2y^3 + 5xy^4$$
$$= -40x^4y - 50x^3y^2 - 5x^2y^3 + 5xy^4$$

77. See answer section in the text.

To obtain the second row: take 12 divided by the numbers in the top row.

This gives $\frac{12}{1}, \frac{12}{2}, \frac{12}{3}, \frac{12}{4}, \frac{12}{6}$ which are then reduced.

To obtain the third row: add the numbers in the first and second rows.

81. See answer section in the text.   Start by distributing:

$$-6.7ab(3.5a^2) + 6.7ab(2.6ab) + 6.7ab(5.1b^2)$$ We leave the rest to you.

Problem Set 6.6

1. $(m + 3)(m + 4)$
$$= (m)(m) + (m)(4) + (3)(m) + (3)(4)$$
$$= m^2 + 7m + 12$$

5. $(v - 8)(v - 9)$
$$= (v)(v) + (v)(-9) + (-8)(v)$$
$$\qquad\qquad\qquad + (-8)(-9)$$
$$= v^2 - 17v + 72$$

9. $(5x + 7)(4x - 3)$
$$= (5x)(4x) + (5x)(-3) + (7)(4x)$$
$$\qquad\qquad + (7)(-3)$$
$$= 20x^2 + 13x - 21$$

13. $(4w - 1)(2 - 3w)$
$$= 8w - 12w^2 - 2 + 3w$$
$$= -12w^2 + 11w - 2$$

17. $(3a - b)(4a + 7b)$
$$= 12a^2 + 21ab - 4ab - 7b^2$$
$$= 12a^2 + 17ab - 7b^2$$

21. $(6x - 7y)(7x - 6y)$
$$= 42x^2 - 36xy - 49yx + 42y^2$$
$$= 42x^2 - 85xy + 42y^2$$

25. $(m + 4)^2 = m^2 + 2(m)(4) + 4^2$
$= m^2 + 8m + 16$

29. $(3t + 1)^2 = (3t)^2 + 2(3t)(1) + 1^2$
$= 9t^2 + 6t + 1$

33. $(4x + 5y)^2$
$= (4x)^2 + 2(4x)(5y) + (5y)^2$
$= 16x^2 + 40xy + 25y^2$

37. $(v + 5)(v - 5)$
$= v^2 - 5^2$
$= v^2 - 25$

41. $(3x + y)(3x - y)$
$= (3x)^2 - y^2$
$= 9x^2 - y^2$

45. $(9a + 4b)(9a - 4b)$
$= (9a)^2 - (4b)^2$
$= 81a^2 - 16b^2$

49. $(x^2 - y^2)(x^2 + y^2)$
$= (x^2)^2 - (y^2)^2$
$= x^4 - y^4$

53. $(100 - 1)^2$ *note form* $(a - b)^2$
$= 100^2 - 2(100)(1) + 1^2$
$= 10000 - 200 + 1$
$= 9800 + 1$
$= 9801$

57.   area of the larger square: $(x + 1)(x + 1) = x^2 + 2x + 1$.
area of the smaller square: x^2
area of the unshaded region: $(x^2 + 2x + 1) - x^2 = 2x + 1$

61. $\left(\dfrac{x}{2} - \dfrac{y}{3}\right)^2$

$= \left(\dfrac{x}{2}\right)^2 - 2\left(\dfrac{x}{2}\right)\left(\dfrac{y}{3}\right) + \left(\dfrac{y}{3}\right)^2$

$= \dfrac{x^2}{4} - \dfrac{xy}{3} + \dfrac{y^2}{9}$

65. $(2x + y)^2 - \left((2x)^2 + y^2\right)$

$= 4x^2 + 4xy + y^2 - 4x^2 - y^2$

$= 4xy$

69. $(a - b)^2 - (b - a)^2$
$= a^2 - 2ab + b^2 - (b^2 - 2ab + a^2)$
$= a^2 - 2ab + b^2 - b^2 + 2ab - a^2$
$= 0$

Notice: $a - b$ and $b - a$ are just opposite in sign. Therefore, their squares must be equal, which is why the difference turns out to be 0.

73. $(x + 8)^2 - (x - 6)^2$
$= x^2 + 16x + 64 - (x^2 - 12x + 36)$
$= x^2 + 16x + 64 - x^2 + 12x - 36$
$= 28x + 28$

77.   See answer section in the text.   Start with:
$(5.83)^2x^2 - 2(5.83)(7.1) x + (7.1)^2$

Problem Set 6.7

1. $\dfrac{18x^3}{6x} = 3x^{3-1} = 3x^2$

5. $\dfrac{15a^2 - 20a}{5a} = \dfrac{15a^2}{5a} - \dfrac{20a}{5a}$

   $= 3a - 4$

9. $\dfrac{63v^4 - 21v^3}{-3v^2} = \dfrac{63v^4}{-3v^2} - \dfrac{21v^3}{-3v^2}$

   $= -21v^2 + 7v$

13. $\dfrac{21b^3 - 14b^2 + 63b}{-7b^2}$

   $= \dfrac{21b^3}{-7b^2} - \dfrac{14b^2}{-7b^2} + \dfrac{63b}{-7b^2}$

   $= -3b + 2 - \dfrac{9}{b}$

17. $\dfrac{36m^3n^3 - 40m^2n^4}{-4mn^2}$

   $= \dfrac{36m^3n^3}{-4mn^2} - \dfrac{40m^2n^4}{-4mn^2}$

   $= -9m^2n + 10mn^2$

21. $\dfrac{30x^2 - 45x^3 + 10x + 35x^4}{5x}$

   $= \dfrac{30x^2}{5x} - \dfrac{45x^3}{5x} + \dfrac{10x}{5x} + \dfrac{35x^4}{5x}$

   $= 6x - 9x^2 + 2 + 7x^3$

   $= 7x^3 - 9x^2 + 6x + 2$

25. $\dfrac{35v^3w - 7v^2w - 28vw^2}{7vw}$

   $= \dfrac{35v^3w}{7vw} - \dfrac{7v^2w}{7vw} - \dfrac{28vw^2}{7vw}$

   $= 5v^2 - v - 4w$

29.
$$
\begin{array}{r}
x + 7 \\
x + 2 \enclose{longdiv}{x^2 + 9x + 14} \\
\underline{x^2 + 2x} \\
7x + 14 \\
\underline{7x + 14} \\
0
\end{array}
$$

answer: $x + 7$

33.
$$
\begin{array}{r}
3w - 4 \\
2w + 3 \enclose{longdiv}{6w^2 + w - 12} \\
\underline{6w^2 + 9w} \\
-8w - 12 \\
\underline{-8w - 12} \\
0
\end{array}
$$

answer: $3w - 4$

37.
$$
\begin{array}{r}
x - 7 \\
x - 5 \enclose{longdiv}{x^2 - 12x + 38} \\
\underline{x^2 - 5x} \\
-7x + 38 \\
\underline{-7x + 35} \\
3
\end{array}
$$

answer: $x - 7 + \dfrac{3}{x - 5}$

41.

$$
\begin{array}{r}
2v^2 \;-\; v \;-\; 3 \\
v+1 \;\big|\; \overline{2v^3 + v^2 - 4v - 3} \\
\underline{2v^3 + 2v^2} \\
-v^2 - 4v \\
\underline{-v^2 - v} \\
-3v - 3 \\
\underline{-3v - 3} \\
0
\end{array}
$$

answer: $2v^2 - v - 3$

45.

$$
\begin{array}{r}
3v^2 + 2v - 4 \\
v+2 \;\big|\; \overline{3v^3 + 8v^2 \qquad - 8} \\
\underline{3v^3 + 6v^2} \\
2v^2 \\
\underline{2v^2 + 4v} \\
-4v - 8 \\
\underline{-4v - 8} \\
0
\end{array}
$$

answer: $3v^2 + 2v - 4$

49.

$$
\begin{array}{r}
4y^2 - 7y - 9 \\
5y-6 \;\big|\; \overline{20y^3 - 59y^2 - 3y + 54} \\
\underline{20y^3 - 24y^2} \\
-35y^2 - 3y \\
\underline{-35y^2 + 42y} \\
-45y + 54 \\
\underline{-45y + 54} \\
0
\end{array}
$$

answer: $4y^2 - 7y - 9$

53.

$$
\begin{array}{r}
y^3 - 3y^2 + 9y - 27 \\
y+3 \;\big|\; \overline{y^4 + 0y^3 + 0y^2 + 0y + 81} \\
\underline{y^4 + 3y^3} \\
-3y^3 + 0y^2 \\
\underline{-3y^3 - 9y^2} \\
9y^2 + 0y \\
\underline{9y^2 + 27y} \\
-27y + 81 \\
\underline{-27y - 81} \\
162
\end{array}
$$

answer: $y^3 - 3y^2 + 9y - 27 + \dfrac{162}{y + 3}$

57.

$$
\begin{array}{r}
x + 2y \\
x-y \;\big|\; \overline{x^2 + xy - 20y^2} \\
\underline{x^2 - xy} \\
2xy - 20y^2 \\
\underline{2xy - 2y^2} \\
-18y^2
\end{array}
$$

answer: $x + 2y - \dfrac{18y^2}{x - y}$

61. Estimate by using $\dfrac{6x^6}{2x^2} = 3x^4$. Answer: D

65. $(x - 5)(x + 5) + \dfrac{6x^3 + 10x}{2x}$

$= x^2 - 25 + \dfrac{6x^3}{2x} + \dfrac{10x}{2x}$

$= x^2 - 25 + 3x^2 + 5$

$= 4x^2 - 20$

69. Let P represent the polynomial.

$\dfrac{6x^2 - 2x - 20}{P} = 2x - 4$

$6x^2 - 2x - 20 = P(2x - 4)$

$P = (6x^2 - 2x - 20) \div (2x - 4)$

$$
\begin{array}{r}
3x + 5 \\
2x - 4 \;\big|\; \overline{6x^2 - 2x - 20} \\
\underline{6x^2 - 12x} \\
10x - 20 \\
\underline{10x - 20} \\
0
\end{array}
$$

answer: $P = 3x + 5$

73. See answer section in the text.   Start with $\dfrac{31.54212}{7.6932}\,x^5$ and use your calculator to do the division.

Solutions To All Review Exercises – Chapter Six

1. $\dfrac{8^7}{8^5} = 8^{7-5} = 8^2 = 64$

2. $(7 - 5)^2 = 2^2 = 4$

3. $7^2 - 5^2 = 49 - 25 = 24$

4. $7^{15} \cdot 7^{-13} = 7^{15-13} = 7^2 = 49$

5. $-1^{100} = -(1^{100}) = -1$
Notice that 1 is raised to the 100th power, *not −1*.

6. $(-1)^{100} = 1$
Notice that -1 is raised to the 100th power. Contrast this with #5.

7. $100^{-1} = \dfrac{1}{100}$

8. $2^5 + 5^2 = 32 + 25 = 57$

9. $\left(\dfrac{3}{5}\right)^{-3} = \left(\dfrac{5}{3}\right)^3 = \dfrac{5^3}{3^3} = \dfrac{125}{27}$

10. $2^{-1} + 3^{-1} = \dfrac{1}{2} + \dfrac{1}{3} = \dfrac{5}{6}$

11. $(2 + 3)^{-1} = 5^{-1} = \dfrac{1}{5}$

12. $\left(\dfrac{1}{3}\right)^{-1} + \left(\dfrac{1}{4}\right)^{-1} = 3 + 4 = 7$

13. $\left(\dfrac{1}{3} + \dfrac{1}{4}\right)^{-1} = \left(\dfrac{4+3}{12}\right)^{-1} = \left(\dfrac{7}{12}\right)^{-1} = \dfrac{12}{7}$

14. $7x^0 = 7(1) = 7$

15. $(7x)^0 = 1$

16. $(-x)^0 = 1$ (for $x \neq 0$)
Remember: 0^0 is undefined.

17. $(-1)x^0 = (-1)(1) = -1$

18. $(2x + 3y)^0 = 1$

19. $(2x)^0 + (3y)^0 = 1 + 1 = 2$

20. $2x^0 + 3y^0 = 2(1) + 3(1) = 2 + 3 = 5$

21. $3x^{-1} = 3\left(\frac{1}{x}\right) = \frac{3}{x}$

22. $(3x)^{-1} = \frac{1}{3x}$

23. $(10)^6 = 1,000,000$

24. $(0.1)^4 = (0.1)(0.1)(0.1)(0.1) = 0.0001$

    or: $(0.1)^4 = \left(\frac{1}{10}\right)^4 = \frac{1}{10,000} = 0.0001$

25. $10^{-3} = \frac{1}{10^3} = \frac{1}{1000}$

26. $(0.1)^{-7} = \left(\frac{1}{10}\right)^{-7} = 10^7 = 10,000,000$

27. $10^2 + 10^3 = 100 + 1000 = 1100$

28. $(10^2)(10^3) = 10^5 = 100,000$

29. $(3^{-2})^{-1} = 3^{(-2)(-1)} = 3^2 = 9$

30. $(3 + 7)^{-2} = 10^{-2} = \frac{1}{100}$ or 0.01

31. $(2x^5)(5x^7) = 10x^{5+7} = 10x^{12}$

32. $\frac{12v^{18}}{6v^6} = \left(\frac{12}{6}\right)v^{18-6} = 2v^{12}$

33. $(2x^3)^4 = 2^4(x^3)^4 = 16x^{12}$

34. $\left(\frac{2v}{5w^2}\right)^3 = \frac{(2v)^3}{(5w^2)^3} = \frac{2^3v^3}{5^3(w^2)^3} = \frac{8v^3}{125w^6}$

35. $(4v^3w^5)^3 = 4^3(v^3)^3(w^5)^3 = 64v^9w^{15}$

36. $\left(\frac{12m^3}{4n^2}\right)^{-2} = \left(\frac{4n^2}{12m^3}\right)^2 = \left(\frac{n^2}{3m^3}\right)^2 = \frac{n^4}{9m^6}$

37. $(3x^2)(4x^3)(5x^{11})$
    $= (3)(4)(5)x^2x^3x^{11}$
    $= 60x^{16}$

38. $\frac{(3x^2)^4}{9x^5} = \frac{81x^8}{9x^5} = 9x^{8-5} = 9x^3$

39. $(-2x^5)^3(-5x^3)^2$
    $= (-2)^3(x^5)^3(-5)^2(x^3)^2$
    $= (-8)(25)\,x^{15}x^6$
    $= -200x^{21}$

40. $(7y^8)(8y^{-6})$
    $= 56y^{8-6}$
    $= 56y^2$

41. $\frac{16w^{-4}}{8w^{-6}} = 2w^{-4-(-6)} = 2w^2$

42. $(5m^{-6})^{-4} = 5^{-4}m^{24} = \frac{m^{24}}{5^4} = \frac{m^{24}}{625}$

Chapter Six Review

43. $(6x)^{-3}\left(\dfrac{x}{6}\right)^{-4}$

$= \dfrac{1}{(6x)^3} \cdot \left(\dfrac{6}{x}\right)^4$

$= \dfrac{6^4}{6^3 x^3 x^4}$

$= \dfrac{6}{x^7}$

44. 1. Any nonzero quantity raised to the zero power is one.

45. $\dfrac{(2a^3 b^3)(8a^2 b^5)}{(4ab^2)^2} = \dfrac{16a^5 b^8}{16a^2 b^4} = a^3 b^4$

46. $1.0 \times 10^{-13} = 0.000\ 000\ 000\ 0001$ watt

47. $2.9979 \times 10^8 = 299{,}790{,}000$ meters per sec

48. $9{,}578{,}420{,}000 = 9.57842 \times 10^9$

49. $0.00437 = 4.37 \times 10^{-3}$
Remember: when using scientific notation, be careful with the sign of the exponent.

Rule of Thumb:

positive exponent	*standard decimal form of the number is greater than 10*
negative exponent	*standard decimal form of the number is less than 1*

50. Since $0.000\ 000\ 000\ 1989$ may be approximated using $0.000\ 000\ 000\ 2000$, we use the following as an estimate: $(4 \times 10^4)(2 \times 10^{-10}) = 8 \times 10^{-6}$
Answer: B

51. See answer section in the text. We suggest that you review the definition of "standard form" as given in the textbook, section 6-4. The terms are arranged so that the term with the largest exponent is written first. The remaining terms are arranged so that the exponents decrease from left to right.

52. $-7x^5$. Review the definition of "monomial" as given in the textbook, section 6-4. A fifth degree monomial in x must have the form ax^5. Choose -7 for a to give the correct coefficient.

53. Monomial, degree zero. The symbol π represents the irrational number pi, which is a constant.

54. Binomial, degree six. If you had difficulty with this, review section 6-4.

55. Four terms, degree three.

56. Trinomial, degree three.

57. $7x^2 - 9x + 13 + 4x^2 + 6x - 11$
$= 7x^2 + 4x^2 - 9x + 6x + 13 - 11$
$= 11x^2 - 3x + 2$

58. $9x^3 - 5x^2 - 7 - 4x^3 - 8x + 11$
$= 9x^3 - 4x^3 - 5x^2 - 8x - 7 + 11$
$= 5x^3 - 5x^2 - 8x + 4$

59. This problem is conveniently arranged with like terms under like terms. All you need to do is add coefficients going down. See answer section in the text.

Chapter Six Review

60. One way to deal with this is to change the signs on the polynomial to be subtracted (the lower one) and then *add*. This gives:

$$\begin{array}{r} 7x^5 \qquad\quad + 9x^3 \qquad + 6x - 3 \\ -4x^5 + 3x^4 + 7x^3 + x^2 + x - 8 \\ \hline 3x^5 + 3x^4 + 16x^3 + x^2 + 7x - 11 \end{array}$$

61. $x^2 - 8x + 7 - 2x^2 - 7x - 11 + 3x^2 + 4x - 8$
$= x^2 - 2x^2 + 3x^2 - 8x - 7x + 4x + 7 - 11 - 8$
$= 2x^2 - 11x - 12$

62. $5x^2(7x^3) + 5x^2(-9x^2) + 5x^2(3x) + 5x^2(1)$
$= 35x^5 - 45x^4 + 15x^3 + 5x^2$

63. $(5v + 1)(7v - 1)$ *Use FOIL.*
$= 35v^2 - 5v + 7v - 1$
$= 35v^2 + 2v - 1$

64. $(5y - 7)^2$ *Note form: $(a - b)^2$.*
$= 25y^2 - 2(5y)(7) + 7^2$
$= 25y^2 - 70y + 49$

65. $(9y + 5)^2$ *Note form: $(a + b)^2$*
$= (9y)^2 + 2(9y)(5) + 5^2$
$= 81y^2 + 90y + 25$

66. $(3a + 5b)(3a - 5b)$ *Note form $(a+b)(a-b)$.*
$= (3a)^2 - (5b)^2$
$= 9a^2 - 25b^2$

67. $(3m + 5)(2m^2 - 6m + 7)$
$= 3m(2m^2 - 6m + 7) + 5(2m^2 - 6m + 7)$
$= 6m^3 - 18m^2 + 21m + 10m^2 - 30m + 35$
$= 6m^3 - 8m^2 - 9m + 35$

68. $(2x + 3y)^2 - (2x - 3y)^2$
$= 4x^2 + 12xy + 9y^2 - 4x^2 + 12xy - 9y^2$
$= 24xy$

69. $\dfrac{-36a^3b^7}{9a^4b^4} = -4\,a^{3-4}\,b^{7-4} = -\dfrac{4b^3}{a}$

70. When the denominator is a monomial (one term), a good way to proceed to break up the original problem into a sum of fractions, each with denominator $3m^2n^3$:

$$\dfrac{15m^5n^4}{3m^2n^3} - \dfrac{21m^4n^5}{3m^2n^3} - \dfrac{3m^3n^6}{3m^2n^3}$$

    You are 'undoing' the addition of fractions. Now reduce each fraction.

$$= 5m^3n - 7m^2n^2 - mn^3$$

Chapter Six Review

71.
$$
\begin{array}{r}
v - 2 \\
v - 4 \overline{\big)\ v^2 - 6v + 8} \\
\underline{v^2 - 4v} \\
-2v + 8 \\
\underline{-2v + 8} \\
0
\end{array}
$$

72.
$$
\begin{array}{r}
7w + 3 \\
3w - 7 \overline{\big)\ 21w^2 - 40w - 21} \\
\underline{21w^2 - 49w} \\
9w - 21 \\
\underline{9w - 21} \\
0
\end{array}
$$

73. $\dfrac{(a + b)^7}{(a + b)^6} = (a + b)^{7-6} = a + b$

74. $(3x + 4y)^2 - (3x + 4y)(3x - 4y)$ *Use the forms $(a + b)^2 = a^2 + 2ab + b^2$*

$= 9x^2 + 24xy + 16y^2 - (9x^2 - 16y^2)$ *and* $(a + b)(a - b) = a^2 - b^2$.

$= 9x^2 + 24xy + 16y^2 - 9x^2 + 16y^2$

$= 32y^2 + 24xy$

75. Work on each fraction separately, using long division.   The procedure is the same as used in exercises #71 and #72; also see section 6-7 in your textbook.

$\dfrac{6x^2 + x - 2}{2x - 1}$ will give $3x + 2$; $\dfrac{8x^2 + 18x - 35}{4x - 5}$ will give $2x + 7$.

The original problem is then equivalent to $(3x + 2) - (2x + 7) = 3x + 2 - 2x - 7 = x - 5.$

76. The perimeter of a figure is the distance around it. For a rectangle, this is:
perimeter = 2(width) + 2(length)
Calling the perimeter P, we have:

$P = 2(x^2 - x + 7) + 2(2x + 3)$
$P = 2x^2 - 2x + 14 + 4x + 6$
$P = 2x^2 + 2x + 20$ This polynomial represents the perimeter of the rectangle.

77. *The area of a rectangle is the product of the length and width.* Calling the area A, we have:
$A = (x^2 - x + 7)(2x + 3)$ The multiplication can be done vertically or horizontally.
If it is done vertically, we have:

$$
\begin{array}{r}
x^2 - x + 7 \\
(\times)\ \underline{2x + 3} \\
2x^3 - 2x^2 + 14x \\
\underline{3x^2 - 3x + 21} \\
2x^3 + x^2 + 11x + 21
\end{array}
$$
   This polynomial represents the area of the rectangle.

Chapter Six Review

78. The perimeter of a triangle is the distance around it; that is, the sum of the lengths of the three sides. We want to find the length of the base, so call this b. Start by setting up an equation for the perimeter of the triangle:

$12x^2 + x + 16 = b + (3x^2 + 2x + 1) + (4x^2 + 7)$ Now simplify the right side.

$12x^2 + x + 16 = b + 7x^2 + 2x + 8$ Solve for b.

$b = 12x^2 - 7x^2 + x - 2x + 16 - 8$

$b = 5x^2 - x + 8$ This polynomial represents the length of the base.

79. The area of a rectangle is the product of the width and length. We want to find the width, so call this quantity w. Start by setting up an equation for the area of the rectangle:

$12x^2 + 11x + 2 = w(4x + 1)$ Now solve for w.

$w = \dfrac{12x^2 + 11x + 2}{4x + 1}$ You will need to use long division, as in #71. We will leave the details to you.

$w = 3x + 2$

80. *The area of a triangle is one-half the product of the base and height.* We want to find the height, so call this quantity h. Start by setting up an equation for the area of the triangle:

$10x^2 + 7x + 1 = (\frac{1}{2})(5x + 1) h$ Multiply both sides by 2 to clear fractions.

$2(10x^2 + 7x + 1) = 2 [(\frac{1}{2})(5x + 1) h]$

$20x^2 + 14x + 2 = (5x + 1) h$ Now solve for h.

$h = \dfrac{20x^2 + 14x + 2}{5x + 1}$ You will need to use long division, as in #71. We will leave the details to you.

$h = 4x + 2$ or $2(2x + 1)$ This polynomial represents the height.

81. This is worked out in the answer section in your textbook.

Solutions To All Mastery Test Problems – Chapter Six

1a. $x^5 x^8 = x^{13}$

1b. $(5y^4)(8y^3) = 40y^7$

1c. $-v^4 v^5 v^6 = -v^{15}$

1d. $(mn^2)(m^3 n) = (m\, m^3)(n^2\, n) = m^4 n^3$

2a. $(w^5)^8 = w^{40}$

2b. $(2y^4)^5 = 2^5 (y^4)^5 = 32y^{20}$

2c. $(-v^2 w^3)^4$

$= (-1)^4 (v^2)^4 (w^3)^4$

$= v^8 w^{12}$

2d. $\left(\dfrac{-3m}{2n}\right)^3$

$= \dfrac{(-3m)^3}{(2n)^3} = \dfrac{(-3)^3 m^3}{2^3 n^3} = -\dfrac{27m^3}{8n^3}$

Chapter Six Mastery Test

3a. $\dfrac{z^8}{z^2} = z^6$

3b. $\dfrac{4x^9}{2x^3} = \left(\dfrac{4}{2}\right)\left(\dfrac{x^9}{x^3}\right) = 2x^6$

3c. $-\dfrac{v^{23}}{v^{23}} = -\left(\dfrac{v^{23}}{v^{23}}\right) = -1$

3d. $\dfrac{-6a^4b^6}{3a^2b^2} = \left(\dfrac{-6}{3}\right)\left(\dfrac{a^4}{a^2}\right)\left(\dfrac{b^6}{b^2}\right) = -2a^2b^4$

4a. $1^0 = 1$

4b. $\left(\dfrac{2}{5}\right)^0 = 1$

4c. $(3x + 5y)^0 = 1$

4d. $3x^0 + 5y^0 = 3(1) + 5(1) = 3 + 5 = 8$

5a. $\left(x^3y^5\right)\left(x^2y^6\right)$

$= (x^3x^2)(y^5y^6)$

$= x^5y^{11}$

5b. $\left(\dfrac{x^2y^3}{xy}\right)^3$

$= \left(xy^2\right)^3$

$= x^3(y^2)^3$

$= x^3y^6$

5c. $\dfrac{(6x^2)^2}{(3x^3)^3} = \dfrac{6^2(x^2)^2}{3^3(x^3)^3}$

$= \dfrac{36x^4}{27x^9} = \dfrac{4}{3x^5}$

5d. $[(2x^2)(3x^4)]^2$

$= (6x^6)^2$

$= 6^2(x^6)^2$

$= 36x^{12}$

6a. $3^{-1} = \dfrac{1}{3}$

6b. $7^{-2} = \dfrac{1}{7^2} = \dfrac{1}{49}$

6c. $\left(\dfrac{2}{3}\right)^{-2} = \left(\dfrac{3}{2}\right)^2 = \dfrac{9}{4}$

6d. $\left(\dfrac{1}{3}\right)^{-1} + \left(\dfrac{1}{6}\right)^{-1} = 3 + 6 = 9$

7a. i. 3.57×10^5
$= 357,000$

7a. ii. 7.35×10^{-5}
$= 0.000\ 0735$

7b. i. 0.000509
$= 5.09 \times 10^{-4}$

7b. ii. $93,050,000$
$= 9.305 \times 10^7$

8a. binomial; degree 3

8b. monomial; degree 0

8c. trinomial; degree 2

8d. polynomial; degree 5

9a. $(4x - 9y) + (3x + 8y)$
$= 4x - 9y + 3x + 8y$
$= 4x + 3x + (-9y) + 8y$
$= 7x - y$

9b. $(2x^2 - 3x + 7) - (5x^2 - 9x - 11)$
$= 2x^2 - 3x + 7 - 5x^2 + 9x + 11$
$= -3x^2 + 6x + 18$

9c.
$$5x^4 - 9x^3 + 7x^2 \qquad + 13$$
$$(+) \quad \underline{4x^4 \qquad\quad + 12x^2 + 9x - 5}$$
$$9x^4 - 9x^3 + 19x^2 + 9x + 8$$

9d.
$$3x^3 \qquad\quad - 7x + 1$$
$$(-) \quad \underline{\qquad 12x^2 + 9x - 5}$$
$$3x^3 - 12x^2 - 16x + 6$$

10a. $(5x^2y^3)(11x^4y^7)$
$= 55x^2x^4y^3y^7$
$= 55x^6y^{10}$

10b. $-3x^2(5x^3 - 2x^2 + 7x - 9)$
$= -15x^2x^3 + 6x^2x^2 - 21x^2x + 27x^2$
$= -15x^5 + 6x^4 - 21x^3 + 27x^2$

10c.
$$x^2 + 3x + 1$$
$$(\times) \qquad\quad x + 5$$
$$\overline{}$$
$$x^3 + 3x^2 + \quad x$$
$$\underline{\qquad 5x^2 + 15x + 5}$$
$$x^3 + 8x^2 + 16x + 5$$

10d.
$$x^2 + xy + y^2$$
$$(\times) \qquad x - 3y$$
$$\overline{}$$
$$x^3 + x^2y + xy^2$$
$$\underline{-3yx^2 - \quad 3xy^2 - 3y^3}$$
$$x^3 - 2x^2y - 2xy^2 - 3y^3$$

11 a. $(x + 2)(x + 3)$
$= x^2 + 3x + 2x + 6$
$= x^2 + 5x + 6$

11b. $(x - 3)(x + 5)$
$= x^2 + 5x - 3x - 15$
$= x^2 + 2x - 15$

11c. $(2x + 5)(6x - 7)$
$= 12x^2 - 14x + 30x - 35$
$= 12x^2 + 16x - 35$

11d. $(3x - y)(x + 4y)$
$= 3x^2 + 12xy - yx - 4y^2$
$= 3x^2 + 11xy - 4y^2$

Chapter Six Mastery Test

12a.
$$(x - 8)^2$$
$$= x^2 - 2(8x) + 8^2$$
$$= x^2 - 16x + 64$$

12b.
$$(2x + 3)^2$$
$$= (2x)^2 + 2(6x) + 3^2$$
$$= 4x^2 + 12x + 9$$

12c.
$$(5x + 7)(5x - 7)$$
$$= (5x)^2 - 7^2$$
$$= 25x^2 - 49$$

12d.
$$(3x - 9y)(3x + 9y)$$
$$= (3x)^2 - (9y)^2$$
$$= 9x^2 - 81y^2$$

13a.
$$\frac{15a^4b^2}{5a^2b^2}$$

$$= 3a^2$$

13b.
$$\frac{36m^4n^3 - 48m^2n^5}{12m^2n^2}$$

$$= \left(\frac{36m^4n^3}{12m^2n^2}\right) - \left(\frac{48m^2n^5}{12m^2n^2}\right)$$

$$= 3m^2n - 4n^3$$

13c.
$$\begin{array}{r} x + 11 \\ x - 2 \overline{\smash{\big)}\ x^2 + 9x - 22} \\ \underline{x^2 - 2x} \\ 11x - 22 \\ \underline{11x - 22} \\ 0 \end{array}$$

13d.
$$\begin{array}{r} 2x^2 + 6x + 8 \\ 3x - 4 \overline{\smash{\big)}\ 6x^3 + 10x^2 \qquad - 32} \\ \underline{6x^3 - 8x^2} \\ 18x^2 \\ \underline{18x^2 - 24x} \\ 24x - 32 \\ \underline{24x - 32} \\ 0 \end{array}$$

Chapter Six Mastery Test

1. $(-1)^8 = (-1)(-1)(-1)(-1)(-1)(-1)(-1)(-1) = 1$
    Any negative number raised to an even power will give a positive result.

2. $-1^8 = -1$, because it is the number 1 that is raised to the eighth power, not -1. The last step taken is to multiply by negative one.

3. $0^8 = 0$, because this is a product of eight zeros.   Zero to any positive power yields zero.

4. $8^0 = 1$.   Any nonzero number raised to the zero power is defined to be 1.

5. $8^{-1} = \dfrac{1}{8^1} = \dfrac{1}{8}$

6. $-1^0 = -(1^0) = -(1) = -1$

7. $(-1)^0 = 1$, since any nonzero number raised to the zero power is defined to be 1.

8. $\left(\dfrac{1}{4}\right)^{-1} + \left(\dfrac{1}{5}\right)^{-1} = 4 + 5 = 9$

9. $\left(\dfrac{1}{4} + \dfrac{1}{5}\right)^{-1} = \left(\dfrac{5+4}{20}\right)^{-1} = \left(\dfrac{9}{20}\right)^{-1} = \dfrac{20}{9}$

10. $\dfrac{5^8}{5^6} = 5^2 = 25$

11. $(3 + 7)^5 = 10^5 = 100{,}000$

12. $10^3 + 10^4 = 1000 + 10000 = 11{,}000$

13. $10^3 10^4 = 10^{3+4} = 10^7 = 10{,}000{,}000$

14. $10^{-1} + 10^{-2}$

    $= \dfrac{1}{10} + \dfrac{1}{100}$

    $= \dfrac{10 + 1}{100}$

    $= \dfrac{11}{100}$ or 0.11

15. $10^{-1} 10^{-2}$ Compare with #17.

    $= 10^{-1 + (-2)}$ Notice: add exponents

    $= 10^{-3}$

    $= \dfrac{1}{10^3}$

    $= \dfrac{1}{1000}$ or 0.001

16. $\dfrac{10^{-1}}{10^{-2}}$

    $= 10^{-1-(-2)}$

    $= 10^{-1+2}$

    $= 10^1$

    $= 10$

17. $(10^{-2})^{-3}$ Compare with #15.

    $= 10^{-2(-3)}$ Notice: multiply exponents

    $= 10^6$

    $= 1{,}000{,}000$

Cumulative Review

18. $(3x + 4y + 5)^0 = 1$ A nonzero quantity raised to the zero power is always 1.

19. $(3x)^0 + (4y)^0 + 5^0 = 1 + 1 + 1 = 3.$ Notice that $(3x)^0 = 1$, $(4y)^0 = 1$ and $5^0 = 1$.

20. $3x^0 + 4y^0 + 5^0 = 3(1) + 4(1) + 1 = 3 + 4 + 1 = 8$
Be careful here. Notice in $3x^0$, only x is raised to a power, not 3x. Compare this problem with #19; in *that* problem, the parentheses tell us that the whole expression 3x is raised to the zero power.

21. $12x + 4x$
$= (12+4)x$
$= 16x$

22. $12x - 4x$
$= (12-4)x$
$= 8x$

23. $(12x)(4x)$
$= (12)(4)x^2$
$= 48x^2$

24. $\frac{12x}{4x} = \frac{12}{4} = 3$

25. $(3x^2)(15x^4)$

$= 3(15)x^2x^4$

$= 45x^6$

26. $\frac{15x^4}{3x^2}$

$= \frac{15}{3}x^{4-2}$

$= 5x^2$

27. $(3x^2)^4$

$= 3^4(x^2)^4$

$= 81x^8$

28. $\left(\frac{2}{x}\right)^{-3}$

$= \left(\frac{x}{2}\right)^3$

$= \frac{x^3}{2^3}$

$= \frac{x^3}{8}$ or $\frac{1}{8}x^3$

29. $(2a^2b^3)^4$

$= 2^4(a^2)^4(b^3)^4$

$= 16a^8b^{12}$

30. $\left(\frac{16a^4}{20b}\right)^{-2}$

$= \left(\frac{20b}{16a^4}\right)^2$

$= \left(\frac{5b}{4a^4}\right)^2$

$= \frac{(5b)^2}{(4a^4)^2}$

$= \frac{25b^2}{16a^8}$

Cumulative Review

31. $\dfrac{36x^{-2}y}{24xy^{-3}}$

$= \dfrac{36}{24} \cdot \dfrac{x^{-2}}{x} \cdot \dfrac{y}{y^{-3}}$

$= \dfrac{3}{2} \cdot x^{-2-1} \cdot y^{1-(-3)}$

$= \dfrac{3}{2} \cdot x^{-3} \cdot y^{4}$

$= \dfrac{3y^4}{2x^3}$

32. $\dfrac{(12xy^2)(3x^2y)}{(4xy)^2}$

$= \dfrac{(12)(3)\, x\, x^2 y^2\, y}{4^2\, x^2\, y^2}$

$= \dfrac{3(3)\, x^3\, y^3}{4\, x^2\, y^2}$

$= \dfrac{9}{4}\, xy \quad \text{or} \quad \dfrac{9xy}{4}$

33. $x^2 - 5xy + y^2 - (3x^2 - 7xy - y^2)$
$= x^2 - 5xy + y^2 - 3x^2 + 7xy + y^2$
$= -2x^2 + 2xy + 2y^2$

34. $(4x - 7y)(3x + 5y)$ Use FOIL.
$= (4x)(3x) + (4x)(5y)$
$\qquad\qquad + (-7y)(3x) + (-7y)(5y)$
$= 12x^2 + 20xy - 21xy - 35y^2$
$= 12x^2 - xy - 35y^2$

35. $(x - 5y)^2 - (x + 5y)^2$
$= x^2 - 10xy + 25y^2 - (x^2 + 10xy + 25y^2)$
$= x^2 - 10xy + 25y^2 - x^2 - 10xy - 25y^2$
$= -20xy$

36.
$$\begin{array}{r}
v + 6 \\
v - 5 \,\overline{\big)\, v^2 + v - 30} \\
\underline{v^2 - 5v } \\
6v - 30 \\
\underline{6v - 30} \\
0
\end{array}$$

37. $(a - 2b)(a^2 + 2ab + 4b^2)$
$= a(a^2 + 2ab + 4b^2) - 2b(a^2 + 2ab + 4b^2)$
$= a^3 + 2a^2b + 4ab^2 - 2ba^2 - 4ab^2 - 8b^3$
$= a^3 - 8b^3$

Cumulative Review

38. Use long division.

$$
\begin{array}{r}
m^3 + 3m^2 + 9m + 27 \\
m - 3\ \overline{)\ m^4 + 0m^3 + 0m^2 + 0m - 81} \\
\underline{m^4 - 3m^3} \\
3m^3 + 0m^2 \\
\underline{3m^3 - 9m^2} \\
9m^2 + 0m \\
\underline{9m^2 - 27m} \\
27m - 81 \\
\underline{27m - 81} \\
0
\end{array}
$$

Notice the coefficients of zeros for missing terms.

Be sure to line up like terms before subtracting.

39. $\dfrac{30a^4b^3 - 20a^3b^4}{5a^2b}$

This problem can be worked by long division, but since the denominator consists of only one term, we will use a different method:

$= \dfrac{30a^4b^3}{5a^2b} - \dfrac{20a^3b^4}{5a^2b}$

Break apart into the difference of two fractions, each with a denominator of $5a^2b$. You are reversing fraction addition.

$= 6a^2b^2 - 4ab^3$

Each fraction was then simplified.

40. $[(12x^2)^3(5x)^2]^2$

$= [\,(12)^3x^6\ 5^2\ x^2\,]^2$

$= [\,(1728)(25)\ x^8\,]^2$

$= [43{,}200\ x^8\,]^2$

$= 1{,}866{,}240{,}000\ x^{16}$

41. 2.34×10^7

As a check, note that it could not be 10^{-7}, since 2.34×10^{-7} is a small number.

$(\ 10^{-7}$ is $\dfrac{1}{10{,}000{,}000}$ or $0.0\,000\,001.)$

42. 4.8×10^{-4}

The effect of multiplying by 10^{-4} is to decrease the size of 4.8, since $10^{-4} = \dfrac{1}{10{,}000}$ or 0.0001. Move the decimal point to the left four places to obtain 0.00048.
Remember, a negative exponent in scientific notation means the resulting standard decimal number will be *less than 1*.

43. "Monomial" means that there will be only one term; sixth degree means that x will be raised to the sixth power, and the coefficient is the number placed in front, indicating quantity and sign.
Answer: $-3x^6$

Cumulative Review

44. $x + 3 \geq 8$ Subtract 3 from both sides.
       $x \geq 5$

45. $-3x < 15$ Divide both sides by -3, and reverse the inequality sign.
       $x > -5$

46. $17x + 21 < 8x - 24$ Subtract 8x from both sides.
       $9x + 21 < -24$ Subtract 21 from both sides.
       $9x < -45$ Divide both sides by 9.
       $x < -5$

47. $19 - \frac{5}{7} v \geq 54$ Subtract 19 from both sides.

       $-\frac{5}{7} v \geq 35$ Multiply both sides by $-\frac{7}{5}$ and reverse the inequality sign.

       $v \leq 35 \left(-\frac{7}{5}\right)$

       $v \leq -49$

48. $5(7y - 3) \leq 6(8y - 5) + 2$ Distribute to remove parentheses.
       $35y - 15 \leq 48y - 30 + 2$
       $35y - 15 \leq 48y - 28$ Add 28 to both sides.
       $35y + 13 \leq 48y$ Subtract 35y from both sides.
       $13 \leq 13y$ Divide both sides by 13.
       $1 \leq y$ or $y \geq 1$

49. $-4 \leq \frac{2y - 4}{3}$ Multiply both sides by 3 to clear fractions.

       $-12 \leq 2y - 4$ Add 4 to both sides.
       $-8 \leq 2y$ Divide both sides by 2.
       $-4 \leq y$ or $y \geq -4$

50. $-3 < 5 - 2x \leq 1$ Subtract 5 from each member.
       $-8 < -2x \leq -4$ Divide each member by -2, reversing inequalities.
       $4 > x \geq 2$ This means x is simultaneously greater than or
                             equal to 2 and less than 4.
       $2 \leq x < 4$ This is the preferred way to write it.

51.   The completed table is found in the answer section in the text.
      Notice that in each column the numbers are a multiple of the numbers in the first column.
      In other words, the second column is obtained by multiplying the entries in the first column
      by 2, the third column is obtained by multiplying the entries in the first column by 4, and so

Cumulative Review

forth. Use the first number in a column to tell you what the correct multiple is.

To obtain the column beginning with 1, divide each of the entries in the first column by 3; to obtain the last column, multiply each entry in the previous column by a.

52.   Let the consecutive odd integers be n and n + 2.

$n + n + 2 = 100$
$2n = 98$
$n = 49$ $n + 2 = 51$ The numbers are 49 and 51.

53.   n   the second number
       $3n + 2$ the first number

$n + 3n + 2 = 22$
$4n = 20$
$n = 5$ $3(5) + 2 = 17$ The numbers are 5 and 17.

54.   x   length of the longer piece
       $x - 93$ length of the shorter piece

$x + x - 93 = 500$
$2x = 593$
$x = 296.5$ $296.5 - 93 = 203.5$ The lengths of the pieces are 296.5 cm
                                                          and 203.5 cm. (Check: do they add to 500?)

55.   We will be using the formula $D = R \cdot T$ in each case.

(a) The boats travel in opposite directions.

slower boat   faster boat
    7 km/h   11 km/h
<------------- • --->

Let t be the number of hours before the boats are 9 km apart if they are traveling in opposite directions. This means that the sum of their distances is 9 km.
distance for the slower boat: 7t   } $7t + 11t = 9$
distance for the faster boat: 11t }

$18t = 9$

$t = \frac{1}{2}$ or 0.5   It will take 0.5 hour.

·(b) The boats travel in the same direction. The faster boat goes farther.

          7 km/h
    •-------------->
    •----------------------------------->
                  11 km/h

Cumulative Review

Let t be the number of hours before the boats are 9 km apart if they are traveling in the same direction. This means that the faster boat will be ahead by 9 km, and we will set the *difference* of their distances to be 9.

distance for the slower boat: 7t $\left.\begin{array}{c}\\\\\end{array}\right\}$ $11t - 7t = 9$
distance for the faster boat: 11t

If you are not sure why we subtracted in this order, keep in mind that the faster boat will go a longer distance than the slower boat, so 11t is larger than 7t. We must subtract smaller from larger to get the *positive* difference of 9.

$4t = 9$

$t = \frac{9}{4}$ or $2\frac{1}{4}$ or 2.25   It will take $2\frac{1}{4}$ hours.

56.   The perimeter of a triangle is the sum of the lengths of the sides.

$48 = x + 2x + 1 + x + 7$
$48 = 4x + 8$
$4x = 40$
$x = 10$ The side labeled x measures 10 cm.

57.   The area of a triangle is one-half the product of the base and height.

$30 = \frac{1}{2}(2x)(5)$

$30 = 5x$
$x = 6$

58.   Let ℓ be the number of liters of the 12% solution.

	strength •	liters of solution	= liters of pure antifreeze
12% solution	0.12	ℓ	0.12ℓ
60% solution	0.60	$16 - \ell$	$0.6(16-\ell)$
mixture	0.30	16	$0.3(16)$

$0.12\ell + 0.6(16 - \ell) = 0.3(16)$ Multiply both sides by 100.
$12\ell + 60(16 - \ell) = 30(16)$
$12\ell + 960 - 60\ell = 480$
$-48\ell = -480$
$\ell = 10$ $16 - 10 = 6$ Use 10 liters of the 12% solution and 6 liters of the 60% solution.

Cumulative Review

59. The ratio is $\dfrac{\text{black sheep}}{\text{total sheep}}$.

Let s be the number of sheep on the ranch.

$\dfrac{3}{375} = \dfrac{60}{s}$

$3s = 22{,}500$
$s = 7{,}500$ There are 7,500 sheep on the ranch.

60. Let s be the score the golfer must achieve on the fourth round to average at most 72.

$\dfrac{72 + 76 + 71 + s}{4} \leq 72$ 'At most' corresponds to *no more than*.

$72 + 76 + 71 + s \leq 288$

$s \leq 69$ The golfer must have a score of at most 69.

Cumulative Review

Problem Set 7.1

1. $12 = 2 \cdot 6 = 2 \cdot 2 \cdot 3 = 2^2 \cdot 3$
   $20 = 2 \cdot 10 = 2 \cdot 2 \cdot 5 = 2^2 \cdot 5$
   $GCF = 2^2 = 4$

5. $6x^3 = 2 \cdot 3x^3$
   $9x^2 = 3^2x^2$
   $GCF = 3x^2$

9. See answer section in the text.

13. See answer section in the text.

17. $32a^3b^4 = 4 \cdot 8\ a^2ab^2b^2$
    $= 4a^2b^2(8ab^2)$

21. $6x + 9y = 3(2x) + 3(3y)$
    $= 3(2x + 3y)$

25. $-33m^4 + 44m^3 - 11m^2$
    $= -11m^2(3m^2) - 11m^2(-4m)$
    $\qquad\qquad\qquad - 11m^2(1)$
    $= -11m^2(3m^2 - 4m + 1)$

29. $x(a - 2b) + 3(a - 2b)$ Let $z = a - 2b$.
    $xz + 3z$ Factor out z.
    $= z(x + 3)$ Now replace z.
    $= (a - 2b)(x + 3)$

COMMENT   *A Substitution Technique*
In #29, we replace the common expression a − 2b with the single letter z in order to simplify the appearance of the expression and make it easier to spot what should be factored out at the next step. Just be sure that the letter you choose, whether it be z or something else, is not already used in the problem.

33. $6x(a-b) = 2 \cdot 3x(a - b)$
    $9y(a - b) = 3^2y(a - b)$
    $GCF = 3(a - b)$

37. $6x^2 + 4x$
    $= (2x)(3x) + (2x)(2)$
    $= 2x(3x + 2)$

41. $35x^2y^2 + 7xy$
    $= (7xy)(5xy) + (7xy)(1)$
    $= 7xy(5xy + 1)$

45. $49x^2 - 63xy + 14y$
    $49x^2 = 7^2x^2$
    $63xy = 3^2 \cdot 7xy$ $\left.\right\}$ $GCF = 7$
    $14y = 2 \cdot 7y$

    $7(7x^2 - 9xy + 2y)$

49. $x(2x - 1) + 3(2x - 1)$
    Let $z = 2x - 1$.

53. $2v(3a - 4c) - 6w(3a - 4c)$
    Let $z = 3a - 4c$.

$xz + 3z$
$= z(x + 3)$
$= (2x - 1)(x + 3)$

$2vz - 6wz$
$= (2z)v - (2z)(3w)$
$= 2z(v - 3w)$
$= 2(3a - 4c)(v - 3w)$

57. $5ax - 2a + 15x - 6$
$a(5x - 2) + 3(5x - 2)$
Let $z = 5x - 2$.
$az + 3z = z(a + 3)$
$= (5x - 2)(a + 3)$

61. Let $z = v - 5$. So far we have:
$v(v - 5) + 7(v - 5)$
$= vz + 7z$ Now factor out z.
$= z(v + 7)$ Replace z with $(v - 5)$.
$= (v - 5)(v + 7)$
or $(v + 7)(v - 5)$

65. Let $z = x - 7$. So far we have:
$x(x - 7) - 3(x - 7)$
$= xz - 3z$ Now factor out z.
$= z(x - 3)$ Replace z with $(x - 7)$.
$= (x - 7)(x - 3)$

69. $(4ax - 4ay) + (-4x + 4y)$
$= 4a(x - y) + (-4)(x - y)$
Let $z = x - y$ and factor.
$4az + (-4)z = 4z(a - 1)$ Replace z.
$= 4(x - y)(a - 1)$

73. Group as follows:
$(ax + ay + a) - (x + y + 1)$
$= a(x + y + 1) - (x + y + 1)$
$= az - z$
$= z(a - 1)$
$= (x + y + 1)(a - 1)$

If you are not sure of the next step, replace
$x + y + 1$ with z and then factor z out.
Replace z with $x + y + 1$.

77. $p + prt$
$= p(1 + rt)$

81. $7.644v^3 - 13.083v$
$= 1.47v(5.2v^2 - 8.9)$
To obtain 5.2, divide 7.644 by 1.47.
To obtain 8.9, divide 13.083 by 1.47.

Problem Set 7.2

1. The product of the constants must be 6 and $(2)(3) = 6$, so we choose $x + 3$.
To make sure it works, multiply out: $(x + 2)(x + 3) = x^2 + 3x + 2x + 6 = x^2 + 5x + 6$.
Always check your factoring by multiplying the factors out to see that you get what you started with.

5. Since $(-3)(-6) = 18$ and $-3 + -6 = -9$, we choose $m - 6$.
Check: $(m - 3)(m - 6) = m^2 - 6m - 3m + 18 = m^2 - 9m + 18$ ✓

9. We need two integers whose sum is 11 and whose product is 18.
$2 \cdot 9 + 18$; $2 + 9 = 11$
Check: $(a + 2)(a + 9) = a^2 + 9a + 2a + 18 = a^2 + 11a + 18$.   ✓

13. $63 = 9 \cdot 7$ This choice doesn't work. To show what happens:

$$(v + 9w)(v - 7w) = v^2 - 7vw + 9vw - 63w^2 = v^2 + 2vw - 63w^2 \qquad \text{(incorrect)}$$
$$(v + 7w)(v - 9w) = v^2 - 9vw + 7vw - 63w^2 = v^2 - 2vw - 63w^2 \qquad \text{(incorrect)}$$

$63 = 3 \cdot 21$ We try this possibility, and find that it multiplies out correctly:
$$(v + 3w)(v - 21w) = v^2 - 21vw + 3vw - 63w^2 = v^2 - 18vw - 63w^2 \quad \text{(correct)} \quad \checkmark$$

17a.   correct blank form: $(m + ?)(m + ?)$
We need two integers whose sum is 7 and whose product is 10.
$10 = 2 \cdot 5$
check: $(m + 2)(m + 5) = m^2 + 5m + 2m + 10 = m^2 + 7m + 10$ $\qquad \checkmark$

17b.   correct blank form: $(m - ?)(m - ?)$
We need two integers whose sum is 7 and whose product is 10.
$10 = 2 \cdot 5$
check: $(m - 2)(m - 5) = m^2 - 5m - 2m + 10 = m^2 - 7m + 10$ $\qquad \checkmark$

Compare #17b with #17a.

17c.   correct blank form: $(m - ?)(m - ?)$
We need two integers whose sum is 11 and whose product is 10.
$10 = 1 \cdot 10$
check: $(m - 1)(m - 10) = m^2 - 10m - m + 10 = m^2 - 11m + 10$ $\qquad \checkmark$

17d.   correct blank form: $(m + ?)(m - ?)$
We need two integers whose difference is 3 and whose product is 10.
$10 = 2 \cdot 5$
check: $(m + 5)(m - 2) = m^2 - 2m + 5m - 10 = m^2 + 3m - 10$ $\qquad \checkmark$

If you had considered $(m + 2)(m - 5)$, the check would have shown it to be incorrect:
$(m + 2)(m - 5) = m^2 - 5m + 2m - 10 = m^2 - 3m - 10$ This is not our original expression.

21.   correct blank form: $(a + ?)(a - ?)$
We need two integers whose difference is 10 and whose product is 11.
$11 = 1 \cdot 11$
Since we need to come out with $-10a$, we will choose to subtract 11 and add 1:
check: $(a + 1)(a - 11) = a^2 - 11a + a - 11 = a^2 - 10a - 11$ $\qquad \checkmark$

> You must **always** check your factorization,
> even if only mentally.

25.   Correct blank form: $(a + ?)(a + ?)$
We need two integers whose sum is 11 and whose product is 10.
$10 = 1 \cdot 10$
check: $(a + 1)(a + 10) = a^2 + 10a + a + 10 = a^2 + 11a + 10$ $\qquad \checkmark$

29. correct blank form: $(x - ?)(x - ?)$
$7 = 7 \cdot 1$
$(x - 7)(x - 1) = x^2 - x - 7x + 8 = x^2 - 8x + 7$ ✓

33. correct blank form: $(p - ?)(p - ?)$
We need two integers whose product is 100 and whose sum is 29.
$100 = 4 \cdot 25$
$(p - 4)(p - 25) = p^2 - 25p - 4p + 100 = p^2 - 29p + 100$ ✓

You should notice that there are many ways to factor 100. Before listing all the factors, try out the familiar ones first.

37. correct blank form: $(n + ?)(n - ?)$.
$36 = 9 \cdot 4$
$(n + 9)(n - 4) = n^2 + 9n - 4n - 36 = n^2 + 5n - 36$ ✓

41. correct blank form: $(x + ?y)(x - ?y)$
$6 = 2 \cdot 3$
$(x + 3y)(x - 2y) = x^2 - 2xy + 3xy - 6y^2 = x^2 + xy - 6y^2$ ✓

45. correct blank form: $(a + ?b)(a - ?b)$
$36 = 6 \cdot 6$
$ = 3 \cdot 12$
$ = 2 \cdot 18$
$(a + 2b)(a - 18b) = a^2 - 18ab + 2ab - 36b^2 = a^2 - 16ab - 36b^2$ ✓

49. correct blank form: $(x - ?)(x - ?)$
You are looking for two integers whose product is 90 and whose sum is 22.
$90 = 2 \cdot 3^2 \cdot 5$
The possible factorizations of 90 are:
$90 = 1 \cdot 90$
$ = 2 \cdot 45$
$ = 3 \cdot 30$ None of these pairs add up to 22, so the polynomial is prime.
$ = 5 \cdot 18$
$ = 6 \cdot 15$
$ = 9 \cdot 10$

53. correct blank form: $(a + ?b)(a - ?b)$
$12 = 1 \cdot 12$ Checking out these combinations, it will be discovered that
$ = 2 \cdot 6$ none of them work. The polynomial is prime.
$ = 3 \cdot 4$

57. $5x^3 + 15x^2 + 10x = 5x(x^2) + 5x(3x) + 5x(2)$
$= 5x(x^2 + 3x + 2)$
$= 5x(x + 2)(x + 1)$

61. $z^2 + 14 + 9z = z^2 + 9z + 14 = (z + 2)(z + 7)$

65. $x^2 + 12xy + 36y^2$ Notice that $36y^2 = (6y)^2$.

The choice $(x + 6y)(x + 6y)$ would certainly give x^2 and $36y^2$; but we need to multiply it out to be sure that the middle term is correct :

$$
\begin{aligned}
\text{check: } (x + 6y)(x + 6y) &= x^2 + x(6y) + 6y(x) + (6y)^2 \\
&= x^2 + 6xy + 6xy + 36y^2 \\
&= x^2 + 12xy + 36y^2
\end{aligned}
$$
✓

69.
$$
\begin{aligned}
&-m^2 - 3m + 18 \\
&= (-1)(m^2 + 3m - 18) \\
&= (-1)(m + 6)(m - 3)
\end{aligned}
$$
First factor out the GCF, -1.
Now factor $m^2 + 3m - 18$.
Notice that the factor (-1) is carried along.

73.
$$
\begin{aligned}
&x^2(a + b) - 6x(a + b) + 8(a + b) \\
&= (a + b)(x^2 - 6x + 8) \\
&= (a + b)(x - 4)(x - 2)
\end{aligned}
$$
Factor out the GCF, $a + b$.
Now concentrate on factoring $x^2 - 6x + 8$.
Notice that $(a - b)$ is carried along.

77. $(3x^2 + 9x + 6) + ax + a$ There are five terms and no GCF. Try grouping.

$$
\begin{aligned}
&= 3(x^2 + 3x + 2) + a(x + 1) \\
&= 3(x + 1)(x + 2) + a(x + 1) \\
&= (x + 1)[\, 3(x + 2) + a\,] \\
&= (x + 1)(3x + 6 + a)
\end{aligned}
$$
Keep going. Factor $x^2 + 3x + 2$.
Factor out the common expression $x + 1$.
If you have trouble with factoring out $x + 1$,
see the COMMENT with #29, section 7-1.

Problem Set 7.3

1. Since $7x^2 = 7x(x)$ and $-1 = (-1)(1)$, the only choice is $x + 1$.
checking: $(7x - 1)(x + 1) = 7x^2 + 7x - x - 1 = 7x^2 + 6x - 1$ ✓

5. The first term must be 3m, since $3m^2 = m(3m)$;
the second term must be 2 since $-8 = -4(2)$.
It appears that $3m + 2$ will work; this can be checked by multiplying the factors out.

Remember: you must be sure that the factors you choose will give not only the correct first and last terms, but a correct middle term as well. Multiply the factors out to check!

9. $45x^2 = 5x(9x)$ and $14y^2 = -y(-14y)$
The factor $9x - 14y$ checks when the factors are multiplied out.

13. $35n^2$ can be factored as $5n(7n)$, which suggests two possible set ups:

$(3m + 7n)(4m - 5n) = 12m^2 - 15mn + 28mn - 35n^2$ (this does not work)

$(3m + 5n)(4m - 7n) = 12m^2 - 21mn + 20mn - 35n^2$ (this checks)   ✓

Always multiply these out (even if only mentally) to eliminate incorrect factorizations.

Suggestion: Read solution to #65 to see how to save time when checking possibilities.

17. $24v^2$ can be factored several ways. You may have to test several set ups before finding the one that works.   Possibilities: $2v(12v)$, $6v(4v)$, $3v(8v)$, $v(24v)$
Only $(3v - 2)(8v - 9)$ yields the correct middle term.

21. $48am^2 - 16am - 84a = 4a(12m^2) + 4a(-4m) + 4a(-21) = 4a(12m^2 - 4m - 21)$
$= 4a(6m + 7)(2m - 3)$

Read solutions to #37 and #65 to see how to save time when checking possibilities.

25.   Because the constant term is positive and the middle term is negative, a correct blank form is:
$(3x - ?)(x - ?)$.
There are now only two possibilities: $(3x - 1)(x - 5)$ or $(3x - 5)(x - 1)$. By multiplying out to check, you will discover that only the second one is correct.

29.   A correct blank form is $(t + ?)(11t + ?)$.

     Possibilities to check: $(t + 9)(11t + 1)$ (correct)   ✓
                                     $(t + 1)(11t + 9)$
                                     $(t + 3)(11t + 3)$

Remember: you must eliminate incorrect products and be absolutely sure of the correct one.   The way to do this is to multiply out each product until the correct one is found.

33.   A correct blank form will have one plus sign, and one minus sign.

     $6v^2 = (2v)(3v)$ or $6v^2 = (v)(6v)$;
     $25 = 5^2$ or $25 = (25)(1)$

     Possibilities to check : $(2v - 5)(3v + 5) = 6v^2 - 5v - 25$ does not match
                                   $(2v + 5)(3v - 5) = 6v^2 + 5v - 25$ correct ✓

There are other combinations, but since we have found the correct one, there is no need to check them.
Read solution to #65 to see how to save time when checking possibilities.

37.   First we will factor out -1:
$(-1)(12x^2 - 20x + 3)$

Now concentrate on factoring $12x^2 - 20x + 3$.   Notice that a correct blank form will have two minus signs.   12 has several factorizations: $12(1)$, $3(4)$ and $6(2)$, which allows for several possible set ups.

$$(3x - 1)(4x - 3) = 12x^2 - 13x + 3 \qquad \text{does not match}$$

$\boxed{TIME\ SAVER}$ *(3x − 3)(4x − 1) need not be tried, since 3x − 3 cannot be one of the factors of $12x^2$ − 20x + 3. Reason: Notice that 3x − 3 = 3(x − 1). If 3x − 3 were a factor, this would imply that 3 also factors out of $12x^2$ − 20x + 3. This is impossible, because 3 is not a factor of 20. This means that we can reject the set up (3x − 3)(4x − 1).*

Expressions of the form ax + a and ax − a can be rejected as factors if the number a is not a factor of the polynomial under consideration.

$$(6x - 1)(2x - 3) = 12x^2 - 20x + 3 \qquad \text{correct } \checkmark$$

Don't forget the factor of -1 in your final answer:
Write $(-1)(6x - 1)(2x - 3)$ or $-(6x - 1)(2x - 3)$, but don't drop the negative one!

41. First we will factor out -1:
$(-1)(12z^2 + 47z + 40)$

Now concentrate on factoring $12z^2 + 47z + 40$. Notice that a correct blank form will have two plus signs. Since 12 and 40 each have several factorizations, there are several possible set ups, two of which are shown here:

$$(3z + 5)(4z +8) = 12z^2 + 44z + 40 \qquad \text{does not match}$$
$$(3z + 8)(4z + 5) = 12z^2 + 47z + 40 \qquad \text{correct } \checkmark$$

$\boxed{TIME\ SAVER}$ *We claim that you can reject (3z + 5)(4z + 8) as a factorization without even multiplying it out to check. Notice that 4z + 8 = 4(z + 2). If 4z + 8 were a factor, this would imply that 4 factors out of $12z^2$ + 47z + 40. This is impossible, because 4 is not a factor of 47. We can reject the set up (3z + 5)(4z + 8).*

The correct factorization is $(-1)(3z + 8)(4z + 5)$ or $-(3z + 8)(4z + 5)$.

Any GCF factored out at the start is still part of the final answer. Don't drop that negative one!

45. The correct blank form will have one plus sign and one minus sign.
There are several possible set ups. $33m^2$ factors as $(3m)(11m)$ or $(33m)(m)$; $6n^2$ factors as $(2n)(3n)$ or $(6n)(n)$. After trial and error, the correct factorization is:

$$(3m + 2n)(11m - 3n) = 33m^2 - 9mn + 22mn - 6n^2 = 33m^2 + 13mn - 6n^2 \quad \checkmark$$

Read solutions to #37 and #65 for hints on how to save time checking the possibilities.

49. The correct blank form will have one plus sign and one minus sign.
There are several possible set ups. $8a^2$ factors as $(4a)(2a)$ or $(8a)(a)$; $9b^2$ factors as $(3b)(3b)$ or $(9b)(b)$.

Since the coefficient of the middle term, -71, is large in magnitude, it makes sense to try a set up that also uses large numbers:

$$(8a + b)(a - 9b) = 8a^2 - 72ab + ab - 9b^2 = 8a^2 - 71ab - 9b^2 \quad \checkmark$$

Read solutions to #37 and #65 for hints on how to save time checking the possibilities.

53.   First we factor out the GCF, $5x^3$:

$5x^3(12x^2 - 29x + 15)$ Now we concentrate on factoring $12x^2 - 29x + 15$.

There are several possibilities, and in each case we must have two minus signs:

$(6x - 5)(2x - 3)$ Multiply this out to see that this will not give the correct middle term.

$(6x - 3)(2x - 5)$ Without even multiplying this out we know this will not work.

>   Reason: $6x - 3 = 3(2x - 1)$, implying 3 would be a factor of $12x^2 - 29x + 15$. This is impossible since 3 is not a factor of 29.   Therefore, $6x - 3$ cannot be one of the factors.   This a great time-saver. See also #37.

$(4x - 3)(3x - 5)$ This does work; multiply it out to see. $\checkmark$

The correct factorization is: $5x^3(4x - 3)(3x - 5)$
Any GCF factored out at the start is still part of the final answer.

57.   First we factor out the GCF:

$7av(30v^2 + 11vw - 30w^2)$

Now concentrate on factoring $(30v^2 + 11vw - 30w^2)$. There will be many possibilities, since 30 factors as $(6)(5)$, $(2)(15)$, $(3)(10)$ or $(30)(1)$.   The correct blank form will have one plus sign and one minus sign.

A suggestion: if you are in doubt as to which set up to try first, and all look equally good to you, choose numbers that are somewhat close together. That is, we will first try a set up with 5 and 6, not 30 and 1.   (This is not a very reliable approach, but it does work more often than it does not.)

$(6v + 5w)(5v - 6w) = 30v^2 - 36vw + 25vw - 30w^2$ sign of middle term is incorrect
$(6v - 5w)(5v + 6w) = 30v^2 + 36vw - 25vw - 30w^2$ correct $\checkmark$

The correct factorization: $7av(6v - 5w)(5v + 6w)$

Any GCF factored out at the start is still part of the final answer.

Read solutions to #37 and #65 for hints on how to save time checking the possibilities.

61. $(x - y)(x^2 + xy + y^2) = x(x^2 + xy + y^2) - y(x^2 + xy + y^2)$
$$= x^3 + x^2y + xy^2 - yx^2 - xy^2 - y^3$$
$$= x^3 - y^3$$

65. The correct blank form will have a plus sign and a minus sign. There are not many possibilities:

$(2y + 1)(11y - 1) = 22y^2 + 9y - 1$ does not match
$(2y - 1)(11y + 1) = 22y^2 - 9y - 1$ does not match
$(22y + 1)(y - 1) = 22y^2 - 21y - 1$ does not match
$(22y - 1)(y + 1) = 22y^2 + 21y - 1$ does not match

The polynomial $22y^2 - 15y - 1$ is prime.

TIME SAVER *Take a look at the last two set ups. We did not have to check both $(22y + 1)(y - 1)$ and $(22y - 1)(y + 1)$. These two are almost the same when expanded; the only difference is a sign change in the middle term.*

$(22y + 1)(y - 1) = 22y^2 - 21y - 1$ If this set up does not give $-15y$.....
                 ↑

$(22y - 1)(y + 1) = 22y^2 + 21y - 1$ then this one won't either. No need to try both!
                 ↑

The same idea applies to the first two set ups and we did not have to expand them both.

69. First rewrite in standard form. As it is, the expression is confusing to deal with.

$$5xy - 3y^2 + 6x^2$$

$$= 6x^2 + 5xy - 3y^2$$ From this, there aren't too many possibilities:

$(3x + y)(2x - 3y) = 6x^2 - 7xy - 3y^2$ middle term does not match

There is no point in checking $(3x - y)(2x + 3y)$, $(3x + 3y)(2x - y)$ or $(3x - 3y)(2x + y)$. Read solutions to #37 and #65 to see why. We go on to the next option, using $6x^2 = (6x)(x)$.

$(6x + y)(x - 3y) = 6x^2 - 17xy - 3y^2$ middle term does not match

We do not need to check $(6x - y)(x + 3y)$, $(6x + 3y)(x - y)$ or $(6x - 3y)(x + y)$. Again, read solutions to #37 and #65 to see why.

This exhausts all the possibilities, and so we conclude that the polynomial is prime.

73. $64a^2 + 0a - 1$ Although $64a^2$ factors in many ways, the one that should easily come to mind is $(8a)(8a)$. We will try this one first.

$(8a + 1)(8a - 1)$ This works. Notice that using 8a in each, in combination with the 1 and -1, gives $64a^2 - 8a + 8a - 1$, and so the middle term drops out.

77.   This problem is the difference of two expressions, each with a factor of $4a^2 - 31ab - 8b^2$. Consider this as if it were a single object, and factor it out. (If it is this step that you are having trouble with, see #29 in section 7-1 for a suggested method.)

$(4a^2 - 31ab - 8b^2)(7m - 5n)$

Now concentrate on factoring $4a^2 - 31ab - 8b^2$. (There are several possibilities, but the list can be reduced somewhat by following the time saving hints given in the solutions to #37 and #65.)

Sometimes it is helpful to take a look at the middle term... since the coefficient of -31 is relatively large in magnitude, it makes sense to first try $4a^2 = (4a)(a)$ and $8b^2 = (8b)(b)$, instead of other factorizations, because these contain the largest numbers.

$4a^2 - 31ab - 8b^2 = (4a + b)(a - 8b)$ Answer: $(4a + b)(a - 8b)(7m - 5n)$

Factoring is sometimes like detective work. You go with your best 'clue', and sometimes it works, sometimes it doesn't. As you practice more with factoring, you will get better in noticing the clues.

Problem Set 7.4

1. $m^2 - 2m + 1$
      $= m^2 - 2(m)(1) + 1^2$
      $= (m - 1)^2$

      Use the form $a^2 - 2ab + b^2 = (a - b)^2$.
      $a = m$ $b = 1$

5. $36v^2 + 12v + 1$
      $= (6v)^2 + 2(6v)(1) + 1^2$
      $= (6v + 1)^2$

      Use the form $a^2 + 2ab + b^2 = (a + b)^2$.
      $a = 6v$ $b = 1$

$\boxed{TIME\ SAVER}$ *Most problems in this section could be worked by the trial-and-error technique used in the last section. But it takes time to check out the various possibilities, so we prefer to use it only when necessary. In this section the idea is to learn to recognize special forms, because they are quicker and easier to factor. It will pay off greatly for you to learn to spot these special forms.*

9. $25x^2 - 20xy + 4y^2$
      $= (5x)^2 - 2(5x)(2y) + (2y)^2$
      $= (5x - 2y)^2$

      Use the form $a^2 - 2ab + b^2 = (a - b)^2$.
      $a = 5x$ $b = 2y$

13. $-x^2 + 16xy - 64y^2$
$= (-1)(x^2 - 16xy + 64y^2)$
$= (-1)[x^2 - 2(x)(8y) + (8y)^2]$
$= (-1)(x - 8y)^2$ or $-(x - 8y)^2$

Factor out -1 first.
Use the form $a^2 - 2ab + b^2 = (a - b)^2$.
$a = x$ $b = 8y$

17. $2x^3 + 16x^2 + 32x$
$= 2x(x^2 + 8x + 16)$
$= 2x(x^2 + 2(x)(4) + 4^2)$
$= 2x(x + 4)^2$

Factor out the GCF first.
Use the form $a^2 + 2ab + b^2 = (a + b)^2$.
$a = x$ $b = 4$

21. $9v^2 - 1$
$= (3v)^2 - 1^2$
$= (3v + 1)(3v - 1)$

Use the form $a^2 - b^2 = (a + b)(a - b)$.
$a = 3v$ $b = 1$

25. $4a^2 - 9b^2$
$= (2a)^2 - (3b)^2$
$= (2a + 3b)(2a - 3b)$

Use the form $a^2 - b^2 = (a + b)(a - b)$.

29. $36 - m^2$
$= (-1)(m^2 - 36)$
$= (-1)(m^2 - 6^2)$
$= -(m + 6)(m - 6)$

Factor out -1 first.
Use the form $a^2 - b^2 = (a + b)(a - b)$.
(The factorization is not complete without the factor of -1; do not remove it.)

33. $25x^2 + 64y^2$
$= (5x)^2 + (8y)^2$

This is a sum of squares, and is prime.

37. $100x^2 - 81y^2$
$= (10x^2) - (9y)^2$
$= (10x + 9y)(10x - 9y)$

Use the form $a^2 - b^2 = (a + b)(a - b)$.

41. $100av^2 - 36aw^2$
$= 4a(25v^2 - 9w^2)$
$= 4a[(5v)^2 - (3w)^2]$
$= 4a(5v + 3w)(5v - 3w)$

Factor out the GCF first.

Use the form $a^2 - b^2 = (a + b)(a - b)$.

45. $x^3 + 27$
$x^3 + 3^3$
$= (x + 3)(x^2 - 3x + 9)$

Use the form $a^3 + b^3 = (a + b)(a^2 - ab + b^2)$.
with x for a and 3 for b.

49. $64a^3 - b^3$
$(4a)^3 - b^3$
$= (4a - b)(16a^2 + 4ab + b^2)$

Use the form $a^3 - b^3 = (a - b)(a^2 + ab + b^2)$,
with 4a for x.

page 136

53. $x^6 - 36$

$= (x^3)^2 - 6^2$

$= (x^3 + 6)(x^3 - 6)$

Use the form $a^2 - b^2 = (a + b)(a - b)$, with x^3 for a and 6 for b.

57. $(a + 2b)^2 - (2a + b)^2$

$\boxed{SUGGESTION}$ *This is a difference of two squares, so the correct form to use is $a^2 - b^2 = (a + b)(a - b)$. This is awkward to apply, since a and b are already being used in the problem. It will help in these cases to rewrite the formula using entirely different variables. Choose any you like, as long as they are not used in the given problem. We will use s and t.*

Apply the formula $s^2 - t^2 = (s + t)(s - t)$, where $s = a + 2b$ and $t = 2a + b$.

$(a + 2b)^2 - (2a + b)^2$

Put in s for a + 2b, and t for 2a + b.

$= s^2 - t^2$

Now factor using the formula.

$= (s + t)(s - t)$

Now replace s with a + 2b and t with 2a + b.

$= (a + 2b + 2a + b)(a + 2b - (2a + b))$

Simplify... be careful with the subtraction.

$= (3a + 3b)(-a + b)$

You can factor 3 out of the first factor.

$= 3(a + b)(-a + b)$

You now have two choices:

(1) Factor -1 out of the second factor. This gives: $-3(a + b)(a - b)$,

or (2) Rewrite $(-a + b)$ as $(b - a)$, and leave the final answer as $3(a + b)(b - a)$.

Some instructors consider the first best, and some will say either is fine. Your instructor may have a preference, so it would be a good idea to ask.

61. $125x^3 + 8y^3$

$= (5x)^3 + (2y)^3$

$= (5x + 2y)(\ (5x)^2 - (5x)(2y) + (2y)^2\ )$

$= (5x + 2y)(25x^2 - 10xy + 4y^2)$

Use the form $a^3 + b^3 = (a + b)(a^2 - ab + b^2)$, with 5x for a, and 2y for b.

65. $\pi r_1{}^2h - \pi r_2{}^2h$

$= \pi h(r_1{}^2 - r_2{}^2)$

$= \pi h(r_1 + r_2)(r_1 - r_2)$

First, factor out the GCF, πh.
Use the form $a^2 - b^2 = (a + b)(a - b)$, with r_1 for a, and r_2 for b.
Be sure to carry along πh. It doesn't go away!

69. $(x^2 - y^2)a^2 - (x^2 - y^2)b^2$

$za^2 - zb^2$

$= z(a^2 - b^2)$

This is a mess! But notice the common expression $x^2 - y^2$. Replace $x^2 - y^2$ with z.

Now you can spot the GCF. Factor z out.

Now use the form $a^2 - b^2 = (a + b)(a - b)$.

$$= z(a + b)(a - b)$$ Replace z with $x^2 - y^2$.

$$= (x^2 - y^2)(a + b)(a - b)$$ Use the form $a^2 - b^2 = (a + b)(a - b)$ to

$$= (x + y)(x - y)(a + b)(a - b)$$ factor $x^2 - y^2$.

Reread this solution. Notice that the idea is to concentrate on one task at a time, and then go on and see what else needs to be done.

73. $343v^3 + w^3$ Notice the cubes.

     $= (7v)^3 + w^3$ Use the form $a^3 + b^3 = (a + b)(a^2 - ab + b^2)$.

     $= (7v + w)(49v^2 - 7vw + w^2)$

Note: You probably wouldn't have recognized that $343 = 7^3$. This is something you would have to discover with your calculator and a little trial and error.

77. $3.64v^2 - 91$ Factor out 3.64 . Divide 91 by 3.64.

     $= 3.64(v^2 - 25)$ Use the form $a^2 - b^2 = (a + b)(a - b)$.

     $= 3.64(v + 5)(v - 5)$

Problem Set 7.5

1a. $x^2 - 16$ Use the form $a^2 - b^2 = (a + b)(a - b)$.

     $= (x + 4)(x - 4)$

1b. $x^2 + 16$ Prime.

1c. $x^2 - 8x + 16$ Use the form $a^2 - 2ab + b^2 = (a - b)^2$, with x for a, and 4 for b.

     $= x^2 - 2(x)(4) + 4^2$

     $= (x - 4)^2$

1d. $x^2 - 17x + 16$ x^2 and 16 are perfect squares, but 17 is not even so this is not a perfect square trinomial. Use trial and error.

     $(x - 16)(x - 1)$ You need two minus signs, and 17 is rather large, so choose $16 = (16)(1)$ instead of $16 = (4)(4)$.
     Always multiply out to check.

5a. $5v - 5w$ Just factor out the GCF.

     $= 5(v - w)$

5b. $5vw - 10v$ The GCF is 5v.

     $= 5v(w - 2)$

5c. $5v^2 - 80$ Factor out the GCF.
       $= 5(v^2 - 16)$ Use the form $a^2 - b^2 = (a + b)(a - b)$, with v for a, and 4 for b.
       $= 5(v + 4)(v - 4)$

5d. $5ax - 5ay + 15a$ Factor out the GCF.
       $= 5a(x - y + 3)$

9a. $7bx^2 + 14bxy - 7bx$ 9b. $x^2 + 8x + 16$
       $= 7bx(x + 2y - 1)$ $= x^2 + 2(4x) + 4^2$
                                       $= (x + 4)^2$ *Multiply out to check!*

13. $36v^2 + 60vw + 25w^2$
       $= (6v)^2 + 2(6v)(5w) + (5w)^2$ Use the form $a^2 + 2ab + b^2 = (a + b)^2$.
       $= (6v + 5w)^2$

17. $36v^2 - 65vw + 25w^2$ (This is not a perfect square trinomial because
       $= (4v - 5w)(9v - 5w)$ 65 is not an even number.) Use trial-and-error.

21. $6a^2 + 5ab - 21b^2$ Use trial-and-error.
       $= (2a - 3b)(3a + 7b)$

25. $10ab^3 - 28ab^2 - 6ab$ Factor out the GCF first.
       $= 2ab(5b^2 - 14b - 3)$ Apply trial-and-error.
       $= 2ab(5b + 1)(b - 3)$

29. $7x^2 - 14x + 7$ Factor out the GCF first.
       $= 7(x^2 - 2x + 1)$ Use the form $a^2 - 2ab + b^2 = (a - b)^2$.
       $= 7(x - 1)^2$

33. $13m^2 + 52n^2$ Factor out the GCF first.
       $= 13(m^2 + 4n^2)$ $m^2 + 4n^2$ is a sum of squares and is prime.

37. $2x(a - b) + (a - b)$ Notice the common expression $a - b$.
       Let $z = a - b$.
       $2xz + z$
       $= z(2x + 1)$
       $= (a - b)(2x + 1)$ Be sure to replace z with $a - b$.

41. $ax^2 + bx^2 - 9a - 9b$ When there are four or more terms, try grouping.
       $= x^2(a + b) + (-9)(a + b)$ Notice the common expression $a + b$.
       $= x^2z + (-9)z$ We have replaced $a + b$ with z. Factor out z.
       $= z(x^2 - 9)$ Replace z with $a + b$. (This could be done later.)
       $= (a + b)(x^2 - 9)$ You're not done yet, since $x^2 - 9 = x^2 - 3^2$.
       $= (a + b)(x + 3)(x - 3)$

45. $-x^2 + x + 6$

    $= -(x^2 - x - 6)$
    $= -(x - 3)(x + 2)$

It is easier to factor if the leading coefficient is positive. Factor out -1.
Now use trial and error. Carry along -1.

49. $-77 + x^2 - 4x$
    $= x^2 - 4x - 77$
    $= (x - 11)(x + 7)$

First rewrite in standard form.
Use trial and error.

53. $x^2(a + b) + 2x(a + b) + (a + b)$
    $= x^2 z + 2xz + z$
    $= z(x^2 + 2x + 1)$
    $= z(x + 1)^2$
    $= (a + b)(x + 1)^2$

Notice the common expression. Let $z = a + b$.
Factor out z.
$x^2 + 2x + 1$ is a perfect square trinomial.
Now replace z with $(a + b)$.

57. $-z^2 - 1$
    $-(z^2 + 1)$

Factor out -1.
$z^2 + 1$ is prime, so we can go no further.

61. $x^2 - 9a^2 + 6a - 1$

This looks unusual. There are four terms, so suppose we try grouping. Let's group the terms in pairs, just as they appear:

$(x^2 - 9a^2) + (6a - 1)$
$= (x + 3a)(x - 3a) + (6a - 1)$

This grouping is not helpful, since we do not see a common expression to factor out. This does not mean that grouping was the wrong approach; maybe we should try a more creative arrangement. Go back to the original expression and group them as shown:

$x^2 - 9a^2 + 6a - 1$
$= x^2 - (9a^2 - 6a + 1)$
$= x^2 - (3a - 1)^2$
$= (x + (3a - 1))\ (x - (3a - 1))$
$= (x + 3a - 1)(x - 3a + 1)$

Use the form $a^2 - 2ab + b^2 = (a - b)^2$.
Use the form $a^2 - b^2 = (a + b)(a - b)$.
Now simplify.

It was the presence of $9a^2$, $6a$ and 1 that suggested this grouping, because they remind us of a perfect square trinomial.

65. $60m^4 + 35m^3 - 50m^2$
    $= 5m^2(12m^2 + 7m - 10)$
    $= 5m^2(3m - 2)(4m + 5)$

Factor out the GCF first.
Use trial-and-error.

69. $2a^3 + 2$
    $= 2(a^3 + 1)$
    $= 2(a + 1)(a^2 - a + 1)$

Factor out the GCF first.
Use the form $a^3 + b^3 = (a + b)(a^2 - ab + b^2)$.

73. $320x^4y - 5xy^4$
     $= 5xy(64x^3 - y^3)$
     $= 5xy[\,(4x)^3 - y^3\,]$
     $= 5xy(4x - y)[\,(4x)^2 + 4xy + y^2\,]$
     $= 5xy(4x - y)(16x^2 + 4xy + y^2)$

Factor out the GCF first.
Notice the cubes.
Use the form $a^3 - b^3 = (a - b)(a^2 + ab + b^2)$.

Problem Set 7.6

1. $4x^2 - 9x + 8 = 0$ See answer section in the text.

5. $(3x - 1)(2x + 1) = 4$
    $6x^2 + 3x - 2x - 1 = 4$
    $6x^2 + x - 5 = 0$ See answer section in the text.

9. $5x = x^2$
    $x^2 - 5x = 0$ See answer section in the text.

13. $(v + 8)(v - 4) = 0$

$v + 8 = 0 \qquad v - 4 = 0$

$v = -8 \qquad v = 4$

17. $(5t + 8)(3t - 5) = 0$

$5t + 8 = 0 \qquad 3t - 5 = 0$

$5t = -8 \qquad 3t = 5$

$t = -\dfrac{8}{5} \qquad t = \dfrac{5}{3}$

21. $x(x - 3)(x + 4) = 0$
     $x = 0, \quad x - 3 = 0, \quad x + 4 = 0$
     $x = 0, \quad x = 3, \quad x = -4$

25. $m(m - 1)(m + 2)(m - 3) = 0$
     $m = 0, \quad m-1 = 0, \quad m+2 = 0, \quad m-3 = 0$
     $m = 0, \quad m = 1, \quad m = -2, \quad m = 3$

29. $m^2 - 6m - 16 = 0$
     $(m - 8)(m + 2) = 0$
     $m - 8 = 0 \qquad m + 2 = 0$
     $m = 8 \qquad m = -2$

33. $2r^2 + 9r - 35 = 0$
     $(2r - 5)(r + 7) = 0$
     $2r - 5 = 0 \qquad r + 7 = 0$
     $r = \dfrac{5}{2} \qquad r = -7$

37. $y^2 = 8y + 33$
     $y^2 - 8y - 33 = 0$
     $(y - 11)(y + 3) = 0$
     $y - 11 = 0 \qquad y + 3 = 0$
     $y = 11 \qquad y = -3$

41. $19w^2 = 5w$
     $19w^2 - 5w = 0$
     $w(19w - 5) = 0$
     $w = 0 \qquad 19w - 5 = 0$
     $\qquad\qquad w = \dfrac{5}{19}$

45. $z^2 = 0$
     $z \cdot z = 0$
     $z = 0 \quad$ (double root)

49. $(a - 1)(a - 2) = 0$
     $a - 1 = 0 \qquad a - 2 = 0$
     $a = 1 \qquad a = 2$

53. $(y + 1)(y + 2) = 12$
$y^2 + 2y + y + 2 = 12$
$y^2 + 3y - 10 = 0$
$(y + 5)(y - 2) = 0$
$y + 5 = 0 \qquad y - 2 = 0$
$y = -5 \qquad y = 2$

57. $2z(2z^2 - 3z - 14) = 0$
$2z(2z - 7)(z + 2) = 0$
$2z = 0 \qquad 2z - 7 = 0 \qquad z + 2 = 0$
$z = 0 \qquad z = \frac{7}{2} \qquad z = -2$

61a. $6x^2 + 7x + 2$
$= (3x + 2)(2x + 1)$

61b. $6x^2 + 7x + 2 = 0$
$(3x + 2)(2x + 1) = 0$
$x = -\frac{2}{3}, \, x = -\frac{1}{2}$

65a. $(3x + 5)(5x - 2)$
$= 15x^2 + 19x - 10$

65b. $(3x + 5)(5x - 2) = 0$
$3x + 5 = 0, \, 5x - 2 = 0$
$x = -\frac{5}{3}, \, x = \frac{2}{5}$

69. $5x(x + 7) + 8(x + 7) = 0$
$5xz + 8z = 0$
$z(5x + 8) = 0$
$(x + 7)(5x + 8) = 0$
$x + 7 = 0 \qquad 5x + 8 = 0$
$x = -7 \qquad x = -\frac{8}{5}$

Let $z = x + 7$ to aid in factoring.

Replace z with $x + 7$.

73. Set each factor equal to zero, and solve for x:
$3.56x - 14.24 = 0 \qquad\qquad 5.87x + 29.35 = 0$
$3.56x = 14.24 \qquad\qquad\quad 5.87x = -29.35$
$x = \frac{14.24}{3.56} = 4 \qquad\qquad x = -\frac{29.35}{5.87} = -5$

77. Let one number be n, and the other $2n + 7$.
$n(2n + 7) = 85$
$2n^2 + 7n - 85 = 0$
$(2n + 17)(n - 5) = 0$
$2n + 17 = 0 \qquad n - 5 = 0$
$n = -\frac{17}{2} \qquad\qquad n = 5$

These are two values for n, not the two numbers asked for in the problem.
For each n the other number required is $2n + 7$:

$n = -\frac{17}{2} \, ; \, 2n + 7 = 2(-\frac{17}{2}) + 7 = -10$

$n = 5 \, ; \, 2n + 7 = 2(5) + 7 = 17$
Answer: $-\frac{17}{2}$ and -10 or 5 and 17.

Problem Set 7.7

1. Let n = the first integer and
   n + 1 = the second integer.

   $n(n + 1) = 56$
   $n^2 + n = 56$
   $n^2 + n - 56 = 0$
   $(n + 8)(n - 7) = 0$
   $n + 8 = 0 \qquad n - 7 = 0$
   $n = -8 \qquad n = 7$

   For n = −8, n + 1 = −7.
   For n = 7, n + 1 = 8.
   The consecutive integers are −8 and −7
   or 7 and 8.

5. Let n = the first odd integer and
   n + 2 = the second odd integer.

   $n(n + 2) = 99$
   $n^2 + 2n = 99$
   $n^2 + 2n - 99 = 0$
   $(n + 11)(n - 9) = 0$
   $n + 11 = 0 \qquad n - 9 = 0$
   $n = -11 \qquad n = 9$

   For n = −11, n + 2 = −9.
   For n = 9, n + 2 = 11.
   The consecutive odd integers
   are −11 and −9 or 9 and 11.

9. Let n = the first integer and
   n + 1 = the second integer.

   $n^2 + (n + 1)^2 = 113$
   $n^2 + n^2 + 2n + 1 = 113$
   $2n^2 + 2n - 112 = 0 \qquad$ *Divide both sides by 2.*
   $2(n^2 + n - 56) = 0$
   $n^2 + n - 56 = 0$
   $(n + 8)(n - 7) = 0$
   $n + 8 = 0 \qquad n - 7 = 0$
   $n = -8 \qquad n = 7$

   For n = −8, n + 1 = −7.
   For n = 7, n + 1 = 8.
   The consecutive integers are −8 and −7
   or 7 and 8.

13. Let n = the first even integer and
    n + 2 = the second even integer.

    $n^2 + (n + 2)^2 = 340$
    $n^2 + n^2 + 4n + 4 = 340$
    $2n^2 + 4n - 336 = 0$
    $2(n^2 + 2n - 168) = 0$
    $n^2 + 2n - 168 = 0$
    $(n + 14)(n - 12) = 0$
    $n + 14 = 0 \qquad n - 12 = 0$
    $n = -14 \qquad n = 12$

    For n = −14, n + 2 = −12.
    For n = 12, n + 2 = 14.
    The consecutive even integers are
    −14 and −12 or 12 and 14.

17. Let n = the first even integer and
    n + 2 = the second even integer.

    their product: $n(n + 2)$
    their sum: $n + n + 2$
    four times their sum: $4(n + n + 2)$
    8 more than that: $8 + 4(n + n + 2)$
    $n(n + 2) = 8 + 4(n + n + 2)$
    $n^2 + 2n = 8 + 4(2n + 2)$
    $n^2 + 2n = 8 + 8n + 8$
    $n^2 - 6n - 16 = 0$
    $(n - 8)(n + 2) = 0$
    $n - 8 = 0 \qquad n + 2 = 0$
    $n = 8 \qquad n = -2$
    If n = 8, then n + 2 = 10.
    If n = −2, then n + 2 = 0.
    The consecutive even integers are 8 and 10 or −2 and 0.

21. Let w = the width of the rectangle, and
    6 + w = the length.

    The area of a rectangle is the product of
    the length and width: $w(6 + w)$
    $w(6 + w) = 91 \qquad$ This is not standard form.
    $6w + w^2 = 91$
    $w^2 + 6w - 91 = 0$
    $(w + 13)(w - 7) = 0$
    $w + 13 = 0 \qquad w - 7 = 0$
    $w = -13 \qquad w = 7$
    (We reject w = −13 because width is not
    negative.)
    The width is 7 cm.

25. $6^2 + 8^2 = c^2$
    $36 + 64 = c^2$
    $c^2 = 100; c = 10$ m

    Use the Pythagorean Theorem : $a^2 + b^2 = c^2$.
    The mathematics says that c = 10 or c = −10,
    but we reject c = − 10 because c represents length.

29. Let d = the distance along the floor from the ladder to the wall, and d + 14 = the distance along the wall from the floor to the top of the ladder.

    $d^2 + (d + 14)^2 = 26^2$ Use the Pythagorean Theorem : $a^2 + b^2 = c^2$.
    $d^2 + d^2 + 28d + 196 = 676$
    $2d^2 + 28d − 480 = 0$ Divide both sides by 2. Notice $\frac{0}{2} = 0$.
    $d^2 + 14d − 240 = 0$
    $(d − 10)(d + 24) = 0$
    $d − 10 = 0 \qquad d + 24 = 0$
    $d = 10 \qquad\qquad d = −24$ *Reject d = −24 because distance is nonnegative.*
    The distance is 10 feet.

33. Let w = the length of the shorter side of the rectangle and w + 17 = the length of the longer side. These two sides and the diagonal give us a right triangle.

    $w^2 + (w + 17)^2 = 25^2$ Use the Pythagorean Theorem : $a^2 + b^2 = c^2$.
    $w^2 + w^2 + 34w + 289 = 625$
    $2w^2 + 34w − 336 = 0$ Divide both sides by 2. Notice that $\frac{0}{2} = 0$.
    $w^2 + 17w − 168 = 0$
    $(w + 24)(w − 7) = 0$
    $w + 24 = 0 \qquad w − 7 = 0$
    $w = −24 \qquad\quad w = 7$ The short side is 7 cm and the long side is
                                        7+17=24 cm.

37. The segments with lengths 20, 15 and x represent the three sides of a right triangle, so it is appropriate to use the Pythagorean Theorem.   The side labeled x is the hypotenuse.

    $x^2 = 20^2 + 15^2$
    $x^2 = 400 + 225$
    $x^2 = 625$
    $x = 25$

41. Call the length of the dashed line segment d (for distance). This dashed line is the hypotenuse of a right triangle, with legs of length 20 feet and 21 feet. We will use the Pythagorean Theorem.

    $d^2 = 20^2 + 21^2$
    $d^2 = 400 + 441$
    $d^2 = 841$ You'll need your calculator for this one.
    $d = 29$ Answer: 29 feet. Be sure to include units with answers.

45. Let w be the width of the shaded square. We are given that the area of the unshaded portion is 65 cm^2, so we will write an equation using this fact.

Notice that the area of the unshaded region is just the area of the larger square minus the area of the smaller square. Start by writing an equation, in words:

(area of larger square) − (area of smaller square) = 65

Now the area of the shaded square is w^2. The larger square has side $2w + 1$ (reread the first sentence of the problem) and so the area of the larger square is $(2w + 1)^2$. We now have:

$(2w + 1)^2 - w^2 = 65$	The left side requires some simplifying.
$4w^2 + 4w + 1 - w^2 = 65$	Collect all terms to the left side.
$3w^2 + 4w - 64 = 0$	You cannot begin factoring unless one side is zero.
$(3w + 16)(w - 4) = 0$	
$w = -\dfrac{16}{3}$ or $w = 4$	We reject $w = -\dfrac{16}{3}$ because width is nonnegative. The width of the shaded square is 4 cm.

Solutions to All Review Exercises − Chapter Seven

1. $12 = 2 \cdot 6 = 2 \cdot 2 \cdot 3 = \quad 2^2 \cdot 3$
    $18 = 2 \cdot 9 = \quad 2 \cdot 3^2$ $\Big\}$ The GCF is $2 \cdot 3 = 6$.
    $30 = 2 \cdot 15 = \quad 2 \cdot 3 \cdot 5$

2. $42 = 2 \cdot 21 = 2 \cdot 3 \cdot 7$
    $56 = 8 \cdot 7 = 2^3 \cdot 7$
    $70 = 7 \cdot 10 = 2 \cdot 5 \cdot 7$ The GCF is $2 \cdot 7 = 14$
    $98 = 2 \cdot 49 = 2 \cdot 7^2$

3. $15x^3y - 25x^2y^2 + 35xy^3$

    $15x^3y = \quad 3(5) x^3y$
    $25x^2y^2 = \quad 5^2x^2y^2$ $\Big\}$ The GCF is $5xy$.
    $35xy^3 = \quad 5(7) xy^3$

4. $10x(3a - b) - 18y(3a - b)$

    $10x(3a - b) = \quad 2(5) x (3a - b)$
    $18y(3a - b) = \quad 2(3^2) y (3a - b)$ The GCF is $2(3a - b)$.

5. $22x - 33 = 11(2x) - 11(3) = 11(2x - 3)$

6. $81m^2 - 1 = (9m)^2 - 1 = (9m + 1)(9m - 1)$ Use the form $a^2 - b^2 = (a + b)(a - b)$.

7. $4m^2 + 9$ This is a sum of squares, and is prime.

8. $x^2 + 10xy + 25y^2$
   $= x^2 + 2(5)xy + (5y)^2$
   $= (x + 5y)^2$

   Use the form $(a + b)^2 = a^2 + 2ab + b^2$.

9. $12m^2 - mn - n^2$
   $= (3m - n)(4m + n)$

   The correct sign pattern will require different signs because of $-n^2$.

10. $64ax^2 - 16ax$
    $= 16ax(4x - 1)$

11. $36a^2 - 12ab + b^2$
    $= (6a)^2 - 2(6)ab + b^2$
    $= (6a - b)^2$

    Use the form $(a - b)^2 = a^2 - 2ab + b^2$.

12. $35v^2 - 11vw - 6w^2$
    $= (5v - 3w)(7v + 2w)$

    Use trial and error.

13. $4m^2 - 9$
    $= (2m + 3)(2m - 3)$

    Use the form $a^2 - b^2 = (a + b)(a - b)$.

14. $a(7b - 2) - 4(7b - 2)$
    $= (7b - 2)(a - 4)$

    Notice the common factor $(7b - 2)$.

15. $22x^3y - 77x^2y$
    $= 11x^2y(2x - 7)$

16. $100t^2 - 220t + 121$
    $= (10t)^2 - 2(10)(11)t + 11^2$
    $= (10t - 11)^2$

    Use the form $(a - b)^2 = a^2 - 2ab + b^2$.

17. $10t^2 + 17t - 11$
    $= (2t - 1)(5t + 11)$

    The correct sign pattern will require different signs because of -11. Trial and error.

18. $8m^2 + 9mn - 14n^2$
    $= (3m - 2n)(6m + 7n)$

    Use trial and error.

19. $2x(3y + 5) - (3y + 5)$
    $= (3y + 5)(2x - 1)$

    Factor out $(3y + 5)$.

20. $12x^5y^2 - 18x^4y^2 + 30x^3y^2 - 24x^2y^2$

    $= 6x^2y^2(2x^3 - 3x^2 + 5x - 4)$

    $= 6x^2y^2(x - 1)(2x^2 - x + 4)$

21. $x^3y - xy^3$

    $= xy(x^2 - y^2)$

    $= xy(x + y)(x - y)$

    First factor out the GCF.

    Use the form $a^2 - b^2 = (a + b)(a - b)$.

22. $4a^2x - b^2x$

    $= x(4a^2 - b^2)$

    $= x(2a + b)(2a - b)$

    Use the form $a^2 - b^2 = (a + b)(a - b)$.

23. $2am - an + 4bm - 2bn$

    $= a(2m - n) + 2b(2m - n)$

    $= (2m - n)(a + 2b)$

    When there are four or more terms, try grouping.

24. $7a^2x^3 + 63b^2x^3$

    $= 7x^3(a^2 + 9b^2)$

    Factor out the GCF.

25. $36av^2 + 180avw + 225aw^2$

    $= 3a(12v^2 + 60vw + 75w^2)$

    $= 3a(3)(4v^2 + 20vw + 25w^2)$

    $= 9a(4v^2 + 20vw + 25w^2)$

    $= 9a[(2v)^2 + 20vw + (5w)^2]$

    $= 9a[(2v)^2 + 2(2v)(5w) + (5w)^2]$

    $= 9a(2v + 5w)^2$

    36, 180 and 225 are all divisible by 3. 12, 60 and 75 are also divisible by 3.

    Notice that $4v^2$ and $25w^2$ are both squares.

    Use the form $a^2 + 2ab + b^2 = (a + b)^2$.

26. $-54a^3b + 252a^2b^2 - 294ab^3$

    $= -6ab(9a^2 - 42ab + 49b^2)$

    $= -6ab(3a - 7b)^2$

    Factor out the GCF.

    Use the form $(a - b)^2 = a^2 - 2ab + b^2$.

27. $-12x^2 + 8xy + 15y^2$

    $= -(12x^2 - 8xy - 5y^2)$

    $= -(2x - 3y)(6x + 5y)$

    Factor out -1.

    Use trial and error.

28. $-36x^2 + 60xy - 25y^2$

    $= -(36x^2 - 60xy + 25y^2)$

    $= -(6x - 5y)^2$

    Factor out -1.

    Use the form $(a - b)^2 = a^2 - 2ab + b^2$.

Chapter Seven Review

29. $mn - 6n^2 + 40m^2$
   $= 40m^2 + mm - 6n^2$
   $= (5m + 2n)(8m - 3n)$

   Write the expression in standard form.
   m comes before n
   The correct sign pattern will require different signs because of $-6n^2$.

30. $19t - 12 + 18t^2$
   $= 18t^2 + 19t - 12$
   $= (2t + 3)(9t - 4)$

   First rewrite in standard form.

31. $av^2 - 9a - bv^2 + 9b$
   $= a(v^2 - 9) - b(v^2 - 9)$
   $= (v^2 - 9)(a - b)$
   $= (v + 3)(v - 3)(a - b)$

   When four or more terms are present, try grouping.
   Use the form $a^2 - b^2 = (a + b)(a - b)$.

32. $16y^4 - 1$
   $= (4y^2) - 1^2$
   $= (4y^2 + 1)(4y^2 - 1)$
   $= (4y^2 + 1)(2y + 1)(2y - 1)$

   Use the form $a^2 - b^2 = (a + b)(a - b)$.

   Use the form $a^2 - b^2 = (a + b)(a - b)$.

33. $2ax^2 - 2ax - 12a + bx^2 - bx - 6b$

   When you have four or more terms, try grouping. Notice that the first three have a common factor of 2a and the last three have a common factor of b. This suggests grouping the first three together and the last three together.

   $(2ax^2 - 2ax - 12a) + (bx^2 - bx - 6b)$
   $= 2a(x^2 - x - 6) + b(x^2 - x - 6)$
   Let $z = x^2 - x - 6$.
   $2az + bz$
   $= z(2a + b)$
   $= (x^2 - x - 6)(2a + b)$
   $= (x - 3)(x + 2)(2a + b)$

   Now replace z with what it stood for.

   Not done yet...

34. $ax^2 + 25a + bx^2 + 25b$
   $= a(x^2 + 25) + b(x^2 + 25)$
   $= (x^2 + 25)(a + b)$

   Use grouping.

35. $v^4 - 625$
   $= (v^2)^2 - (25)^2$
   $= (v^2 + 5)(v^2 - 5)$
   $= (v^2 + 5)(v + 5)(v - 5)$

   This is a difference of squares.
   Use the form $a^2 - b^2 = (a + b)(a - b)$.
   Use the form $a^2 - b^2 = (a + b)(a - b)$.

36. $2w^5 - 32w$
   $= 2w(w^4 - 16)$
   $= 2w(w^2 + 4)(w^2 - 4)$
   $= 2w(w^2 + 4)(w + 2)(w - 2)$

   $w^4 - 16 = (w^2)^2 - 4^2$

Chapter Seven Review

37. $(2x - 3)(x + 1) - (x - 9)(x + 1)$ Notice the common factor of $x + 1$.
    Let $z = x + 1$.
    $(2x - 3)z - (x - 9)z$
    $= z[(2x - 3) - (x - 9)]$ Be careful with subtraction.
    $= z(2x - 3 - x + 9)$
    $= z(x + 6)$ Replace z with $x + 1$.
    $= (x + 1)(x + 6)$

38. $(3v + 4)(v - 2) - (v - 2)(v - 5)$ Notice the common factor $(v - 2)$.
    $= (v - 2)[(3v + 4) - (v - 5)]$
    $= (v - 2)[3v + 4 - v + 5]$
    $= (v - 2)(2v + 9)$

39. $x^2(a^2 + b^2) - 4(a^2 + b^2)$ Notice the common factor $a^2 + b^2$.
    $= (a^2 + b^2)(x^2 - 4)$ Use the form $a^2 - b^2 = (a + b)(a - b)$.
    $= (a^2 + b^2)(x + 2)(x - 2)$

40. $x^2(x^2 - 5x - 6) - 9(x^2 - 5x - 6)$ Factor out $(x^2 - 5x - 6)$.
    $= (x^2 - 5x - 6)(x^2 - 9)$ Each factor can be factored.
    $= (x - 6)(x + 1)(x + 3)(x - 3)$

41a. $64x^2 - 1$
    $= (8x + 1)(8x - 1)$

41b. $64x^2 + 1$ is prime.

41c. $64x^3 - 1$ Use the form $a^3 - b^3 = (a - b)(a^2 + ab + b^2)$.
    $= (4x)^3 - 1$
    $= (4x - 1)(16x^2 + 4x + 1)$

41d. $64x^3 + 1$ Use the form $a^3 + b^3 = (a + b)(a^2 - ab + b^2)$.
    $= (4x)^3 + 1$
    $= (4x + 1)(16x^2 - 4x + 1)$

42. $(a + b)(a^2 - ab + b^2)$

    $= a(a^2 - ab + b^2)$
    $\quad + b(a^2 - ab + b^2)$

    $= a^3 - a^2b + ab^2$
    $\quad + ba^2 - ab^2 + b^3$

    $= a^3 + b^3$

43a. $(6x + 3y)(6x - 3y)$ Use FOIL.

$$= (6x)^2 - (6x)(3y) + (3y)(6x) - (3y)^2$$
$$= 36x^2 - 18xy + 18xy - 9y^2$$
$$= 36x^2 - 9y^2$$

43b. $9(4x^2 - y^2)$

$$= 9(4x^2) - 9(y^2)$$
$$= 36x^2 - 9y^2$$

43c.   This is explained in the answer section in your book.

44.   Two can be factored out of each factor:
$$(2x + 2)(2x - 2) = 2(x + 1)2(x - 1) = 4(x + 1)(x - 1)$$

45.   Because $v^2 - 9$ can be factored as $(v + 3)(v - 3)$.
The error is in saying that $(v^2 + 9)(v^2 - 9)$ is a <u>complete</u> factorization of $v^4 - 81$.

46. $20x^2 + 35xy + 12y^2$ (prime)

47. $(v + 7)(2v - 9) = 0$
$v + 7 = 0$ $2v - 9 = 0$
$v = -7$ $v = \dfrac{9}{2}$

48. $y(y - 4)(2y + 5) = 0$
$y = 0 \quad y = 4 \quad y = -\dfrac{5}{2}$

49. $6w^2 - 5w - 21 = 0$
$(3w - 7)(2w + 3) = 0$
$3w - 7 = 0$ $2w + 3 = 0$
$w = \dfrac{7}{3}$ $w = -\dfrac{3}{2}$

50. $4m^2 = 28m - 49$ Write in standard form.
$4m^2 - 28m + 49 = 0$ Use the form $(a - b)^2 = a^2 - 2ab + b^2$.
$(2m - 7)^2 = 0$
$2m - 7 = 0$
$m = \dfrac{7}{2}$ (double root)

51. $(2n + 1)(3n - 2) = 20$ Write in standard form.
$6n^2 - 4n + 3n - 2 = 20$
$6n^2 - n - 22 = 0$
$(6n + 11)(n - 2) = 0$
$6n + 11 = 0 \qquad n - 2 = 0$
$n = -\dfrac{11}{6}, \qquad n = 2$

52. $(3x + 4)(2x - 5) = (x - 4)(2x + 1)$ Write in standard form.
$6x^2 - 15x + 8x - 20 = 2x^2 + x - 8x - 4$
$6x^2 - 7x - 20 = 2x^2 - 7x - 4$
$4x^2 - 16 = 0$
$4(x^2 - 4) = 0$
$x^2 - 4 = 0$
$(x + 2)(x - 2) = 0$
$x = -2, \; x = 2$

53. $5t(t^2 - 4) = 6(t^2 - 4)$
$5t(t^2 - 4) - 6(t^2 - 4) = 0$
Let $z = t^2 - 4$.
$5tz - 6z = 0$
$z(5t - 6) = 0$

$(t^2 - 4)(5t - 6) = 0$
$(t + 2)(t - 2)(5t - 6) = 0$
$t + 2 = 0 \qquad t - 2 = 0 \qquad 5t - 6 = 0$
$t = -2 \qquad\quad t = 2 \qquad\quad t = \dfrac{6}{5}$

54. $x^3 = 9x$ Bring all terms to the left.
$x^3 - 9x = 0$
$x(x^2 - 9) = 0$
$x(x + 3)(x - 3) = 0$
$x = 0 \qquad x + 3 = 0 \qquad x - 3 = 0$
$x = 0 \qquad x = -3 \qquad x = 3$

55a. $6x^2 - 13x + 6$
$= (3x - 2)(2x - 3)$

55b. $6x^2 - 13x + 6 = 0$
$(3x - 2)(2x - 3) = 0$
$3x - 2 = 0 \qquad 2x - 3 = 0$
$x = \dfrac{2}{3} \qquad\qquad x = \dfrac{3}{2}$

56a. $14x^2 - x - 15$
$= (14x - 15)(x + 1)$

56b. $14x^2 - x - 15 = 0$
$(14x - 15)(x + 1) = 0$
$14x - 15 = 0 \qquad x + 1 = 0$
$x = \dfrac{15}{4} \qquad\qquad x = -1$

Chapter Seven Review

57. Let the integers be n and n + 2.
$$n(n + 2) = 6 + 7n$$
$$n^2 + 2n = 6 + 7n$$
$$n^2 - 5n - 6 = 0$$
$$(n - 6)(n + 1) = 0$$
$$n = 6 \qquad n = -1$$

(We reject n = −1 since it is not even)
The consecutive even integers are 6 and 8.

58. width   w
length   2w − 4
area of a rectangle = (width)(length)

$$48 = w(2w - 4)$$
$$48 = 2w^2 - 4w$$
$$2w^2 - 4w - 48 = 0 \qquad \text{Divide both sides by 2.}$$
$$w^2 - 2w - 24 = 0$$
$$(w - 6)(w + 4) = 0$$
$$w = 6 \qquad w = -4 \qquad \text{We reject } w = -4, \text{ because width is nonnegative.}$$

The width is 6 cm and the length is 2(6) −4 = 8 cm.

59. Let c = the length of the wire.

$$7^2 + 24^2 = c^2$$
$$49 + 576 = c^2$$
$$625 = c^2$$
$$c = 25 \qquad \text{The length of the wire is 25 m.}$$

60. width   w
length   2w − 1
diagonal   2w + 1

Apply the Pythagorean Theorem:

$$w^2 + (2w - 1)^2 = (2w + 1)^2$$
$$w^2 + 4w^2 - 4w + 1 = 4w^2 + 4w + 1$$
$$w^2 - 8w = 0$$
$$w(w - 8) = 0$$
$$w = 0, \ w = 8 \qquad \text{The width is 8 m.}$$

61. This is worked out in the answer section in your book.

62. $$ax + ay + bx + by + x + y$$
$$= a(x + y) + b(x + y) + 1(x + y)$$
$$= (x + y)(a + b + 1)$$

1a. $18x^3 - 12x^2 + 30x$
$= 6x(3x^2) + 6x(-2x) + 6x(5)$
$= 6x(3x^2 - 2x + 5)$

1b. $35ax^3y^2 + 15ax^2y^3$
$= 5ax^2y^2(7x + 3y)$

1c. $-6y^4 + 10y^3 - 14y^2 + 38y$
$= 2y(-3y^3 + 5y^2 - 7y + 19)$

1d. $15v(2a - b) - 18w(2a - b)$
Let $z = 2a - b$.
$15vz - 18wz$
$= 3z(5v - 6w)$
$= 3(2a - b)(5v - 6w)$

2a. $ax - 3a + 5x - 15$
$= a(x - 3) + 5(x - 3)$
$= (x - 3)(a + 5)$

2b. $x^2 - 2x + 7x - 14$
$= x(x - 2) + 7(x - 2)$
$= (x - 2)(x + 7)$

2c. $2bx - 3by - 2cx + 3cy$
$= b(2x - 3y) - c(2x - 3y)$
$= (2x - 3y)(b - c)$

2d. $ax + ay + az + bx + by + bz$
$= a(x + y + z) + b(x + y + z)$
$= (x + y + z)(a + b)$

3a. $x^2 - 2x - 3$
$= (x - 3)(x + 1)$

3b. $y^2 - 11y + 18$
$= (y - 9)(y - 2)$

3c. $v^2 + 13vw + 36w^2$
$= (v + 9w)(v + 4w)$

3d. $a^2 + 7ab - 144b^2$
$= (a - 9b)(a + 16b)$

4a. $11t^2 + 56t + 5$
$= (11t + 1)(t + 5)$

4b. $15r^2 + r - 2$
$= (5r + 2)(3r - 1)$

4c. $18x^2 - 23xy - 6y^2$
$= (9x + 2y)(2x - 3y)$

4d. $12a^2 - 25ab + 12b^2$
$= (3a - 4b)(4a - 3b)$

5a. $11t^2 - 56t + 5$
$= 11t^2 - 55t - t + 5$
$= 11t(t - 5) - (t - 5)$
$= (11t - 1)(t - 5)$

5b. $15r^2 - r - 2$
$= 15r^2 + 5r - 6r - 2$
$= 5r(3r + 1) - 2(3r + 1)$
$= (5r - 2)(3r + 1)$

5c. $18x^2 + 23xy - 6y^2$
$= 18x^2 + 27xy - 4xy - 6y^2$
$= 9x(2x + 3y) - 2y(2x + 3y)$
$= (9x - 2y)(2x + 3y)$

5d. $15a^2 + 8ab - 12b^2$
$= 15a^2 - 10ab + 18ab - 12b^2$
$= 5a(3a - 2b) + 6b(3a - 2b)$
$= (5a + 6b)(3a - 2b)$

Chapter Seven Mastery Test

6a. $49b^2 + 42b + 9$
$= (7b + 3)^2$

6b. $25v^2 - 90v + 81$
$= (5v - 9)^2$

6c. $16m^2 - 88mn + 121n^2$
$= (4m - 11n)^2$

6d. $36x^2 + 156xy + 169y^2$
$= (6x + 13y)^2$

7a. $m^2 - 81$
$= (m + 9)(m - 9)$

7b. $100a^2 - 1$
$= (10a + 1)(10a - 1)$

7c. $25x^2 - 121y^2$
$= (5x + 11y)(5x - 11y)$

7d. $144m^2 - 49n^2$
$= (12m + 7n)(12m - 7n)$

8a. $125z^3 - 1$ (difference of cubes)
$= (5z - 1)(25z^2 + 5z + 1)$

8b. $z^3 + 125$ (sum of cubes)
$= (z + 5)(z^2 - 5z + 25)$

8c. $27x^3 - 64$ (difference of cubes)
$= (3x - 4)(9x^2 + 12x + 16)$

8d. $8m^3 + 125b^3$ (sum of cubes)
$= (2m + 5b)(4m^2 - 10mb + 25b^2)$

9a. $112x^3 - 63x$
$= 7x(16x^2 - 9)$
$= 7x(4x + 3)(4x - 3)$

9b. $175ax^2 + 140axy + 28ay^2$
$= 7a(25x^2 + 20xy + 4y^2)$
$= 7a(5x + 2y)^2$

9c. $75ax^2 - 75axy + 12ay^2$
$= 3a(25x^2 - 25xy + 4y^2)$
$= 3a(5x - y)(5x - 4y)$

9d. $2ax^2 + bx^2 - 2ay^2 - by^2$
$= (2ax^2 - 2ay^2) + (bx^2 - by^2)$
$= 2a(x^2 - y^2) + b(x^2 - y^2)$
$= (x^2 - y^2)(2a + b)$
$= (x + y)(x - y)(2a + b)$

10a. $x^2 - x - 72 = 0$
$(x - 9)(x + 8) = 0$
$x = 9 \quad x = -8$

10b. $2x^2 = 13x + 7$
$2x^2 - 13x - 7 = 0$
$(2x + 1)(x - 7) = 0$
$x = -\frac{1}{2} \quad x = 7$

10c. $(x + 1)(3x - 1) = 4$
$3x^2 + 2x - 1 - 4 = 0$
$3x^2 + 2x - 5 = 0$
$(3x + 5)(x - 1) = 0$
$x = -\frac{5}{3} \quad x = 1$

10d. $(2x - 3)(x + 8) = x(x + 3)$
$2x^2 + 13x - 24 = x^2 + 3x$
$x^2 + 10x - 24 = 0$
$(x - 2)(x + 12) = 0$
$x = 2 \quad x = -12$

11a.　　Let the two consecutive odd integers be n and n + 2.

$$n(n + 2) = n + 56$$
$$n^2 + 2n = n + 56$$
$$n^2 + n - 56 = 0$$
$$(n + 8)(n - 7) = 0$$
$$n = -8 \qquad n = 7 \qquad\qquad \text{Answer: } -8 \text{ and } -6 \text{ or } 7 \text{ and } 9$$

11b.　　Let h = the height of the building and h + 5 = the base. The area of a triangle equals one-half the product of the base and height.

$$42 = \tfrac{1}{2}(h + 5)(h)$$

$$84 = h^2 + 5h$$
$$h^2 + 5h - 84 = 0$$
$$(h + 12)(h - 7) = 0$$
$$h = -12 \qquad h = 7$$

Answer: The height is 7 m.

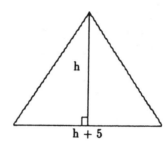

12a.　　Length of the longer leg:　　x
　　　　Length of the shorter leg:　x − 7
　　　　Length of the hypotenuse: x + 1

By the Pythagorean Theorem:

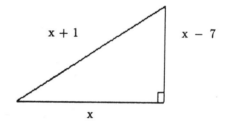

$$x^2 + (x - 7)^2 = (x + 1)^2$$
$$x^2 + x^2 - 14x + 49 = x^2 + 2x + 1$$
$$x^2 - 16x + 48 = 0$$
$$(x - 12)(x - 4) = 0$$
$$x = 12 \quad x = 4$$

We reject x = 4 because x − 7 would then be negative.
Answer: The triangle has sides of lengths 12 cm, 5 cm and 13 cm.

12b.　　Length of the ladder:　x
　　　　Vertical distance up the wall to the top of the ladder: x − 4
　　　　Horizontal distance from the base of the ladder to the wall:　x − 8

By the Pythagorean Theorem,　$(x - 4)^2 + (x - 8)^2 = x^2$

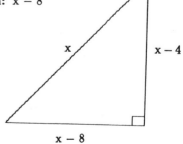

$$x^2 - 8x + 16 + x^2 - 16x + 64 = x^2$$
$$x^2 - 24x + 80 = 0$$
$$(x - 4)(x - 20) = 0$$
$$x = 4 \qquad\qquad x = 20$$

We reject x = 4 because x − 8 would then be negative.
Answer: The ladder is 20 feet long.

Chapter Seven Mastery Test

Problem Set 8.1

1. $\dfrac{65}{8x} = \dfrac{65}{8(-5)} = \dfrac{5(13)}{8(-5)} = -\dfrac{13}{8}$

5. This expression is undefined for $x = -5$ because the denominator would be zero.

9. $\dfrac{5x+4y}{7x-8y} = \dfrac{5(-2)+4(3)}{7(-2)-8(3)} = \dfrac{-10+12}{-14-24}$

$= \dfrac{2}{-38} = -\dfrac{2}{2(19)} = -\dfrac{1}{19}$

13. Set the denominator equal to zero and you immediately have: $x = 0$

17. Set the denominator equal to zero and solve for the variable:
$4v + 5 = 0$

$v = -\dfrac{5}{4}$

21. Set the denominator equal to zero and solve for the variable:
$w^2 - 7w - 8 = 0$
$(w - 8)(w + 1) = 0$
$w - 8 = 0 \qquad w + 1 = 0$
$w = 8 \qquad\quad\ w = -1$

25. $\dfrac{18x^8}{27x^4} = \dfrac{9(2)x^8}{9(3)x^4} = \dfrac{2x^{8-4}}{3} = \dfrac{2x^4}{3}$ or $\dfrac{2}{3}x^4$

29. $\dfrac{7(y - 5)}{y(y - 5)} = \dfrac{7}{y}$

The reason $y - 5$ drops out is because it is a *factor of both the numerator and denominator* .

Notice : $\dfrac{7(y - 5)}{y(y - 5)} = \dfrac{7}{y} \cdot \dfrac{y - 5}{y - 5} = \dfrac{7}{y} \cdot 1 = \dfrac{7}{y}$

33. $\dfrac{3x - 15}{xy - 5y} = \dfrac{3(x - 5)}{y(x - 5)} = \dfrac{3}{y}$

Remember! You must factor the numerator and denominator before doing any canceling. It is only the common factors that cancel.

37. $\dfrac{9m^2 - 16n^2}{27m + 36n} = \dfrac{(3m)^2 - (4n)^2}{9(3m + 4n)} = \dfrac{(3m + 4n)(3m - 4n)}{9(3m + 4n)} = \dfrac{3m - 4n}{9}$

41. $\dfrac{n^2 - 2n - 15}{n^2 - 7n + 10} = \dfrac{(n - 5)(n + 3)}{(n - 2)(n - 5)} = \dfrac{n + 3}{n - 2}$

45. $\dfrac{2x^2 - 3xy - 2y^2}{4x^2 + 4xy + y^2} = \dfrac{(2x + y)(x - 2y)}{(2x)^2 + 4xy + y^2} = \dfrac{(2x + y)(x - 2y)}{(2x + y)^2} = \dfrac{(2x + y)(x - 2y)}{(2x + y)(2x + y)}$

$= \dfrac{x - 2y}{2x + y}$

49. $\dfrac{m - n}{7n - 7m} = \dfrac{m - n}{7(n - m)} = \dfrac{(-1)(n - m)}{7(n - m)} = \dfrac{-1}{7} = -\dfrac{1}{7}$

> *Remember:* $\dfrac{a - b}{b - a} = -1$

> *Remember:* $\dfrac{a}{-b} = \dfrac{-a}{b} = -\dfrac{a}{b}$

53. The fraction $\dfrac{x - 3}{x - 2}$ can be obtained from $\dfrac{3 - x}{2 - x}$ by multiplying numerator and denominator by -1, so response B is correct.

Also, since $\dfrac{3 - x}{2 - x} = \dfrac{3 - x}{-1(x - 2)} = -\dfrac{3 - x}{x - 2}$, response C is also correct.

57. We will use 10 as an approximation to 9.98. The expression becomes:

$\dfrac{10^2 + 25}{10^2 - 75} = \dfrac{100 + 25}{100 - 75} = \dfrac{125}{25} = \dfrac{25(5)}{25(1)} = 5$ The correct response is C.

61. $\dfrac{ax - bx + cx}{-5a + 5b - 5c} = \dfrac{x(a - b + c)}{-5(a - b + c)} = \dfrac{x}{-5} = -\dfrac{x}{5}$

65. $\dfrac{x^2 - 10xy + 25y^2}{2x^2 - 9xy - 5y^2} = \dfrac{(x - 5y)^2}{(2x + y)(x - 5y)} = \dfrac{x - 5y}{2x + y}$

69. $\dfrac{(5.605)^2 - 2(5.605) + 5}{(5.605)^2 + 2(5.605) - 5} = \dfrac{25.206025}{37.626025} \approx 0.670$

73. A rational expression is a quotient of polynomials. Since each real constant is also a polynomial (although a very simple one) ordinary fractions of constants like $\dfrac{5}{6}$ are rational expressions as long as the denominator is not zero.

77. See answer section in the text.

81. First compare the denominators of the fractions. To obtain $3 - 4x$, $4x - 3$ was multiplied by -1. For these two fractions to be equivalent, the numerator of the original fraction must also be multiplied by -1. Therefore the numerator of the second fraction is $(-1)(3x - 4)$ or $4 - 3x$.

Problem Set 8.2

1. $\dfrac{14(55)}{15(42)} = \dfrac{7(2)(5)(11)}{5(3)(7)(6)} = \dfrac{2(11)}{3(6)} = \dfrac{2(11)}{3(3)(2)} = \dfrac{11}{9}$ See the TIME SAVER that follows:

5. $\dfrac{7m(6)}{9(14m^2)} = \dfrac{7m(3)(2)}{3(3)(7)2m^2} = \dfrac{m}{3m^2} = \dfrac{1}{3m}$

9. $\dfrac{2x(x-1)(3)(x+3)}{6(x+3)(x^2)(x+1)} = \dfrac{x-1}{x(x+1)}$

13. $\dfrac{(4x-9)(9x+4)}{(9x-4)(9-4x)} = \dfrac{(4x-9)(9x+4)}{(9-4x)(9x-4)} = (-1)\dfrac{9x+4}{9x-4} = -\dfrac{9x+4}{9x-4}$

17. $\dfrac{6y+54}{12xy^2} \cdot \dfrac{36x^2y}{8y+72} = \dfrac{(6y+54)(36x^2y)}{12xy^2(8y+72)} = \dfrac{6(y+9)(36x^2y)}{12xy^2(8)(y+9)} = \dfrac{6(3)x}{8y} = \dfrac{9x}{4y}$

21. $\dfrac{(v^2-36)(v-4)}{(v^2-16)(v+6)} = \dfrac{(v+6)(v-6)(v-4)}{(v+4)(v-4)(v+6)} = \dfrac{v-6}{v+4}$

25. $\dfrac{(v-2)(v+4)(v-4)}{(v+4)(v-5)(v-4)(v-2)} = \dfrac{1}{v-5}$

29. $\dfrac{(3y-9)(y^2+6y+9)}{(y^2-9)(8y+24)} = \dfrac{3(y-3)(y+3)^2}{(y+3)(y-3)(8)(y+3)} = \dfrac{3}{8}$

33. $\dfrac{(m^2-m-20)(m+3)}{(m^2-25)(m^2+7m+12)} = \dfrac{(m-5)(m+4)(m+3)}{(m+5)(m-5)(m+3)(m+4)} = \dfrac{1}{m+5}$

37. $\dfrac{(w^2+6w-16)(w^2-16w+64)}{(w^2-64)(w^2-10w+16)} = \dfrac{(w+8)(w-2)(w-8)^2}{(w+8)(w-8)(w-8)(w-2)} = 1$

41. $\dfrac{(10x^2-14x-12)(1)}{(7x^2-28)(35x^2+21x)} = \dfrac{2(5x^2-7x-6)}{7(x^2-4)(7x)(5x+3)} = \dfrac{2(5x+3)(x-2)}{7(x+2)(x-2)(7x)(5x+3)}$

$= \dfrac{2}{49x(x+2)}$

45. $\dfrac{2x+2}{x-5} \cdot \dfrac{x-5}{3x+9} \cdot \dfrac{5x+15}{7x+7} = \dfrac{(2x+2)(x-5)(5x+15)}{(x-5)(3x+9)(7x+7)} = \dfrac{2(x+1)5(x+3)}{3(x+3)7(x+1)} = \dfrac{10}{21}$

49. $\dfrac{v-1}{v+1} \div \left(\dfrac{v+1}{v-3} \cdot \dfrac{v^2-1}{v^2-2v-3} \right)$ You should work inside parentheses first.

$= \dfrac{v-1}{v+1} \div \left(\dfrac{(v+1)(v^2-1)}{(v-3)(v^2-2v-3)} \right)$

$= \dfrac{v-1}{v+1} \div \dfrac{(v+1)(v+1)(v-1)}{(v-3)(v-3)(v+1)}$ We wish to reduce as much as possible.

$= \dfrac{v-1}{v+1} \div \dfrac{(v+1)(v-1)}{(v-3)(v-3)}$ Now invert the second fraction and multiply.

$= \dfrac{(v-1)(v-3)(v-3)}{(v+1)(v+1)(v-1)}$

$= \dfrac{(v-3)^2}{(v+1)^2} \quad \text{or} \quad \dfrac{v^2-6v+9}{v^2+2v+1}$ The first answer is much more compact.

53. $\dfrac{ax+bx-ay-by}{x^2-y^2} \cdot \dfrac{x^2+4xy+3y^2}{ax+bx-4ay-4by} = \dfrac{[\,x(a+b)-y(a+b)\,][\,(x+3y)(x+y)\,]}{[\,(x+y)(x-y)\,][\,x(a+b)-4y(a+b)\,]}$

$= \dfrac{(x-y)(a+b)(x+3y)(x+y)}{(x+y)(x-y)(x-4y)(a+b)} = \dfrac{x+3y}{x-4y}$

57. $\dfrac{R}{\left(\dfrac{x-3}{x-5}\right)} = \dfrac{x^2-3x-10}{x^2-4x+3}$ To solve for R, multiply both sides by $\dfrac{x-3}{x-5}$.

$R = \dfrac{x^2-3x-10}{x^2-4x+3} \cdot \left(\dfrac{x-3}{x-5}\right) = \dfrac{(x^2-3x-10)(x-3)}{(x^2-4x+3)(x-5)}$

$= \dfrac{(x-5)(x+2)(x-3)}{(x-3)(x-1)(x-5)} = \dfrac{x+2}{x-1}$

61. $\dfrac{x^3+y^3}{7x^2-7xy+7y^2} \div \dfrac{x^2+2xy+y^2}{x^2-y^2}$ It can be helpful to reduce each fraction as much as possible before doing anything else.

$= \dfrac{(x+y)(x^2-xy+y^2)}{7(x^2-xy+y^2)} \div \dfrac{(x+y)^2}{(x+y)(x-y)}$ All expressions have been factored. Reduce.

$= \dfrac{x+y}{7} \div \dfrac{x+y}{x-y}$ This will be much easier to work with.

$= \dfrac{(x+y)(x-y)}{7(x+y)}$

$= \dfrac{x-y}{7}$

page 159

section 8.2

Problem Set 8.3

1. $\dfrac{13}{40} + \dfrac{9}{40} = \dfrac{13 + 9}{40} = \dfrac{22}{40} = \dfrac{11}{20}$

5. $\dfrac{22}{135} + \dfrac{23}{135} = \dfrac{22 + 23}{135} = \dfrac{45}{135} = \dfrac{1}{3}$

9. $\dfrac{20}{a^3} - \dfrac{17}{a^3} = \dfrac{20 - 17}{a^3} = \dfrac{3}{a^3}$

13. $\dfrac{11v + 7}{6v^2} - \dfrac{5v + 7}{6v^2}$

   $= \dfrac{11v + 7 - 5v - 7}{6v^2}$

   $= \dfrac{6v}{6v^2} = \dfrac{1}{v}$

17. $\dfrac{11w + 7}{5w} + \dfrac{4w + 3}{5w}$

   $= \dfrac{11w + 7 + 4w + 3}{5w}$

   $= \dfrac{15w + 10}{5w}$

   $= \dfrac{5(3w + 2)}{5w}$

   $= \dfrac{3w + 2}{w}$

21. $\dfrac{6}{x + 3} + \dfrac{3}{x + 3}$

   $= \dfrac{6 + 3}{x + 3}$

   $= \dfrac{9}{x + 3}$

25. $\dfrac{b + 1}{b + 3} + \dfrac{2}{b + 3}$

   $= \dfrac{b + 1 + 2}{b + 3}$

   $= \dfrac{b + 3}{b + 3}$

   $= 1$

29. $\dfrac{y}{x - y} - \dfrac{x}{x - y}$

   $= \dfrac{y - x}{x - y}$

   $= -1$

   $\boxed{\textit{Remember: } \dfrac{a - b}{b - a} = -1}$

33. $\dfrac{5v - 1}{6v + 3} - \dfrac{2v + 5}{6v + 3} = \dfrac{5v - 1 - (2v + 5)}{6v + 3} = \dfrac{5v - 1 - 2v - 5}{6v + 3} = \dfrac{3v - 6}{6v + 3}$

   $= \dfrac{3(v - 2)}{3(2v + 1)} = \dfrac{v - 2}{2v + 1}$

37. $\dfrac{8}{3x} + \dfrac{2}{3x} - \dfrac{1}{3x} = \dfrac{9}{3x} = \dfrac{3}{x}$

41. $\dfrac{v}{v^2 - 1} + \dfrac{1}{v^2 - 1} = \dfrac{v + 1}{v^2 - 1} = \dfrac{v + 1}{(v + 1)(v - 1)} = \dfrac{1}{v - 1}$

45. $\dfrac{13b + 69}{b^2 + 5b - 6} - \dfrac{2b + 3}{b^2 + 5b - 6} = \dfrac{13b + 69 - 2b - 3}{b^2 + 5b - 6} = \dfrac{11b + 66}{b^2 + 5b - 6}$

$\qquad = \dfrac{11(b + 6)}{(b + 6)(b - 1)} = \dfrac{11}{b - 1}$

49. $\dfrac{3m^2 + 4m}{15m^2 - 2m - 1} + \dfrac{2m^2 + 2m + 1}{15m^2 - 2m - 1} = \dfrac{3m^2 + 4m + 2m^2 + 2m + 1}{15m^2 - 2m - 1}$

$\qquad = \dfrac{5m^2 + 6m + 1}{15m^2 - 2m - 1} = \dfrac{(5m + 1)(m + 1)}{(5m + 1)(3m - 1)} = \dfrac{m + 1}{3m - 1}$

53. $\dfrac{4v^2 + v - 1}{2v^2 - v} - \dfrac{v^2 + v + 1}{2v^2 - v} = \dfrac{4v^2 + v - 1 - v^2 - v - 1}{2v^2 - v}$

$\qquad = \dfrac{3v^2 - 2}{2v^2 - v}$

57. $R + \dfrac{5}{3x - 7} = \dfrac{2}{3x - 7}$ $\qquad$ Subtract $\dfrac{5}{3x - 7}$ from both sides to solve for R.

$\qquad R = \dfrac{2}{3x - 7} - \dfrac{5}{3x - 7} = \dfrac{-3}{3x - 7}$ or $-\dfrac{3x}{3x - 7}$

61. The missing parts are -5 and 2.

First of all, $\dfrac{5}{y - 2x}$ can be written with the denominator $2x - y$ if we multiply both numerator and denominator by -1. Therefore,

$\qquad \dfrac{5}{y - 2x} = \dfrac{(-1)5}{(-1)(y - 2x)} = \dfrac{-5}{2x - y}$

We now have the sum $\quad \dfrac{7}{2x - y} + \dfrac{-5}{2x - y} = \dfrac{7 - 5}{2x - y} = \dfrac{2}{2x - y}$.

Problem Set 8.4

1. $\dfrac{3}{4} = \dfrac{3(3)}{4(3)} = \dfrac{9}{12}$.

5. $\dfrac{2}{3m} = \dfrac{2(7m^2)}{3m(7m^2)} = \dfrac{14m^2}{21m^3}$

9. $\dfrac{6}{3v + 2} = \dfrac{6(v)}{(3v + 2)(v)} = \dfrac{6v}{3v^2 + 2v}$

13. $\dfrac{m}{m + 1} = \dfrac{m(m - 1)}{(m + 1)(m - 1)}$ or $\dfrac{m^2 - m}{m^2 - 1}$

17. $\dfrac{2n + 1}{2n + 3} = \dfrac{(2n + 1)(3n - 2)}{(2n + 3)(3n - 2)}$ or $\dfrac{6n^2 - n - 2}{6n^2 + 5n - 6}$

21. $\begin{aligned}12 &= 2 \cdot 6 = 2 \cdot 2 \cdot 3 = \\ 30 &= 2 \cdot 15 = \end{aligned}$ $\left.\begin{aligned} & 2^2 \cdot 3 \\ & 2 \ \cdot 3 \cdot 5\end{aligned}\right\}$ LCD $= 2^2 \cdot 3 \cdot 5 = 60$

25. $\begin{aligned}15x^2 &= \\ 27x^3 &= \end{aligned}$ $\left.\begin{aligned} & 3 \ \cdot 5 \cdot x^2 \\ & 3^3 \cdot x^3\end{aligned}\right\}$ LCD $= 3^3 \cdot 5 \cdot x^3 = 135x^3$

29. $\begin{aligned}m^2 - n^2 &= \\ 5m - 5n &= \end{aligned}$ $\left.\begin{aligned} & (m + n)(m - n) \\ & 5 \ (m - n)\end{aligned}\right\}$ LCD $= 5(m + n)(m - n)$

33. $\begin{aligned}w^2 + w - 2 &= \\ w^2 - w - 6 &= \\ w^2 - 4w + 3 &= \end{aligned}$ $\left.\begin{aligned} & (w + 2)(w - 1) \\ & (w - 3)(w + 2) \\ & (w - 3)(w - 1)\end{aligned}\right\}$ LCD $= (w + 2)(w - 1)(w - 3)$

37. $\dfrac{3x + y}{26} + \dfrac{4y - x}{65}$ $\qquad \left.\begin{aligned} 26 &= 2 \cdot 13 \\ 65 &= 5 \cdot 13 \end{aligned}\right\}$ LCD $= 2 \cdot 5 \cdot 13$

$\dfrac{(3x + y)(5)}{2(13)(5)} + \dfrac{(4y - x)(2)}{5(13)(2)} = \dfrac{(3x + y)(5) + (4y - x)(2)}{2(5)(13)} = \dfrac{15x + 5y + 8y - 2x}{2(5)(13)}$

$= \dfrac{13x + 13y}{2(5)(13)} = \dfrac{13(x + y)}{2(5)(13)} = \dfrac{x + y}{10}$

$\boxed{TIME\ SAVER}$ *As you go along, leave the denominator as $2 \cdot 5 \cdot 13$ (and not 130). Although it is not incorrect to write 130, it is very inefficient. The factored form $2 \cdot 5 \cdot 13$ is exactly what you will need when you reduce the fraction. Notice that the common factor of 13 divides out. Only when the fraction is completely reduced do we multiply together what's left: $2(5) = 10$.*

41. $\dfrac{5}{22x^3} - \dfrac{2}{33x^2}$ $\qquad \left.\begin{aligned} 22x^3 &= 2 \cdot 11 \cdot x^3 \\ 33x^2 &= 3 \cdot 11 \cdot x^2 \end{aligned}\right\}$ LCD $= 2 \cdot 3 \cdot 11 \cdot x^3$

$\dfrac{3(5)}{3(2)(11)x^3} - \dfrac{2x(2)}{2x(3)(11)x^2} = \dfrac{15 - 4x}{2(3)(11)x^3} = \dfrac{15 - 4x}{66x^3}$

45. $\dfrac{a + 1}{a^2 b} + \dfrac{1 - b}{ab^2}$ $\qquad$ The LCD is $a^2 b^2$.

$= \dfrac{b(a + 1)}{a^2 b^2} + \dfrac{a(1 - b)}{a^2 b^2} = \dfrac{b(a + 1) + a(1 - b)}{a^2 b^2} = \dfrac{ba + b + a - ab}{a^2 b^2} = \dfrac{a + b}{a^2 b^2}$

49. $\dfrac{3}{a+2} - \dfrac{2}{a+5}$ As $a+2$ and $a+5$ are both prime, the LCD is $(a+2)(a+5)$.

$$\dfrac{3(a+5)}{(a+2)(a+5)} - \dfrac{2(a+2)}{(a+2)(a+5)} = \dfrac{3a+15-2a-4}{(a+2)(a+5)} = \dfrac{a+11}{(a+5)(a+2)}$$

53. $\dfrac{2}{5-y} + \dfrac{4y}{y^2-25}$ Notice that $\dfrac{2}{5-y} = \dfrac{2}{(-1)(y-5)} = \dfrac{-2}{y-5}$ and so we may rewrite the

problem as $\dfrac{-2}{y-5} + \dfrac{4y}{y^2-25}$.

$$\left.\begin{array}{l} y-5 = \\ y^2-25 = \end{array}\right. \quad \begin{array}{l} y-5 \\ (y+5)(y-5) \end{array} \left.\right\} \quad \text{The LCD is } (y+5)(y-5).$$

$$\dfrac{-2(y+5)}{(y+5)(y-5)} + \dfrac{4y}{(y+5)(y-5)} = \dfrac{-2y-10+4y}{(y+5)(y-5)} = \dfrac{2y-10}{(y+5)(y-5)} = \dfrac{2(y-5)}{(y+5)(y-5)} = \dfrac{2}{y+5}$$

> $\boxed{\textit{TIME SAVER}}$ *Notice that the denominator $(y+5)(y-5)$ was not at any time multiplied out. There is no need to multiply it out; you will need the factored form when you reduce.*

57. $\dfrac{v}{v^2-36} - \dfrac{1}{2v+12}$ $\begin{array}{l} v^2-36 = (v+6)(v-6) \\ 2v+12 = 2(v+6) \end{array} \left.\right\}$ The LCD is $2(v+6)(v-6)$.

$$\dfrac{v(2)}{2(v+6)(v-6)} - \dfrac{1(v-6)}{2(v+6)(v-6)} = \dfrac{2v-v+6}{2(v+6)(v-6)} = \dfrac{v+6}{2(v+6)(v-6)} = \dfrac{1}{2(v-6)}$$

61. $5 - \dfrac{2}{x-1}$ The LCD is $x-1$.

$$= \dfrac{5(x-1)}{x-1} - \dfrac{2}{x-1} = \dfrac{5x-5-2}{x-1} = \dfrac{5x-7}{x-1}$$

65. $\dfrac{t-1}{3t^2+7t-6} + \dfrac{t-1}{6t^2-19t+10}$

$\begin{array}{l} 3t^2+7t-6 = (3t-2)(t+3) \\ 6t^2-19t+10 = (2t-5)(3t-2) \end{array} \left.\right\}$ The LCD is $(3t-2)(t+3)(2t-5)$.

$$\dfrac{(t-1)(2t-5)}{(3t-2)(t+3)(2t-5)} + \dfrac{(t-1)(t+3)}{(2t-5)(3t-2)(t+3)} = \dfrac{2t^2-7t+5+t^2+2t-3}{(2t-5)(3t-2)(t+3)}$$

$$= \dfrac{3t^2-5t+2}{(2t-5)(3t-2)(t+3)} = \dfrac{(3t-2)(t-1)}{(2t-5)(3t-2)(t+3)} = \dfrac{t-1}{(2t-5)(t+3)}$$

69.

$$\frac{5}{2m^2 - 3m - 2} + \frac{m}{6m^2 + 7m + 2}$$

$$\left. \begin{array}{l} 2m^2 - 3m - 2 = (2m + 1)(m - 2) \\ 6m^2 + 7m + 2 = (2m + 1)(3m + 2) \end{array} \right\} \text{ The LCD is } (2m + 1)(m - 2)(3m + 2)$$

$$\frac{5(3m + 2)}{(2m + 1)(m - 2)(3m + 2)} + \frac{m(m - 2)}{(2m + 1)(3m + 2)(m - 2)} = \frac{5(3m + 2) + m(m - 2)}{(2m + 1)(3m + 2)(m - 2)}$$

$$= \frac{15m + 10 + m^2 - 2m}{(2m + 1)(3m + 2)(m - 2)} = \frac{m^2 + 13m + 10}{(2m + 1)(3m + 2)(m - 2)} = \frac{m^2 + 13m + 10}{(2m + 1)(3m + 2)(m - 2)}$$

73.

$$R + \frac{2m - 40}{m^2 - 25} = \frac{5}{m + 5} \qquad\qquad \text{Subtract } \frac{2m - 40}{m^2 - 25} \text{ from both sides to solve for R.}$$

$$R = \frac{5}{m + 5} - \frac{2m - 40}{m^2 - 25}$$

$$\left. \begin{array}{l} m + 5 = m + 5 \\ m^2 - 25 = (m + 5)(m - 5) \end{array} \right\} \text{ The LCD is } (m + 5)(m - 5).$$

$$R = \frac{5(m - 5)}{(m + 5)(m - 5)} - \frac{2m - 40}{(m + 5)(m - 5)} = \frac{5m - 25 - 2m + 40}{(m + 5)(m - 5)} = \frac{3m + 15}{(m + 5)(m - 5)}$$

$$= \frac{3(m + 5)}{(m + 5)(m - 5)} = \frac{3}{m - 5}$$

Problem Set 8.5

1a. $\quad \dfrac{\frac{2}{3}}{4} = \frac{2}{3} \div 4 = \frac{2}{3} \cdot \frac{1}{4} = \frac{2}{3(4)} = \frac{1}{6}$

1b. $\quad \dfrac{2}{\frac{3}{4}} = \frac{2}{1} \times \frac{4}{3} = \frac{2(4)}{1(3)} = \frac{8}{3}$

5. $\quad \dfrac{a^2 - 3a - 4}{1 - \frac{4}{a}} = \dfrac{a^2 - 3a - 4}{\frac{a}{a} - \frac{4}{a}} = \dfrac{a^2 - 3a - 4}{\frac{a - 4}{a}} = (a^2 - 3a - 4) \div \frac{a - 4}{a}$

$$= \frac{a^2 - 3a - 4}{1} \cdot \frac{a}{a - 4} = \frac{(a^2 - 3a - 4)(a)}{a - 4} = \frac{(a - 4)(a + 1)(a)}{a - 4} = (a + 1)(a)$$

9.
$$\frac{\frac{1}{4} - \frac{1}{m^2}}{\frac{1}{2} + \frac{1}{m}} = \frac{\left(\frac{1}{4} - \frac{1}{m^2}\right)4m^2}{\left(\frac{1}{2} + \frac{1}{m}\right)4m^2}$$

Multiply numerator and denominator by the LCD of the fractions.

$$= \frac{\frac{1}{4}(4m^2) - \frac{1}{m^2}(4m^2)}{\frac{1}{2}(4m^2) + \frac{1}{m}(4m^2)} = \frac{m^2 - 4}{2m^2 + 4m} = \frac{(m + 2)(m - 2)}{2m(m + 2)} = \frac{m - 2}{2m}$$

13.
$$\frac{\frac{3x}{4y^2}}{\frac{9x^2}{8y}} = \frac{\left(\frac{3x}{4y^2}\right)8y^2}{\left(\frac{9x^2}{8y}\right)8y^2} = \frac{\frac{(3x)8y^2}{4y^2}}{\frac{(9x^2)8y^2}{8y}} = \frac{6x}{9x^2y} = \frac{2}{3xy}$$

17.
$$\frac{3 - \frac{1}{v}}{9 - \frac{1}{v^2}} = \frac{\left(3 - \frac{1}{v}\right)v^2}{\left(9 - \frac{1}{v^2}\right)v^2} = \frac{3v^2 - \frac{1}{v}(v^2)}{9v^2 - \left(\frac{1}{v^2}\right)v^2} = \frac{3v^2 - v}{9v^2 - 1} = \frac{v(3v - 1)}{(3v + 1)(3v - 1)} = \frac{v}{3v + 1}$$

21.
$$\frac{\frac{2}{w} + 6}{w + \frac{1}{6}} = \frac{\left(\frac{2}{w} + 6\right)6w}{\left(w + \frac{1}{6}\right)6w} = \frac{\left(\frac{2}{w}\right)6w + 6(6w)}{w(6w) + \left(\frac{1}{6}\right)(6w)} = \frac{12 + 36w}{6w^2 + w} = \frac{12(1 + 3w)}{w(6w + 1)}$$

25.
$$\frac{\frac{1}{ab^3} + \frac{2}{a^2b^2}}{\frac{1}{a^2b^2} - \frac{2}{a^3b}} = \frac{\left(\frac{1}{ab^3} + \frac{2}{a^2b^2}\right)a^3b^3}{\left(\frac{1}{a^2b^2} - \frac{2}{a^3b}\right)a^3b^3}$$

Multiply numerator and denominator by the LCD of the fractions to clear them.

$$= \frac{\left(\frac{1}{ab^3}\right)a^3b^3 + \left(\frac{2}{a^2b^2}\right)a^3b^3}{\left(\frac{1}{a^2b^2}\right)a^3b^3 - \left(\frac{2}{a^3b}\right)a^3b^3} = \frac{a^2 + 2ab}{ab - 2b^2} = \frac{a(a + 2b)}{b(a - 2b)} \qquad \text{(no common factors to reduce)}$$

29.
$$\frac{x + 3 - \frac{12}{x - 1}}{x + 2 - \frac{10}{x - 1}} = \frac{\left(x + 3 - \frac{12}{x - 1}\right)(x - 1)}{\left(x + 2 - \frac{10}{x - 1}\right)(x - 1)} = \frac{(x + 3)(x - 1) - \left(\frac{12}{x - 1}\right)(x - 1)}{(x + 2)(x - 1) - \left(\frac{10}{x - 1}\right)(x - 1)}$$

$$= \frac{x^2 - x + 3x - 3 - 12}{x^2 - x + 2x - 2 - 10} = \frac{x^2 + 2x - 15}{x^2 + x - 12} = \frac{(x + 5)(x - 3)}{(x + 4)(x - 3)} = \frac{x + 5}{x + 4}$$

section 8.5

33. $\dfrac{v}{v-4} - \dfrac{2v}{v+3} \div \dfrac{v^2-4v}{2v+6}$ Order of operations: division before subtraction!

First: $\dfrac{2v}{v+3} \cdot \dfrac{2v+6}{v^2-4v} = \dfrac{2v(2v+6)}{(v+3)(v^2-4v)} = \dfrac{2v(2)(v+3)}{(v+3)(v)(v-4)} = \dfrac{4}{v-4}$

Then: $\dfrac{v}{v-4} - \dfrac{4}{v-4} = \dfrac{v-4}{v-4} = 1.$

37. $\left(y + \dfrac{3y+4}{y+2}\right)\left(y - \dfrac{3y+2}{y+4}\right) = \left(\dfrac{y(y+2)}{y+2} + \dfrac{3y+4}{y+2}\right)\left(\dfrac{y(y+4)}{y+4} - \dfrac{3y+2}{y+4}\right)$

$= \left(\dfrac{y^2+2y+3y+4}{y+2}\right)\left(\dfrac{y^2+4y-3y-2}{y+4}\right) = \left(\dfrac{y^2+5y+4}{y+2}\right)\left(\dfrac{y^2+y-2}{y+4}\right)$

$= \dfrac{(y+4)(y+1)(y+2)(y-1)}{(y+2)(y+4)} = (y+1)(y-1)$

41. $\dfrac{211}{429}$ can be approximated by $\dfrac{200}{400} = \dfrac{1}{2}$;

$\dfrac{107}{325}$ can be approximated by $\dfrac{100}{300} = \dfrac{1}{3}$.

$\dfrac{1}{2} \times \dfrac{3}{1} = \dfrac{3}{2}$, and so the best response is C.

45. $\dfrac{2 + \frac{1}{2}}{1 + \frac{1}{1 + \frac{1}{2}}}$ First: $\dfrac{1}{1 + \frac{1}{2}} = \dfrac{1}{\frac{2+1}{2}} = \dfrac{1}{\frac{3}{2}} = \dfrac{2}{3}$

$\dfrac{2 + \frac{1}{2}}{1 + \frac{2}{3}} = \dfrac{\frac{4+1}{2}}{\frac{3+2}{3}} = \dfrac{\frac{5}{2}}{\frac{5}{3}} = \dfrac{5}{2} \cdot \dfrac{3}{5} = \dfrac{3}{2}$

49. $\dfrac{\frac{3}{x+3} + \frac{1}{x-1}}{\frac{1}{x-1} + \frac{2}{x+2}} = \dfrac{\frac{3(x-1) + 1(x+3)}{(x+3)(x-1)}}{\frac{1(x+2) + 2(x-1)}{(x-1)(x+2)}}$

$= \dfrac{\frac{3x-3+x+3}{(x+3)(x-1)}}{\frac{x+2+2x-2}{(x-1)(x+2)}} = \dfrac{4x}{(x+3)(x-1)} \div \dfrac{3x}{(x-1)(x+2)}$

$= \dfrac{4x(x-1)(x+2)}{(x+3)(x-1)3x} = \dfrac{4(x+2)}{3(x+3)}$

53. If you do not have a graphics calculator, but one of your classmates does, you might want to take a look at what it can do. They are not hard to learn to use, and the display shows everything you type in, *so you can check your input before the final calculation is made.*
If you do have one and are not sure how to proceed with this problem, ask your instructor. Not all calculators are the same, so we will not try to give you directions here.

You can also analyze this problem without a calculator. The same order of operations that were covered earlier in this book will be followed by the calculator. What is in parentheses will be done first, and so on. Without a calculator, you might proceed like this:

$$\frac{1.5 + \frac{1}{1.5}}{1.5 - \frac{1}{1.5}} = \frac{\frac{3}{2} + \frac{2}{3}}{\frac{3}{2} - \frac{2}{3}} = \frac{\frac{13}{6}}{\frac{5}{6}} = \frac{13}{6} \times \frac{6}{5} = \frac{13}{5} \text{ or } 2.6$$

57. $\left(\dfrac{1}{x + 2} + \dfrac{1}{x + 3}\right)\dfrac{2x + 4}{2x + 5}$

First: $\dfrac{1}{x + 2} + \dfrac{1}{x + 3} = \dfrac{x + 3 + x + 2}{(x + 2)(x + 3)} = \dfrac{2x + 5}{(x + 2)(x + 3)}$

Then: $\dfrac{2x + 5}{(x + 2)(x + 3)} \cdot \dfrac{2x + 4}{2x + 5} = \dfrac{(2x + 5)(2)(x + 2)}{(x + 2)(x + 3)(2x + 5)} = \dfrac{2}{x + 3}$

Problem Set 8.6

1. Set the denominator equal to zero and solve for the variable: $v - 4 = 0$; $v = 4$.

5. $m - 1 = 0$ $\qquad m - 6 = 0$
$m = 1$ $\qquad\quad m = 6$

9. $\dfrac{x - 3}{2} - \dfrac{x}{5} = \dfrac{x}{4}$ $\quad$ LCD $= 20$

$20\left(\dfrac{x - 3}{2} - \dfrac{x}{5}\right) = 20\left(\dfrac{x}{4}\right)$

$20\left(\dfrac{x - 3}{2}\right) - 20\left(\dfrac{x}{5}\right) = 20\left(\dfrac{x}{4}\right)$

$10(x - 3) - 4x = 5x$
$10x - 30 - 4x = 5x$
$6x - 30 = 5x$
$x = 30$

13. $\dfrac{6}{n} - \dfrac{5}{n} = 2$

$\dfrac{6 - 5}{n} = 2$

$\dfrac{1}{n} = 2$

$1 = 2n$
$n = \dfrac{1}{2}$

17. $\dfrac{3}{v - 1} = 6$

$(v - 1)\left(\dfrac{3}{v - 1}\right) = (v - 1)6$

$3 = 6v - 6$
$9 = 6v$
$v = \dfrac{9}{6} = \dfrac{3}{2}$

21. $\dfrac{x}{x+3} = \dfrac{1}{x+3} + \dfrac{1}{3}$ $\qquad\qquad$ The LCD is $3(x+3)$.

$$3(x+3)\left(\dfrac{x}{x+3}\right) = 3(x+3)\left(\dfrac{1}{x+3}\right) + 3(x+3)\left(\dfrac{1}{3}\right)$$

$3x = 3 + x + 3$

$2x = 6$

$x = 3$

25. $1 - \dfrac{1}{v-1} = \dfrac{3}{1-v}$ $\qquad\qquad$ Remember: $1 - v = -(v-1)$

$1 - \dfrac{1}{v-1} = -\dfrac{3}{v-1}$ $\qquad\qquad$ The LCD is $v-1$.

$$(v-1)\left(1 - \dfrac{1}{v-1}\right) = (v-1)\left(-\dfrac{3}{v-1}\right)$$

$v - 1 - 1 = -3$

$v - 2 = -3$

$v = -1$

29. $\dfrac{6}{y-4} = \dfrac{9}{2y-3}$ $\qquad\qquad$ You have two choices:
$\qquad\qquad\qquad\qquad\qquad\qquad\qquad$ (1) Cross multiply or
$\qquad\qquad\qquad\qquad\qquad\qquad\qquad$ (2) Multiply both sides by the LCD

We will cross multiply:

$6(2y - 3) = 9(y - 4)$

$12y - 18 = 9y - 36$

$3y = -18$

$y = -6$

33. $\dfrac{3}{v+4} - \dfrac{6}{v^2 - 16} = 1$ $\qquad\qquad$ Since $v^2 - 16 = (v+4)(v-4)$, this is the LCD.

$$(v^2 - 16)\left(\dfrac{3}{v+4} - \dfrac{6}{v^2 - 16}\right) = (v^2 - 16)(1)$$

$3(v - 4) - 6 = v^2 - 16$

$3v - 12 - 6 = v^2 - 16$

$v^2 - 3v + 2 = 0$

$(v - 2)(v - 1) = 0$

$v = 2 \quad v = 1$

37. $a^2 - 25 = (a + 5)(a - 5)$, so the LCD will be $(a + 5)(a - 5)$.

$$(a + 5)(a - 5) \left(\frac{10}{a^2 - 25} + \frac{1}{a + 5} \right) = (a + 5)(a - 5) \left(\frac{3}{a - 5} \right)$$

$10 + a - 5 = 3(a + 5)$

$5 + a = 3a + 15$

$-10 = 2a$

$a = -5$ *This is an excluded value, so there is no solution.*

41. $a = \frac{bc}{d}$ Multiply both sides by d.

$d(a) = d\left(\frac{bc}{d}\right)$

$da = bc$ Divide both sides by c.

$\frac{da}{c} = \frac{bc}{c}$

$b = \frac{da}{c}$ or $\frac{ad}{c}$

45a.   When <u>simplifying</u>, we find the LCD and use it as the common denominator. Each fraction is <u>rewritten</u> using this new denominator. The LCD for these fractions is $(v - 2)(v - 3)$.

$\frac{12}{v - 2}$ becomes $\frac{12(v - 3)}{(v - 2)(v - 3)}$

$\frac{9}{v - 3}$ becomes $\frac{9(v - 2)}{(v - 3)(v - 2)}$

We now add the 'new' fractions:

$$\frac{12(v - 3)}{(v - 2)(v - 3)} + \frac{9(v - 2)}{(v - 3)(v - 2)} = \frac{12(v - 3) + 9(v - 2)}{(v - 2)(v - 3)} = \frac{21v - 54}{(v - 2)(v - 3)}$$

or $\frac{3(7v - 18)}{(v - 2)(v - 3)}$

    $\boxed{WARNING!}$ *When simplifying a sum or difference of fractions, you use the LCD as a common denominator, and rewrite the fractions using this denominator. You DO NOT "multiply both sides" by the LCD and clear the fractions, because there aren't any "sides"; we are not dealing with an equation. Notice that the final expression is still a fraction. Contrast this with what is done in #45b.*

45b.   When <u>solving an equation</u> involving fractions, we still find the LCD, but we can use it to clear the fractions. The resulting equation is easier to solve. The LCD is $(v - 3)(v - 2)$.

Multiply both sides of the equation by $(v - 3)(v - 2)$:

$$(v - 3)(v - 2) \left(\frac{12}{v - 2} \right) = (v - 3)(v - 2) \left(\frac{9}{v - 3} \right)$$

$12(v - 3) = 9(v - 2)$

$12v - 36 = 9v - 18$

$3v = 18$

$v = 6$

Notice that we no longer have fractions. That was the purpose of multiplying by the LCD.

49. $\dfrac{2 - 6x}{x^2 - x - 6} = \dfrac{x + 1}{x - 3} + \dfrac{x - 1}{x + 2}$

$\dfrac{2 - 6x}{(x - 3)(x + 2)} = \dfrac{x + 1}{x - 3} + \dfrac{x - 1}{x + 2}$

The LCD is $(x - 3)(x + 2)$.

$(x - 3)(x + 2)\left(\dfrac{2 - 6x}{(x - 3)(x + 2)}\right) = (x - 3)(x + 2)\left(\dfrac{x + 1}{x - 3}\right) + (x - 3)(x + 2)\left(\dfrac{x - 1}{x + 2}\right)$

$2 - 6x = (x + 2)(x + 1) + (x - 3)(x - 1)$

$2 - 6x = x^2 + 3x + 2 + x^2 - 4x + 3$

$0 = 2x^2 + 5x + 3$

$0 = (2x + 3)(x + 1)$

$2x + 3 = 0 \qquad\qquad x + 1 = 0$

$x = -\dfrac{3}{2} \qquad\qquad x = -1$

53. $\dfrac{n - 20}{n^2 - 5n - 6} + \dfrac{12}{2n^2 - 13n + 6} = \dfrac{7n - 2}{2n^2 + n - 1}$

$\dfrac{n - 20}{(n - 6)(n + 1)} + \dfrac{12}{(2n - 1)(n - 6)} = \dfrac{7n - 2}{(2n - 1)(n + 1)}$

The LCD is $(n - 6)(n + 1)(2n - 1)$.

$(n - 6)(n + 1)(2n - 1)\left(\dfrac{n - 20}{(n - 6)(n + 1)} + \dfrac{12}{(2n - 1)(n - 6)}\right)$

$\qquad\qquad = (n - 6)(n + 1)(2n - 1)\left(\dfrac{7n - 2}{(2n - 1)(n + 1)}\right)$

$(2n - 1)(n - 20) + (n + 1)12 = (n - 6)(7n - 2)$

$2n^2 - 41n + 20 + 12n + 12 = 7n^2 - 44n + 12$

$0 = 5n^2 - 15n - 20$

Divide both sides by 5.

$0 = n^2 - 3n - 4$

$0 = (n - 4)(n + 1)$

$n = 4 \qquad\qquad n = -1$

n = −1 is an excluded value. n = 4 is the only solution.

57. $\dfrac{5}{a-2} = \dfrac{3}{b+2}$

$5(b + 2) = (a - 2)(3)$

$5b + 10 = 3a - 6$ Isolate b on one side by subtracting 10 from both sides.

$5b = 3a - 16$ Divide both sides by 5.

$b = \dfrac{3a - 16}{5}$

Problem Set 8.7

1. Let x be the number. Five times its reciprocal is $5\left(\dfrac{1}{x}\right)$ or $\dfrac{5}{x}$.

$x + \dfrac{5}{x} = \dfrac{46}{3}$ The LCD is 3x.

$3x\left(x + \dfrac{5}{x}\right) = 3x\left(\dfrac{46}{3}\right)$

$3x^2 + 15 = 46x$

$3x^2 - 46x + 15 = 0$

$(3x - 1)(x - 15) = 0$

$3x - 1 = 0 \qquad\qquad x - 15 = 0$

$x = \dfrac{1}{3} \qquad\qquad x = 15$ Two solutions: $\dfrac{1}{3}$, 15

5. Let x be one of the numbers and $68 - x$ be the other.

$\dfrac{x}{68 - x} = \dfrac{8}{9}$ The LCD is $9(68 - x)$.

$9(68 - x)\left(\dfrac{x}{68 - x}\right) = 9(68 - x)\left(\dfrac{8}{9}\right)$

$9x = 8(68 - x)$

$9x = 544 - 8x$

$17x = 544$

$x = 32$

$68 - x = 68 - 32 = 36$ The two numbers are 32 and 36.

> You may also solve this type by cross-multiplying: $9x = 8(68 - x)$
> $9x = 544 - 8x$
> $17x = 544$
> $x = 32$

9. Let x be the number.

$\dfrac{17 + x}{23 + x} = \dfrac{4}{5}$ You may cross-multiply. (Remember: If $\dfrac{a}{b} = \dfrac{c}{d}$ then $ad = bc$.)

$(17 + x)(5) = (23 + x)(4)$

$85 + 5x = 92 + 4x$

$x = 7$ The number is 7.

13. Let x be the number; then its reciprocal is $\frac{1}{x}$ and one-half of that is $\frac{1}{2}\left(\frac{1}{x}\right)$ or $\frac{1}{2x}$.

$x + \frac{1}{2x} = \frac{11}{6}$ The LCD is 6x.

$6x\left(x + \frac{1}{2x}\right) = 6x\left(\frac{11}{6}\right)$

$6x^2 + 3 = 11x$

$6x^2 - 11x + 3 = 0$

$(3x - 1)(2x - 3) = 0$

$3x - 1 = 0 \qquad 2x - 3 = 0$

$x = \frac{1}{3} \qquad x = \frac{3}{2}$ The number could be either $\frac{1}{3}$ or $\frac{3}{2}$.

17. $y = \frac{k}{x}$ (First find k.)

$4 = \frac{k}{6} \; ; \; k = 24$

$y = \frac{24}{x}$

$y = \frac{24}{8} = 3$

21a. Let t be the time required before they are 56 km apart.

	R.	·	T	=	D	(Rate · Time = Distance)
upstream	12		t		12t	
downstream	20		t		20t	

The sum of their distances is 56 :

$12t + 20t = 56$

$32t = 56$

$t = \frac{56}{32} = \frac{7}{4} = 1\frac{3}{4}$ The time needed is $1\frac{3}{4}$ hours.

21b. Let r be the speed of the boat in still water.

The speed of the boat upstream (against the current) is $r - 7$;
the speed of the boat downstream (with the current) is $r + 7$.

	R	$\cdot$ T	= D	(Rate · Time = Distance)
upstream	$r - 7$		9	
downstream	$r + 7$		30	

COMMENT *Tables like these are used for many types of word problems. It often happens that one of the columns can be immediately filled out with given information (in this case, the distance column) and one of the columns contains the variable expressions that you set up (in this case the rate column). The remaining column may require you to use the formula at the top.*

In this case, we need expressions for time.

Since $R \cdot T = D$, dividing both sides by R gives $T = \dfrac{D}{R}$.

Going upstream, $T = \dfrac{9}{r - 7}$.　　Going downstream, $T = \dfrac{30}{r + 7}$.

Now finish filling out the table:

	R	$\cdot$ T	= D
upstream	$r - 7$	$\dfrac{9}{r - 7}$	9
downstream	$r + 7$	$\dfrac{30}{r + 7}$	30

The time required going upstream is the same as that going downstream, so

$$\frac{9}{r - 7} = \frac{30}{r + 7}$$

$$9(r + 7) = 30(r - 7)$$

$$9r + 63 = 30r - 210$$

$$273 = 21r$$

$$r = \frac{273}{21} = 13$$　　　　The speed of the boat in still water is 13 km/h.

25a. Let t be the number of hours required to fill the tank. Notice that t is also the number of hours that each pipe will work; they start together and finish when the tank is filled.

$\dfrac{1}{10}$ = the rate of the larger pipe

$\dfrac{1}{15}$ = the rate of the smaller pipe

$$Rate = \frac{1}{time\ required\ alone}$$

The chart records what happens *while they operate together to do the job.*

	R	·	T	=	W
larger pipe	$\dfrac{1}{10}$		t		$\dfrac{t}{10}$
smaller pipe	$\dfrac{1}{15}$		t		$\dfrac{t}{15}$

$\Big($work done by the larger pipe$\Big) + \Big($work done by the smaller pipe$\Big)$ = one whole tank filled

$\dfrac{t}{10} + \dfrac{t}{15} = 1$ .... The LCD is 30.

$30\left(\dfrac{t}{10} + \dfrac{t}{15}\right) = 30$

$3t + 2t = 30$
$5t = 30$
$t = 6$ .... It will take 6 hours to fill the tank.

25b. Let t be the number of hours required by the larger pipe alone;
$t + 6$ = the time required by the smaller pipe alone.

$\dfrac{1}{t}$ = the rate of the larger pipe

$\dfrac{1}{t + 6}$ = the rate of the smaller pipe

$$Rate = \frac{1}{time\ required\ alone}$$

The chart records what happens *while they operate together to do the job.*

	R	·	T	=	W
larger pipe	$\dfrac{1}{t}$		4		$4\left(\dfrac{1}{t}\right)$ or $\dfrac{4}{t}$
smaller pipe	$\dfrac{1}{t + 6}$		4		$4\left(\dfrac{1}{t + 6}\right)$ or $\dfrac{4}{t + 6}$

$\Big($work done by the larger pipe$\Big) + \Big($work done by the smaller pipe$\Big)$ = one whole tank filled

page 174

section 8.7

$\frac{4}{t} + \frac{4}{t + 6} = 1$ The LCD is $t(t + 6)$.

$t(t + 6)\left(\frac{4}{t} + \frac{4}{t + 6}\right) = t(t + 6)(1)$

$4(t + 6) + 4t = t^2 + 6t$
$4t + 24 + 4t = t^2 + 6t$
$0 = t^2 - 2t - 24$
$0 = (t - 6)(t + 4)$
$t - 6 = 0 \qquad t + 4 = 0$
$t = 6 \qquad\;\; t = -4$ We reject $t = -4$ because t represents time.

The larger pipe would take 6 hours to fill the tank by itself.

29.   As with all word problems, start by identifying the important quantities.

the integers	their reciprocals
n	$\frac{1}{n}$
n + 1	$\frac{1}{n + 1}$

The sum of these reciprocals is $\frac{1}{n} + \frac{1}{n + 1}$.

The problem says that the sum of the reciprocals will be 11 times the <u>reciprocal of their product</u>.

Since their product is $n(n + 1)$, the reciprocal is $\frac{1}{n(n + 1)}$.

Now that we have all the pieces to the equation, let's set it up:

The sum of the reciprocals   is   11 times the reciprocal of their product

$\frac{1}{n} \; + \; \frac{1}{n + 1} \qquad\qquad = \qquad\qquad \frac{11}{n(n + 1)}$ The LCD is $n(n + 1)$.

$n(n + 1)\left(\frac{1}{n} \; + \; \frac{1}{n + 1}\right) \; = \; n(n + 1)\left(\frac{11}{n(n + 1)}\right)$

$(n + 1) + n = 11$
$2n = 10$
$n = 5$ The first integer is 5, the next consecutive one is 6.

33.   We have lots of information here, so you may want to use a chart.

Let r be the *monthly* interest rate for the 30-year mortgage. (We will later multiply by 12 to give the *annual* interest rate, which is what the problem is asking for.)

We are using r as the monthly rate because the dollar amounts of interest ($700 and $850) are only for the first month.

r = monthly interest rate on the 30-year mortgage

r − 0.00125 = monthly interest rate for the 15-year mortgage

(Remember, we must express 1.5% as 0.015 in calculations; then the monthly rate is $\frac{0.015}{12} = 0.00125$.)

The formula   Principal · Rate · Time = Interest will be applied, with Time = 1 (month).

	P ·	R	=	I
30-year		r		850
15-year		r − 0.00125		700

We do not know the principal, however from P · R = I, we have $P = \frac{I}{R}$.

The principal for the 30-year mortgage is: $\frac{850}{r}$

The principal for the 15-year mortgage is: $\frac{700}{r - 0.00125}$

Since these quantities represent the same amount (what will be borrowed), we have:

$\frac{850}{r} = \frac{700}{r - 0.00125}$ We may cross-multiply.

$850(r - 0.00125) = 700r$

$850r - 1.0625 = 700r$

$150r = 1.0625$

$r = \frac{1.0625}{150} = 0.00708333....$

The annual rate is $12(\frac{1.0625}{150}) = 0.085$, or 8.5%.   The annual rate on the 15-year mortgage was given to be 1.5% less, so it is 7%.

37.   If w varies directly as x, then there is a constant of variation k_1 such that $w = k_1 x$.

If x varies inversely as y, then there is a constant of variation k_2 such that $x = \frac{k_2}{y}$.

Putting $x = \frac{k_2}{y}$ into the first equation gives $w = k_1\left(\frac{k_2}{y}\right) = \frac{k_1 k_2}{y}$.

If we let $k_1 k_2 = k_3$, then $w = \frac{k_3}{y}$, which says w varies inversely as y.

41a.   Let t be the number of hours required for the tank to be filled.

$\frac{1}{14}$ = the rate of the inlet pipe

$-\frac{1}{18}$ = the rate of the outlet pipe

$$Rate = \frac{1}{time\ required\ alone}$$

We express the rate of the outlet pipe as a negative quantity because it is emptying the tank, and water is lost. This is the opposite of filling the tank.

The chart records what happens *while they operate together to do the job.*

	R	·	T	=	W
inlet pipe	$\frac{1}{14}$		t		$\frac{t}{14}$
outlet pipe	$-\frac{1}{18}$		t		$-\frac{t}{18}$

$\left(\text{work done by the inlet pipe}\right) + \left(\text{work done by the outlet pipe}\right) = \text{one whole tank filled}$

$\frac{t}{14} - \frac{t}{18} = 1$ $\qquad$ The LCD is 126.

$126\left(\frac{t}{14} - \frac{t}{18}\right) = 126$

$9t - 7t = 126$
$2t = 126$
$t = 63$ $\qquad$ It will take 63 hours to fill the tank.

41b. Let t be the number of hours required for the inlet pipe to fill the tank.
The number of hours needed for the outlet pipe to empty it is: $t + 3$

$\frac{1}{t}$ = the rate of the inlet pipe

$-\frac{1}{t+3}$ = the rate of the outlet pipe $\qquad$ $\boxed{Rate = \dfrac{1}{time\ required\ alone}}$

We express the rate of the outlet pipe as a negative quantity because it is emptying the tank, which is the opposite of filling the tank.

The chart records what happens *while they operate together to do the job.*

	R	·	T	=	W
inlet pipe	$\frac{1}{t}$		60		$\frac{60}{t}$
outlet pipe	$-\frac{1}{t+3}$		60		$-\frac{60}{t+3}$

$\left(\text{work done by the inlet pipe}\right) + \left(\text{work done by the outlet pipe}\right) = \text{one whole tank filled}$

$\frac{60}{t} - \frac{60}{t+3} = 1$ $\qquad$ The LCD is $t(t+3)$.

$t(t+3)\left(\frac{60}{t} - \frac{60}{t+3}\right) = t(t+3)$

$60(t+3) - 60t = t(t+3)$

$$60t + 180 - 60t = t^2 + 3t$$
$$t^2 + 3t - 180 = 0$$
$$(t + 15)(t - 12) = 0$$
$$t = -15 \quad t = 12$$

It would take 12 hours for the inlet pipe to fill the tank if the outlet pipe is closed.

Solutions to All Review Exercises – Chapter Eight

1. $\dfrac{24}{-5 - 7} = \dfrac{24}{-12} = -2$

2. $\dfrac{4(-5) + 9}{2(-5) - 1} = \dfrac{-11}{-11} = 1$

3. $\dfrac{-5 + 5}{-5 - 5} = \dfrac{0}{-10} = 0$

4. $\dfrac{-5 + 12}{(-5)^2 - 25} = \dfrac{7}{25 - 25} = \dfrac{7}{0}$ undefined

5. A rational expression cannot contain roots of the variable. Answer: D

6. B. Using x = 3 causes division by zero.

7. $x = 0$

8. $x = -3$

9. $(x - 3)(x + 4) = 0$
   $x - 3 = 0 \qquad x + 4 = 0$
   $x = 3 \qquad\quad x = -4$

10. $x^2 - 7x + 6 = 0$
    $(x - 6)(x - 1) = 0$
    $x = 6, x = 1$

11. $\dfrac{56x^3y^4}{63x^2y^2} = \dfrac{56}{63} \cdot \dfrac{x^3}{x^2} \cdot \dfrac{y^4}{y^2} = \dfrac{8}{9} \cdot x \cdot y^2 = \dfrac{8}{9}xy^2$ or $\dfrac{8xy^2}{9}$

12. $\dfrac{12y - 36}{20y - 60} = \dfrac{12(y - 3)}{20(y - 3)} = \dfrac{12}{20} = \dfrac{3}{5}$

13. $\dfrac{15x^2 - 15}{25x + 25} = \dfrac{15(x + 1)(x - 1)}{25(x + 1)} = \dfrac{3(x - 1)}{5}$

14. $\dfrac{3 - 7m}{14m^2 - 6m} = \dfrac{3 - 7m}{2m(7m - 3)} = \dfrac{(-1)(7m - 3)}{2m(7m - 3)} = -\dfrac{1}{2m}$

15. $\dfrac{v^2 - 49}{v^2 + 12v + 35} = \dfrac{(v + 7)(v - 7)}{(v + 5)(v + 7)} = \dfrac{v - 7}{v + 5}$

16. $\dfrac{10x^2 + 29xy + 10y^2}{6x^2 + 13xy - 5y^2} = \dfrac{(5x + 2y)(2x + 5y)}{(2x + 5y)(3x - y)} = \dfrac{5x + 2y}{3x - y}$

17. $\dfrac{3}{7} = \dfrac{3(x + 3)}{7(x + 3)} = \dfrac{3x + 9}{7x + 21}$

18. $\dfrac{5x - y}{x - 7} = \dfrac{(-1)(5x - y)}{(-1)(x - 7)} = \dfrac{y - 5x}{7 - x}$

19. $\dfrac{2v - 1}{v + 3} = \dfrac{(2v - 1)(2v - 1)}{(v + 3)(2v - 1)} = \dfrac{4v^2 - 4v + 1}{2v^2 + 5v - 3}$

20. $\dfrac{3m + n}{4m - n} = \dfrac{(3m + n)(2m + n)}{(4m - n)(2m + n)} = \dfrac{6m^2 + 5mn + n^2}{8m^2 + 2mn - n^2}$

21. $\left.\begin{array}{l} 24a^2b^3 = 2^3 \cdot 3a^2b^3 \\ 56ab^5 \ \ = 2^3 \cdot 7ab^5 \end{array}\right\}$ $\quad$ LCD $= 2^3 \cdot 3 \cdot 7a^2b^5 = 168a^2b^5$

22. $\left.\begin{array}{l} 15x + 20y = 5(3x + 4y) \\ 18x + 24y = 6(3x + 4y) = 2(3)(3x + 4y) \end{array}\right\}$ $\quad$ LCD $= 30(3x + 4y)$

23. $10m^2 - 10n^2 = 10(m^2 - n^2) = 10(m + n)(m - n) = 2(5)(m + n)(m - n)$
    $15m + 15n = 15(m + n) = 3(5)(m + n)$
    LCD $= 30(m + n)(m - n)$

24. $\left.\begin{array}{l} v^2 - 9w^2 = (v + 3w)(v - 3w) \\ 7v^2 + 20vw - 3w^2 = (v + 3w)(7v - w) \end{array}\right\}$ $\quad$ LCD $= (v + 3w)(v - 3w)(7v - w)$

25. $\dfrac{2x - 3}{3x - 5} + \dfrac{4x - 7}{3x - 5} = \dfrac{2x - 3 + 4x - 7}{3x - 5} = \dfrac{6x - 10}{3x - 5} = \dfrac{2(3x - 5)}{3x - 5} = 2$

26. $\dfrac{2y + 1}{y + 3} - \dfrac{y - 2}{y + 3} = \dfrac{2y + 1 - (y - 2)}{y + 3} = \dfrac{y + 3}{y + 3} = 1$

27. $\dfrac{3v + 5}{(v + 8)(v - 9)} - \dfrac{2v - 3}{(v + 8)(v - 9)} = \dfrac{3v + 5 - 2v + 3}{(v + 8)(v - 9)} = \dfrac{v + 8}{(v + 8)(v - 9)} = \dfrac{1}{v - 9}$

28. $\dfrac{3x + 1}{2x^2 - 9x - 12} - \dfrac{4x - 5}{2x^2 - 9x - 12} = \dfrac{3x + 1 - 4x + 5}{2x^2 - 9x - 12} = \dfrac{-x + 6}{2x^2 - 9x - 12}$

29. $\dfrac{12w^2}{18w - 45} \cdot \dfrac{2w^3 - 5w^2}{15w^3} = \dfrac{12w^2(2w^3 - 5w^2)}{(18w - 45)15w^3} = \dfrac{12w^2(w^2)(2w - 5)}{9(2w - 5)15w^3} = \dfrac{4w}{45}$

30. $\dfrac{x^2 - 4y^2}{4y - x} \div \dfrac{2x + 4y}{x - 4y} = \dfrac{(x^2 - 4y^2)(x - 4y)}{(4y - x)(2x + 4y)} = -\dfrac{(x + 2y)(x - 2y)}{2(x + 2y)} = -\dfrac{x - 2y}{2}$

31. $\dfrac{m^2 - 2m - 15}{m^2 + 5m + 6} \div \dfrac{m^2 - 25}{m^2 - 2m - 8} = \dfrac{(m^2 - 2m - 15)(m^2 - 2m - 8)}{(m^2 + 5m + 6)(m^2 - 25)}$

    $= \dfrac{(m - 5)(m + 3)(m - 4)(m + 2)}{(m + 2)(m + 3)(m + 5)(m - 5)} = \dfrac{m - 4}{m + 5}$

32.
$$\frac{(a^2 + 2ab + b^2)(a^2 - 2ab + b^2)}{(a^2 - b^2)(a^2 + b^2)} = \frac{(a + b)^2(a - b)^2}{(a + b)(a - b)(a^2 + b^2)}$$

$$= \frac{(a + b)(a - b)}{a^2 + b^2} \text{ or } \frac{a^2 - b^2}{a^2 + b^2}$$

33.
$$\frac{\left(\dfrac{m^2 - n^2}{m^3 n^3}\right)}{\left(\dfrac{m + n}{mn}\right)} = \frac{m^2 - n^2}{m^3 n^3} \cdot \frac{mn}{m + n} = \frac{(m + n)(m - n)mn}{m^3 n^3 (m + n)} = \frac{(m - n)}{m^2 n^2}$$

34.
$$\frac{1 + \dfrac{2}{m}}{\dfrac{6}{m} + 3} = \frac{m\left(1 + \dfrac{2}{m}\right)}{m\left(\dfrac{6}{m} + 3\right)} = \frac{m + 2}{6 + 3m} = \frac{m + 2}{3(m + 2)} = \frac{1}{3}$$

35.
$$\frac{\dfrac{28x^2 y^3}{15w^2 y}}{\dfrac{16w^3 y}{25wz^2}} = \frac{(28x^2 y^3)(25wz^2)}{(15w^2 y)(16w^3 y)} = \frac{4(7)(5^2)x^2 y^3 wz^2}{3(5)(4^2)w^5 y^2} = \frac{35x^2 yz^2}{12w^4}$$

36.
$$\frac{1 + \dfrac{11}{v} + \dfrac{24}{v^2}}{1 + \dfrac{9}{v} + \dfrac{18}{v^2}} = \frac{v^2\left(1 + \dfrac{11}{v} + \dfrac{24}{v^2}\right)}{v^2\left(1 + \dfrac{9}{v} + \dfrac{18}{v^2}\right)} = \frac{v^2 + 11v + 24}{v^2 + 9v + 18} = \frac{(v + 8)(v + 3)}{(v + 3)(v + 6)} = \frac{v + 8}{v + 6}$$

37.
$$\frac{7}{2a - 5} + \frac{4}{5 - 2a} = \frac{7}{2a - 5} + \frac{4}{(-1)(2a - 5)} = \frac{7}{2a - 5} - \frac{4}{2a - 5} = \frac{3}{2a - 5}$$

38.
$$\frac{5}{a + 2} + \frac{4}{a - 3} = \frac{5(a - 3) + 4(a + 2)}{(x + 2)(a - 3)} = \frac{5a - 15 + 4a + 8}{(a + 2)(a - 3)} = \frac{9a - 7}{(a + 2)(a - 3)}$$

39.
$$\frac{3}{x^2 + 4x} + \frac{3}{4x + 16}$$

$$\left. \begin{array}{l} x^2 + 4x = x(x + 4) \\ 4x + 16 = 4(x + 4) \end{array} \right\} \quad \text{LCD} = 4x(x + 4)$$

$$\frac{3(4)}{x(x + 4)(4)} + \frac{3x}{4(x + 4)(x)} = \frac{12 + 3x}{4x(x + 4)} = \frac{3(x + 4)}{4x(x + 4)} = \frac{3}{4x}$$

40.

$$\frac{x+1}{2x^2+11x+15} + \frac{x+1}{2x^2+9x+10}$$

$$\left.\begin{array}{l} 2x^2 + 11x + 15 = (2x+5)(x+3) \\ 2x^2 + 9x + 10 = (2x+5)(x+2) \end{array}\right\} \quad \text{LCD} = (2x+5)(x+3)(x+2)$$

$$\frac{(x+1)(x+2)}{(2x+5)(x+3)(x+2)} + \frac{(x+1)(x+3)}{(2x+5)(x+2)(x+3)} = \frac{x^2+3x+2+x^2+4x+3}{(2x+5)(x+3)(x+2)}$$

$$= \frac{2x^2+7x+5}{(2x+5)(x+3)(x+2)} = \frac{(2x+5)(x+1)}{(2x+5)(x+3)(x+2)} = \frac{x+1}{(x+3)(x+2)}$$

41.

$$\frac{m}{m-1} - \frac{m+1}{3m-3} \cdot \frac{6}{2m+2}$$

First: $$\frac{(m+1)(6)}{(3m-3)(2m+2)} = \frac{(m+1)(6)}{3(m-1)(2)(m+1)} = \frac{1}{m-1}$$

Then: $$\frac{m}{m-1} - \frac{1}{m-1} = \frac{m-1}{m-1} = 1$$

42.

$$\left(\frac{m}{m-1} - \frac{1}{m+1}\right)\left(\frac{3m+3}{5m^2+5}\right)$$

First: $$\frac{m}{m-1} - \frac{1}{m+1} = \frac{m(m+1)-(m-1)}{(m-1)(m+1)} = \frac{m^2+m-m+1}{(m-1)(m+1)}$$

$$= \frac{m^2+1}{(m-1)(m+1)}$$

Then: $$\left(\frac{m^2+1}{(m-1)(m+1)}\right)\left(\frac{3m+3}{5m^2+5}\right) = \frac{(m^2+1)\,3(m+1)}{(m-1)(m+1)(5)(m^2+1)} = \frac{3}{5(m-1)}$$

43.

$$\left(\frac{1}{x} - \frac{1}{y}\right)\left(\frac{3xy}{5x-5y}\right)$$

Within the parentheses, find common denominators and do the subtraction.

$$\left(\frac{y-x}{xy}\right)\left(\frac{3xy}{5x-5y}\right) = \frac{(y-x)(3xy)}{(xy)(5)(x-y)} = \frac{-3xy}{5xy} = -\frac{3}{5}$$

44.

$$\left(\frac{5}{y} - \frac{15}{x}\right)\left(\frac{1}{x-3y} - \frac{1}{x+3y}\right)$$

Within parentheses, find common denominators and subtract the fractions.

$$\left(\frac{5x-15y}{xy}\right)\left(\frac{6y}{(x-3y)(x+3y)}\right) = \frac{5(x-3y)6y}{(xy)(x-3y)(x+3y)} = \frac{30}{x(x+3y)}$$

Chapter Eight Review

45. $\dfrac{(5v - 20)(4v^2 + 12v + 9)(11v)}{(6v + 9)(v^2 - 6v + 8)(50v + 75)} = \dfrac{5(v - 4)(2v + 3)^2(11v)}{3(2v + 3)(v - 4)(v - 2)25(2v + 3)} = \dfrac{11v}{15(v - 2)}$

46. $\dfrac{(2x - y)(18x^2y^2)}{(12xy)(2x - y)^2} + \dfrac{(4x^3y^3)(5)(2x + 3y)}{(2x + 3y)(x - 2y)(40x^2y^2)}$

$= \dfrac{3xy}{2(2x - y)} + \dfrac{xy}{2(x - 2y)}$

$= \dfrac{3xy(x - 2y)}{2(2x - y)(x - 2y)} + \dfrac{xy(2x - y)}{2(x - 2y)(2x - y)}$

$= \dfrac{3x^2y - 6xy^2 + 2x^2y - xy^2}{2(2x - y)(x - 2y)}$

$= \dfrac{5x^2y - 7xy^2}{2(2x - y)(x - 2y)}$ or $\dfrac{xy(5x - 7y)}{2(2x - y)(x - 2y)}$

47. $\dfrac{16}{2w - 1} = 4$

$16 = 4(2w - 1)$
$16 = 8w - 4$
$20 = 8w$

$w = \dfrac{20}{8} = \dfrac{5}{2}$ or $2\dfrac{1}{2}$

48. $\dfrac{3}{5y} - \dfrac{7}{6y} = \dfrac{17}{30}$ The LCD is 30y.

$30y\left(\dfrac{3}{5y} - \dfrac{7}{6y}\right) = 30y\left(\dfrac{17}{30}\right)$

$6(3) - 5(7) = 17y$
$18 - 35 = 17y$
$-17 = 17y$
$\quad\; y = -1$

49. $\dfrac{x + 4}{2x - 1} = \dfrac{5}{7}$ Remember: If $\dfrac{a}{b} = \dfrac{c}{d}$, then $ad = bc$.

$(x + 4)7 = (2x - 1)5$
$7x + 28 = 10x - 5$
$33 = 3x$
$x = 11$

50. $\dfrac{v + 1}{v + 3} = \dfrac{3}{v + 3} + 2$ The LCD is $v + 3$.

$(v + 3)\left(\dfrac{v + 1}{v + 3}\right) = (v + 3)\left(\dfrac{3}{v + 3} + 2\right)$

Chapter Eight Review

$$v + 1 = 3 + (v + 3)(2)$$
$$v + 1 = 3 + 2v + 6$$
$$-v = 8$$
$$v = -8$$

51. $\dfrac{m - 1}{3m - 4} - \dfrac{2}{3} = \dfrac{3 - 2m}{3m - 4}$ The LCD is $3(3m - 4)$.

$$3(3m - 4)\left(\dfrac{m - 1}{3m - 4} - \dfrac{2}{3}\right) = 3(3m - 4)\left(\dfrac{3 - 2m}{3m - 4}\right)$$

$$3(m - 1) - 2(3m - 4) = 3(3 - 2m)$$
$$3m - 3 - 6m + 8 = 9 - 6m$$
$$-3m + 5 = 9 - 6m$$
$$3m = 4$$

$$m = \dfrac{4}{3}$$ This is the excluded value! No solution.

52. $\dfrac{6}{n^2 + 7n} + \dfrac{n + 1}{n + 7} = \dfrac{n + 2}{n}$ The LCD is $n(n + 7)$.

$$n(n + 7)\left(\dfrac{6}{n(n + 7)} + \dfrac{n + 1}{n + 7}\right) = n(n + 7)\left(\dfrac{n + 2}{n}\right)$$

$$6 + n(n + 1) = (n + 7)(n + 2)$$

$$6 + n^2 + n = n^2 + 9n + 14$$

$$-8n = 8$$

$$n = -1$$

53. $\dfrac{x - 2}{x + 4} + \dfrac{x + 1}{x + 6} = \dfrac{11x + 32}{x^2 + 10x + 24}$ $x^2 + 10x + 24 = (x + 4)(x + 6)$
The LCD is $(x + 4)(x + 6)$.

$$(x + 4)(x + 6)\left(\dfrac{x - 2}{x + 4} + \dfrac{x + 1}{x + 6}\right) = (x + 4)(x + 6)\left(\dfrac{11x + 32}{(x + 4)(x + 6)}\right)$$

$$(x + 4)(x + 6)\left(\dfrac{x - 2}{x + 4}\right) + (x + 4)(x + 6)\left(\dfrac{x + 1}{x + 6}\right)$$
$$= (x + 4)(x + 6)\left(\dfrac{11x + 32}{(x + 4)(x + 6)}\right)$$

$$(x + 6)(x - 2) + (x + 4)(x + 1) = 11x + 32$$
$$x^2 + 4x - 12 + x^2 + 5x + 4 = 11x + 32$$
$$2x^2 + 9x - 8 = 11x + 32$$
$$2x^2 - 2x - 40 = 0$$ Divide both sides by 2.

$x^2 - x - 20 = 0$

$(x - 5)(x + 4) = 0$

$x - 5 = 0 \qquad x + 4 = 0$

$x = 5 \qquad x = -4$ *We reject x = −4 because it is an excluded value.*

54. $\dfrac{2}{x^2 - 1} + \dfrac{3}{(x + 1)^2} = \dfrac{1}{(x - 1)^2}$ The LCD is $(x + 1)^2(x - 1)^2$.

$(x + 1)^2(x - 1)^2 \left(\dfrac{2}{x^2 - 1} + \dfrac{3}{(x + 1)^2} \right) = (x + 1)^2(x - 1)^2 \left(\dfrac{1}{(x - 1)^2} \right)$

$2(x + 1)(x - 1) + 3(x - 1)^2 = (x + 1)^2$

$2(x^2 - 1) + 3(x - 1)^2 = (x + 1)^2$

$2x^2 - 2 + 3x^2 - 6x + 3 = x^2 + 2x + 1$

$4x^2 - 8x = 0$
$4x(x - 2) = 0$
$x = 0 \quad x = 2$

55. $y = \dfrac{3}{x}$ Multiply both sides by x.

$yx = 3$ Now divide by y.

$x = \dfrac{3}{y}$

56. $y = \dfrac{1}{x + 3}$

$y(x + 3) = 1$
$yx + 3y = 1$
$yx = 1 - 3y$

$x = \dfrac{1 - 3y}{y}$

57. $y = \dfrac{x}{x + 3}$ Multiply both sides by x + 3.

$(x + 3)y = (x + 3)\left(\dfrac{x}{x + 3} \right)$

                 Bring all terms with x to the left side,
$xy + 3y = x$ and all other terms to the right side.
$xy - x = -3y$ Factor out x.
$x(y - 1) = -3y$ Divide by y − 1.

$x = \dfrac{-3y}{y - 1}$ or $-\dfrac{3y}{y - 1}$ $\left[\text{or} : \dfrac{-3y}{y - 1} = \dfrac{(-3y)(-1)}{(y - 1)(-1)} = \dfrac{3y}{1 - y} \right]$

Chapter Eight Review

58. $\dfrac{6}{y} = \dfrac{5}{x} + \dfrac{1}{w}$ The LCD is xyw.

$$xyw\left(\dfrac{6}{y}\right) = xyw\left(\dfrac{5}{x} + \dfrac{1}{w}\right)$$

$$6xw = 5yw + xy$$
$$6xw - xy = 5yw$$
$$x(6w - y) = 5yw$$

$$x = \dfrac{5wy}{6w - y}$$

59. $\dfrac{4x^2 - 4x - 2}{x^2 - 3x + 2} + R = \dfrac{x + 1}{x - 1}$

$R = \dfrac{x + 1}{x - 1} - \dfrac{4x^2 - 4x - 2}{x^2 - 3x + 2}$ Since $x^2 - 3x + 2 = (x - 2)(x - 1)$, the LCD is $(x - 2)(x - 1)$.

$$= \dfrac{(x + 1)(x - 2) - 4x^2 + 4x + 2}{(x - 1)(x - 2)}$$

$$= \dfrac{x^2 - x - 2 - 4x^2 + 4x + 2}{(x - 1)(x - 2)}$$

$$= \dfrac{-3x^2 + 3x}{(x - 1)(x - 2)}$$

$$= \dfrac{-3x(x - 1)}{(x - 1)(x - 2)}$$

$$= -\dfrac{3x}{x - 2}$$

60. $R \cdot \dfrac{x^2 - 8x + 15}{x^2 - 4x - 12} = \dfrac{7x^2 - 21x}{x^2 + 7x + 10}$

$$R = \dfrac{7x^2 - 21x}{x^2 + 7x + 10} \div \dfrac{x^2 - 8x + 15}{x^2 - 4x - 12}$$

$$= \dfrac{7x(x - 3)(x - 6)(x + 2)}{(x + 2)(x + 5)(x - 5)(x - 3)}$$

$$= \dfrac{7x(x - 6)}{(x + 5)(x - 5)} \text{ or } \dfrac{7x^2 - 42x}{x^2 - 25}$$

61. Let x equal the second number, and $3x - 5$ equal the first number.

Their reciprocals: $\dfrac{1}{x}$ and $\dfrac{1}{3x - 5}$.

$$\frac{1}{x} + \frac{1}{3x - 5} = \frac{3}{10} \qquad \text{The LCD is } 10x(3x - 5).$$

$$10x(3x - 5)\left(\frac{1}{x} + \frac{1}{3x - 5}\right) = 10x(3x - 5)\left(\frac{3}{10}\right)$$

$$10x(3x - 5)\left(\frac{1}{x}\right) + 10x(3x - 5)\left(\frac{1}{3x - 5}\right) = 10x(3x - 5)\left(\frac{3}{10}\right)$$

$$10(3x - 5) + 10x = x(3x - 5)(3)$$
$$30x - 50 + 10x = 9x^2 - 15x$$
$$0 = 9x^2 - 55x + 50$$
$$0 = (9x - 10)(x - 5)$$
$$9x - 10 = 0 \qquad x - 5 = 0$$

$$x = \frac{10}{9} \qquad\qquad x = 5$$

If $x = \frac{10}{9}$, then the first number is $3x - 5 = 3\left(\frac{10}{9}\right) - 5 = \frac{10}{3} - 5 = \frac{10 - 15}{3} = -\frac{5}{3}$.

If $x = 5$, then the first number is $3x - 5 = 3(5) - 5 = 15 - 5 = 10$.

There are two pairs of numbers that solve the problem: $\frac{10}{9}$ and $-\frac{5}{3}$ or 5 and 10.

62. $n \qquad$ first even integer
$\quad n + 2 \quad$ next even integer

$$\frac{1}{n} + \frac{1}{n + 2} = \frac{13}{84} \qquad \text{The LCD is } 84n(n + 2).$$

$$84n(n + 2)\left(\frac{1}{n} + \frac{1}{n + 2}\right) = 84n(n + 2)\left(\frac{13}{84}\right)$$

$$84(n + 2) + 84n = 13n(n + 2)$$
$$84n + 168 + 84n = 13n^2 + 26n$$
$$13n^2 - 142n - 168 = 0$$
$$(n - 12)(13n + 14) = 0$$

$$n = 12 \quad n = -\frac{14}{13} \text{ (not an integer)} \qquad \text{Answer: 12 and 14}$$

63. Let t be the correct time.

$$\frac{23}{24} = \frac{t - 3}{t + 2}$$

$$23(t + 2) = 24(t - 3)$$
$$23t + 46 = 24t - 72$$
$$-t = -118 \qquad\qquad \text{The time is 118 seconds.}$$

64. $y = \frac{k}{x} \quad$ Find k first.

$$3 = \frac{k}{15}$$

$$k = 45$$

Chapter Eight Review

$$y = \frac{45}{x}$$

Now find y when x is 9.

$$y = \frac{45}{9} = 5$$

65. $I = \frac{k}{R}$ I = current; R = resistance

$$2 = \frac{k}{5}; \ k = 10$$

$$I = \frac{10}{R}$$

$$I = \frac{10}{0.5} = 20$$ The current is 20 amps.

66. Let r be the airspeed of the airplane. (This is the speed as if there were no wind.)

	R	·	T	=	D
with the wind	r + 30		$\frac{450}{r + 30}$		450
against the wind	r − 30		$\frac{300}{r - 30}$		300

$$\frac{450}{r + 30} = \frac{300}{r - 30}$$

$450(r - 30) = 300(r + 30)$
$450r - 13500 = 300r + 9000$
$150r = 22500$
$\ \ r = 150$ The airspeed of the plane is 150 mi/h.

67. Let t = the number of hours the passenger train would need.
The freight train would require 3t hours.

	R	·	T	=	D
freight	$\frac{1}{3t}$		3		$\frac{1}{t}$
passenger	$\frac{1}{t}$		3		$\frac{3}{t}$

$$\frac{1}{t} + \frac{3}{t} = 1 \text{ route}$$

$$\frac{4}{t} = 1$$

$t = 4$ The passenger train needs 4 hours.
$3t = 12$ The freight train needs 12 hours.

Chapter Eight Review

68.   Let the integers be n and n + 2.

$$\frac{n^2}{(n + 2)^2} = \frac{9}{16}$$

$$16n^2 = 9(n + 2)^2$$

$$16n^2 = 9n^2 + 36n + 36$$

$$7n^2 - 36n - 36 = 0$$

$$(7n + 6)(n - 6) = 0$$

$n = -\frac{6}{7}$ $n = 6$ (We reject $-\frac{6}{7}$ because it isn't an integer.)

The numbers are 6 and 8.

69a.   Let t equal the number of hours required to fill the tank, using both pipes simultaneously.

Rate of larger pipe: $\frac{1}{20}$ tank per hour

Rate of smaller pipe: $\frac{1}{30}$ tank per hour

$$\boxed{Rate\ of\ work = \frac{1}{time\ required\ alone}}$$

	R	·	T	=	W
larger pipe	$\frac{1}{20}$		t		$\frac{t}{20}$
smaller pipe	$\frac{1}{30}$		t		$\frac{t}{30}$

$\left(\text{work done by the larger pipe}\right) + \left(\text{work done by the smaller pipe}\right) =$ one whole tank filled

$\frac{t}{20} + \frac{t}{30} = 1$ The LCD is 60.

$60\left(\frac{t}{20} + \frac{t}{30}\right) = 60$

$3t + 2t = 60$
$5t = 60$
$t = 12$ It will take 12 hours to fill the tank using both pipes simultaneously.

69b.   Let t equal the number of hours required for the larger pipe to fill the tank if working alone; then t + 7 is the time required by the smaller pipe to fill the tank if working alone.

Larger pipe's rate: $\frac{1}{t}$ tank per hour

Smaller pipe's rate: $\frac{1}{t + 7}$ tank per hour

$$\boxed{Rate\ of\ work = \frac{1}{time\ required\ alone}}$$

Chapter Eight Review

	R	·	T	=	W
larger pipe	$\frac{1}{t}$		12		$\frac{12}{t}$
smaller pipe	$\frac{1}{t+7}$		12		$\frac{12}{t+7}$

$\left(\text{work done by the larger pipe}\right)+\left(\text{work done by the smaller pipe}\right) = \text{one whole tank filled}$

$\frac{12}{t} + \frac{12}{t+7} = 1$ 　　　　　　　The LCD is $t(t + 7)$.

$t(t+7)\left(\frac{12}{t} + \frac{12}{t+7}\right) = t(t+7)$

$t(t+7)\left(\frac{12}{t}\right) + t(t+7)\left(\frac{12}{t+7}\right) = t(t+7)$

$(t+7)(12) + t(12) = t^2 + 7t$

$12t + 84 + 12t = t^2 + 7t$

$0 = t^2 - 17t - 84$
$0 = (t - 21)(t + 4)$
$t - 21 = 0$ 　　　　$t + 4 = 0$
$t = 21$ 　　　　　　$t = -4$ 　　　　*We reject $t = -4$ because t represents time.*

The larger pipe would take 21 hours to fill the tank if working alone.

70a. Let t equal the number of hours required to fill the tank, using both pipes simultaneously.

Rate of outlet pipe: $-\frac{1}{12}$ tank per hour 　　　$\boxed{Rate\ of\ work = \frac{1}{time\ required\ alone}}$

Rate of inlet pipe: $\frac{1}{4}$ tank per hour

We will use a negative rate for the outlet pipe because it *drains* water, which is the opposite of adding water.

	R	·	T	=	W
outlet pipe	$-\frac{1}{12}$		t		$-\frac{t}{12}$
inlet pipe	$\frac{1}{4}$		t		$\frac{t}{4}$

$\left(\text{work done by the outlet pipe}\right)+\left(\text{work done by the inlet pipe}\right) = \text{one whole tank filled}$

$-\frac{t}{12} + \frac{t}{4} = 1$ 　　　　　　　The LCD is 12.

$12\left(-\frac{t}{12} + \frac{t}{4}\right) = 12$

$\boxed{\text{Chapter Eight Review}}$

$-t + 3t = 12$
$2t = 12$
$t = 6$ It will take 6 hours to fill the tank using both pipes simultaneously.

70b.   Let t equal the number of hours required for the inlet pipe to fill the tank if working alone; then $t + 3$ is the time required by the outlet pipe to drain the tank if working alone.

Rate of inlet pipe: $\frac{1}{t}$ tank per hour

Rate of outlet pipe: $-\dfrac{1}{t + 3}$ tank per hour

$$\boxed{Rate\ of\ work = \frac{1}{time\ required\ alone}}$$

We are using a negative rate because we are *losing* water at this rate.

	R $\cdot$	T $=$	W
inlet pipe	$\frac{1}{t}$	60	$\frac{60}{t}$
outlet pipe	$-\dfrac{1}{t + 3}$	60	$-\dfrac{60}{t + 3}$

$\left(\text{work done by the inlet pipe}\right) + \left(\text{work done by the outlet pipe}\right) =$ one whole tank filled

$\dfrac{60}{t} - \dfrac{60}{t + 3} = 1$ The LCD is $t(t + 3)$.

$t(t + 3)\left(\dfrac{60}{t} - \dfrac{60}{t + 3}\right) = t(t + 3)$

$t(t + 3)\left(\dfrac{60}{t}\right) - t(t + 3)\left(\dfrac{60}{t + 3}\right) = t(t + 3)$

$(t + 3)(60) - t(60) = t^2 + 3t$

$60t + 180 - 60t = t^2 + 3t$
$0 = t^2 + 3t - 180$
$0 = (t - 12)(t + 15)$
$t - 12 = 0$ $t + 15 = 0$
$t = 12$ $t = -15$ *We reject t $=-15$ because t represents time.*
                                   It would take 12 hours.

71.   The correct solution is in the answer section in the text.

72.   The error in student B's work is that $x = 1$ is an excluded value. There is no solution.

Solutions To All Mastery Test Problems – Chapter Eight

1a.   0   1b. -5 1c. $\frac{2}{5}$ 1d. $-5, 3$

2a. $\dfrac{36v^2}{81v^5} = \dfrac{9(4)v^2}{9(9)v^5} = \dfrac{4}{9v^3}$

2b. $\dfrac{7x - 14y}{5x - 10y} = \dfrac{7(x - 2y)}{5(x - 2y)} = \dfrac{7}{5}$

2c. $\dfrac{8 - 2m}{m^2 - 16} = \dfrac{2(4 - m)}{(m + 4)(m - 4)}$

$= -\dfrac{2}{m + 4}$ $\boxed{Remember:\ \dfrac{a - b}{b - a} = -1.}$

2d. $\dfrac{4x^2 - 9}{2x^2 + 3x - 9} = \dfrac{(2x + 3)(2x - 3)}{(2x - 3)(x + 3)}$

$= \dfrac{2x + 3}{x + 3}$

3a. $\dfrac{2x}{15y} \cdot \dfrac{5y^2}{8x^3} = \dfrac{2x(5y^2)}{15y(8x^3)}$

$= \dfrac{2(5)xy^2}{2^3 3(5)x^3 y} = \dfrac{y}{12x^2}$

3b. $\dfrac{35}{2v - w} \cdot \dfrac{3w - 6v}{49} = \dfrac{5(7)(3)(w - 2v)}{(2v - w)(7)(7)}$

$= -\dfrac{15}{7}$ or $-2\dfrac{1}{7}$

3c. $\dfrac{3m^3 - 6m^2 n}{m^2 - 4n^2} \cdot \dfrac{m^2 + mn - 2n^2}{3m^2 - 3n^2}$

$= \dfrac{3m^2(m - 2n)(m + 2n)(m - n)}{(m + 2n)(m - 2n)(3)(m + n)(m - n)}$

$= \dfrac{m^2}{m + n}$

3d. $\dfrac{x^2 - 4}{x^2 - 4x - 21} \cdot \dfrac{x^2 - 2x - 35}{x^2 - 7x + 10}$

$= \dfrac{(x + 2)(x - 2)(x - 7)(x + 5)}{(x - 7)(x + 3)(x - 5)(x - 2)}$

$= \dfrac{(x + 2)(x + 5)}{(x + 3)(x - 5)}$

4a. $\dfrac{15a^2 b}{36a^3 b^3} \div \dfrac{45}{ab}$

$= \dfrac{(15a^2 b)(ab)}{(36a^3 b^3)(45)} = \dfrac{15a^3 b^2}{36(45)a^3 b^3}$

$= \dfrac{3(5)}{3(12)(5)(9)b} = \dfrac{1}{108b}$

4b. $\dfrac{7m - 21}{9m + 18} \div \dfrac{m^2 - 9}{m^2 - 4}$

$= \dfrac{7(m - 3)(m + 2)(m - 2)}{9(m + 2)(m + 3)(m - 3)}$

$= \dfrac{7(m - 2)}{9(m + 3)}$

4c. $\dfrac{v + 3w}{3v - w} \div \dfrac{w + 3v}{w - 3v}$

$= \dfrac{(v + 3w)(w - 3v)}{(3v - w)(w + 3v)} = -\dfrac{v + 3w}{w + 3v}$

4d. $\dfrac{15x - 45y}{25x^2 y^2} \div \dfrac{x^2 - 9y^2}{7x + 21y}$

$= \dfrac{15(x - 3y)(7)(x + 3y)}{25x^2 y^2(x + 3y)(x - 3y)}$

$= \dfrac{3(7)}{5x^2 y^2} = \dfrac{21}{5x^2 y^2}$

5a. $\dfrac{2x}{x - 5} - \dfrac{10}{x - 5}$

5b. $\dfrac{y}{y^2 - 9} + \dfrac{3}{y^2 - 9}$

$$= \frac{2x - 10}{x - 5} = \frac{2(x - 5)}{x - 5} = 2 \qquad\qquad = \frac{y + 3}{(y + 3)(y - 3)} = \frac{1}{y - 3}$$

5c. $\dfrac{7}{v - 7} - \dfrac{v}{v - 7}$

$= \dfrac{7 - v}{v - 7} = -1$

5d. $\dfrac{2x - 3}{x^2 - 7x + 10} + \dfrac{x - 3}{x^2 - 7x + 10}$

$= \dfrac{2x - 3 + x - 3}{x^2 - 7x + 10} = \dfrac{3x - 6}{x^2 - 7x + 10}$

$= \dfrac{3(x - 2)}{(x - 2)(x - 5)} = \dfrac{3}{x - 5}$

6a. $\dfrac{5}{6x} = \dfrac{5(3x^2)}{6x(3x^2)} = \dfrac{15x^2}{18x^3}$

6b. $\dfrac{5x}{x - y} = \dfrac{5x(2xy^2)}{(x - y)(2xy^2)} = \dfrac{10x^2y^2}{2x^2y^2 - 2xy^3}$

6c. $\dfrac{m - 2}{m + 4} = \dfrac{(m - 2)(m + 3)}{(m + 4)(m + 3)}$

$= \dfrac{m^2 + m - 6}{m^2 + 7m + 12}$

6d. $\dfrac{a + 3b}{a - 3b} = \dfrac{(a + 3b)(-1)}{(a - 3b)(-1)} = \dfrac{-a - 3b}{3b - a}$

7a. $45 = 3^2(5)$
$60 = 2^2\,(3)(5)$
LCD $= 2^2 3^2\,5$ or 180

7b. $24a^2b^3 = 2^3(3)a^2b^3$
$60a^3b = 2^2(3)(5)a^3b$
LCD $= 2^3(3)(5)a^3b^3$ or $120a^3b^3$

7c. $8v - 8w = 2^3(v - w)$
$4v^2 - 4w^2 = 2^2(v + w)(v - w)$
LCD $= 2^3(v - w)(v + w)$ or $8(v^2 - w^2)$

7d. $m^2 - 9 = (m + 3)(m - 3)$
$m^2 - m - 6 = (m - 3)(m + 2)$
LCD $= (m + 3)(m - 3)(m + 2)$

8a. $\dfrac{5}{2x} + \dfrac{4}{3y}$

$\dfrac{5(3y) + 4(2x)}{(2x)(3y)} = \dfrac{15y + 8x}{6xy}$

8b. $\dfrac{y^2}{y^2 - 36} - \dfrac{3}{y - 6} = \dfrac{y^2 - 3(y + 6)}{(y + 6)(y - 6)}$

$= \dfrac{y^2 - 3y - 18}{(y + 6)(y - 6)} = \dfrac{(y - 6)(y + 3)}{(y + 6)(y - 6)}$

$= \dfrac{y + 3}{y + 6}$

Chapter Eight Mastery Test

8c. $\dfrac{m}{(3m+1)(m-4)} + \dfrac{m}{(3m+1)(2m+5)}$

$= \dfrac{m(2m+5) + m(m-4)}{(3m+1)(m-4)(2m+5)}$

$= \dfrac{2m^2 + 5m + m^2 - 4m}{(3m+1)(m-4)(2m+5)}$

$= \dfrac{m(3m+1)}{(3m+1)(m-4)(2m+5)}$

$= \dfrac{m}{(m-4)(2m+5)}$

8d. $\dfrac{4}{x-4} - \dfrac{1}{x+2} + \dfrac{x+8}{x^2-2x-8}$

$= \dfrac{4(x+2) - (x-4) + x + 8}{(x-4)(x+2)}$

$= \dfrac{4x+20}{(x-4)(x+2)}$ or $\dfrac{4(x+5)}{(x-4)(x+2)}$

9a. $\dfrac{24}{\frac{3}{8}} = 24\left(\dfrac{8}{3}\right) = 64$

9b. $\dfrac{\frac{24}{3}}{8} = \dfrac{8}{8} = 1$

9c. $= \dfrac{m^2\left(1 + \frac{9}{m} + \frac{14}{m^2}\right)}{m^2\left(1 - \frac{5}{m} - \frac{14}{m^2}\right)}$

$= \dfrac{m^2 + 9m + 14}{m^2 - 5m - 14} = \dfrac{(m+2)(m+7)}{(m-7)(m+2)}$

$= \dfrac{m+7}{m-7}$

9d. $= \dfrac{\left(\frac{(x-6)(x+4)+9}{x+4}\right)}{\left(\frac{(x+1)(x-4)-14}{x-4}\right)} = \dfrac{\left(\frac{x^2-2x-15}{x+4}\right)}{\left(\frac{x^2-3x-18}{x-4}\right)}$

$= \dfrac{(x-5)(x+3)(x-4)}{(x+4)(x-6)(x+3)}$

$= \dfrac{(x-5)(x-4)}{(x+4)(x-6)}$

10a. $\dfrac{x}{5} - \dfrac{x^3}{15} \cdot \dfrac{6}{x^2}$

$= \dfrac{x}{5} - \dfrac{2x}{5}$

$= \dfrac{x - 2x}{5} = -\dfrac{x}{5}$

10b. $\dfrac{x}{x-2} + \dfrac{3x+9}{3x-2} \cdot \dfrac{-2}{x+3}$

$= \dfrac{x}{x-2} + \dfrac{-6}{3x-2} = \dfrac{x(3x-2) - 6(x-2)}{(x-2)(3x-2)}$

$= \dfrac{3x^2 - 8x + 12}{(x-2)(3x-2)}$ (It reduces no further.)

10c. $\left(\dfrac{v}{v+2} + \dfrac{1}{v+3}\right) \cdot \dfrac{v^2+6v+8}{v^2+4v+2}$

$= \left(\dfrac{v(v+3) + v + 2}{(v+2)(v+3)}\right)\left(\dfrac{(v+2)(v+4)}{v^2+4v+2}\right)$

$= \dfrac{(v^2+4v+2)(v+4)}{(v+3)(v^2+4v+2)} = \dfrac{v+4}{v+3}$

10d. $\left(\dfrac{6}{x} + \dfrac{9}{x^2}\right) - \left(1 + \dfrac{3}{x}\right)^2$

$= \dfrac{6}{x} + \dfrac{9}{x^2} - 1 - \dfrac{6}{x} - \dfrac{9}{x^2} = -1$

Chapter Eight Mastery Test

11a. $\dfrac{2v + 1}{5v + 5} = \dfrac{3}{8}$

$8(2v + 1) = (5v + 5)(3)$
$16v + 8 = 15v + 15$
$v = 7$

11b. $\dfrac{2w}{2w + 3} = \dfrac{5}{7} - \dfrac{3}{2w + 3}$

Multiply both sides by $7(2w + 3)$:
$7(2w) = 5(2w + 3) - 7(3)$
$14w = 10w + 15 - 21$
$4w = -6$; $w = -\dfrac{3}{2}$ or $-1\dfrac{1}{2}$

This is an excluded value. No solution.

11c. $\dfrac{6}{y + 1} = \dfrac{21}{4y - 1}$

$6(4y - 1) = (y + 1)(21)$
$24y - 6 = 21y + 21$
$3y = 27$
$y = 9$

11d. $\dfrac{x^2 - 11}{x^2 - 4x - 5} = \dfrac{x + 2}{x - 5} + \dfrac{x + 3}{x + 1}$

Multiply both sides by $(x - 5)(x + 1)$:
$x^2 - 11 = (x+2)(x+1) + (x+3)(x-5)$
$x^2 - 11 = 2x^2 + x - 13$
$x^2 + x - 2 = 0$
$(x + 2)(x - 1) = 0$
$x = -2 \qquad\qquad x = 1$

12a. $\dfrac{5}{x + 2} = \dfrac{3}{8}$

$5(8) = (x+2)(3)$
$40 = 3x + 6$

$x = \dfrac{34}{3}$ or $11\dfrac{1}{3}$

12b. $v = \dfrac{w + 5}{x - 3}$

$v(x - 3) = w + 5$
$vx - 3v = w + 5$

$vx = 3v + w + 5$

$x = \dfrac{3v + w + 5}{v}$

12c. $\dfrac{x + 4}{x - 5} = y$

$x + 4 = (x - 5)y$
$x + 4 = xy - 5y$
$xy - x = 4 + 5y$
$x(y - 1) = 4 + 5y$

$x = \dfrac{4 + 5y}{y - 1}$

12d. $\dfrac{2}{y} = \dfrac{3}{x} - \dfrac{5}{z}$

Multiply both sides by xyz to clear fractions:
$2xz = 3yz - 5xy$
$2xz + 5xy = 3yz$
$x(2z + 5y) = 3yz$

$x = \dfrac{3yz}{2z + 5y}$

13a. the second number: n
the first number: $3n + 4$

its reciprocal: $\dfrac{1}{n}$
its reciprocal: $\dfrac{1}{3n + 4}$

$\dfrac{1}{n} + \dfrac{1}{3n + 4} = \dfrac{3}{5}$

Multiply both sides by $5n(3n + 4)$.

$5(3n + 4) + 5n = 3n(3n + 4)$
$15n + 20 + 5n = 9n^2 + 12n$
$9n^2 - 8n - 20 = 0$
$(9n + 10)(n - 2) = 0$

$$n = -\frac{10}{9} \qquad\qquad n = 2 \qquad\qquad\qquad \text{Answer: } -\frac{10}{9} \text{ and } \frac{2}{3} \text{ or } 2 \text{ and } 10$$

$$3n+4 = \frac{2}{3} \qquad\qquad 3n+4 = 10$$

13b. Let r = the airspeed of the planes. (This is the rate as if there were no wind.)

Plane #1 (going into the wind)
rate : $(r - 60)$ mi/h

Plane #2 (going with the wind)
rate : $(r + 60)$ mi/h

distance: 315 miles

distance: 525 miles

time: $\dfrac{\text{distance}}{\text{rate}} = \dfrac{315}{r - 60}$

time: $\dfrac{525}{r + 60}$

$$\frac{315}{r - 60} = \frac{525}{r + 60} \qquad \text{They travel the same amount of time.}$$

$315(r + 60) = 525(r - 60)$
$315r + 18900 = 525r - 31500$
$50400 = 210r$
$r = 240$ mi/h $\qquad$ This is the airspeed of each plane.

14a. $y = \dfrac{k}{x};\ \ \dfrac{2}{3} = \dfrac{k}{\left(\frac{3}{4}\right)}\ ;\ \ k = \dfrac{2}{3}\left(\dfrac{3}{4}\right) = \dfrac{1}{2}$

$y = \left(\dfrac{1}{2}\right)\dfrac{1}{x}\ \text{ or }\ y = \dfrac{1}{2x}$

$y = \dfrac{1}{2\left(\frac{5}{16}\right)} = \dfrac{16}{10} = \dfrac{8}{5}\ \text{ or }\ 1\dfrac{3}{5}$

14b. $P = \dfrac{k}{V}\ ;\ \ 4 = \dfrac{k}{36}\ \ k = 144$

$P = \dfrac{144}{V}$

$P = \dfrac{144}{24} = \dfrac{24(6)}{24} = 6$

The pressure is 6 Newtons per cm^2.

Chapter Eight Mastery Test

Problem Set 9.1

Note: Graphs required in the solutions may be found in the answer section of the text.

1. See answer section in the text.

5. See answer section in the text.

9. See answer section in the text.

13. All points with a negative first coordinate and positive second coordinate lie in the second quadrant.

17. All points with both coordinates negative lie in the third quadrant.

21. All points with first coordinate equal to zero lie on the y-axis.

25. See answer section in the text.

29. $5(10) + 6(24) = 50 + 144 = 194$ (no)

33. $7(-2) = -14$ $2(-7) = -14$ (yes)

37. Point C is $(-4, 0)$.   Substitute $x = -4$ and $y = 0$ into the equation to determine if $(-4,0)$ is a solution: $3(-4) - 0 = -12$ (yes)

41. See answer section in the text.

45. If the y-coordinate is 0, for example $(2,0)$ or $(-3, 0)$, the point must be positioned on the x-axis, so the set of *all* such points would be the entire x-axis.

49. Consider some sample points: $(-3, 0)$, $(-3, 1)$ and $(-3, 2)$. Since the x-coordinates are all the same, the points are all the same distance from the y-axis (3 units) and so this line must be a vertical line.   See answer section in the text.

53. The right-most point has coordinates $(3, -1)$; the left-most point has coordinates $(-3, 2)$.

57. See answer section in the text. The point with x-coordinate of 0 is a point on the y-axis. The point with y-coordinate of 0 is a point on the x-axis.

61a. See answer section in the text.

61b. The area of a triangle is $\frac{1}{2}$(base)(height).
The base is 4 units and height is 3 units, so area $= \frac{1}{2}(4)(3) = 6$ square units.

61c. The perimeter of a figure is the distance around it; or, the sum of the lengths of the sides.   Find the length of the hypotenuse by using the Pythagorean Theorem:

$c^2 = 4^2 + 3^2$
$c^2 = 25$
$c = 5$

We reject $c = -5$, as c is length.
The perimeter is $3 + 4 + 5 = 12$ units.

65. See answer section in the text. You might want to plot a point with negative x- and y-coordinates, for example, $(-2, -4)$, and then note what quadrant it is in.

69. See answer section in the text. Plot some points of your choosing, as long as the y-coordinate is one more than the x-coordinate. For example: $(0,1)$, $(1,2)$, and $(-1, 0)$. Connect them with a line. (If they do not all lie in a line, you have made an error somewhere!)

Problem Set 9.2

Note: Graphs required in the solutions may be found in the answer section of the text.

1. Substitute in $x = 0$ and solve for y:

    The point is $(0, -3)$.

    $3(0) - 7y = 21$
    $0 - 7y = 21$
    $-7y = 21 \; ; y = -3$

5. Substitute in $x = 4$ and solve for y:

    The point is $(4,5)$.

    $-2(4) + 8y = 32$
    $-8 + 8y = 32$
    $8y = 40 \; ; \; y = 5$

9. $x - y = 4$

$\underline{x = 0:}$	$0 - y = 4$	$\underline{y = 0:}$	$x - 0 = 4$	$\underline{x = 1:}$	$1 - y = 4$
	$-y = 4$		$x = 4$		$-y = 3$
	$y = -4$				$y = -3$

x	y
0	−4
4	0
1	−3

    These three points are used to make the graph. See answer section in the text.

13. $-3x + 2y - 6 = 0$

$\underline{x = 0:}$	$-3(0) + 2y - 6 = 0$	$\underline{y = 0:}$	$-3x + 2(0) - 6 = 0$
	$2y - 6 = 0$		$-3x - 6 = 0$
	$2y = 6$		$-3x = 6$
	$y = 3$		$x = -2$

$\underline{x = 1:}$	$-3(1) + 2y - 6 = 0$	x	y
	$-3 + 2y - 6 = 0$	0	3
	$-9 + 2y = 0$	−2	0
	$2y = 9$	1	$\frac{9}{2}$
	$y = \frac{9}{2}$		

page 197

section   9.2

17. $y = 3$ In this case, no matter what the x-coordinate, the y-coordinate
                                 will be 3.   See answer section in the text.

21. $5x + 7y - 35 = 0$ To find the x-intercept, set $y = 0$ and solve for x.
     $5x = 35$
     $x = 7$ $(7,0)$ x-intercept

     $5x + 7y - 35 = 0$ To find the y-intercept, set $x = 0$ and solve for y.
     $7y = 35$
     $y = 5$ $(0,5)$ y-intercept

25. $5x + 7y = 0$ To find the x-intercept, set $y = 0$ and solve for x.
     $5x = 0$
     $x = 0$ $(0,0)$ x-intercept

     $5x + 7y = 0$ To find the y-intercept, set $x = 0$ and solve for y.
     $7y = 0$
     $y = 0$ $(0,0)$ y-intercept

29. $3x = 4$ To find the x-intercept, set $y = 0$ and solve for x.

     $x = \frac{4}{3}$ for all y-values, including $y = 0$.   The x-intercept is $(\frac{4}{3}, 0)$.

     $3x = 4$ To find the y-intercept, set $x = 0$ and solve for y.
     $0 = 4$ No solution; there is no y-intercept.

*Notice that $3x = 4$ is equivalent to $x = \frac{4}{3}$. This means that $x = \frac{4}{3}$ for all points (x,y) on the line
and so this must be a vertical line. (The only vertical line that could have an y-intercept would be
the y-axis.)*

33. $3x - 4y = 7$ To find the x-intercept, set $y = 0$ and solve for x.
     $3x = 7$
     $x = \frac{7}{3}$ $(\frac{7}{3},0)$ x-intercept

     $3x - 4y = 7$ To find the y-intercept, set $x = 0$ and solve for y.
     $-4y = 7$
     $y = -\frac{7}{4}$ $(0, -\frac{7}{4})$ y-intercept

37. $x - y + 4 = 0$ intercepts: $(0,4)$ $(-4,0)$ | YOU MAY WISH TO
                                third point: $(1,5)$ | USE OTHER POINTS.
                                                                     | THESE ARE OFFERED
     COMMENT *A third point can be used as a check.* | AS A SUGGESTION.
     If all three points do not lie in a line, then one of
     them is incorrect.

41. $3x - 5y - 15 = 0$ intercepts: $(0,-3)$ $(5,0)$

45. $x + 2y = 0$ intercept: $(0,0)$ The line goes through the origin.
                          second point: $(2, -1)$

49. $x = -4$ This is a vertical line passing through the x-intercept $(-4,0)$.

53. $y - 1 = 0$
    $y = 1$ This is a horizontal line passing through the y-intercept $(0,1)$.

57. $y = x$ intercept: $(0,0)$
                     second point: $(1,1)$

61. $\dfrac{3x}{4} + \dfrac{2y}{5} = 1$ intercepts: $(\frac{4}{3}, 0)$ $(0, \frac{5}{2})$

65. See answer section in the text.

69. For each point on the line, the x-coordinate is 4. Therefore, the equation is $x = 4$, and the line is a vertical line through $(4,0)$.

73. This is a horizontal line, and every y-coordinate is 5. The equation is $y = 5$.

77. We start by first noticing that the line goes through $(0,0)$, which means that the equation must be of the form $y = ax$. There is more than one way to determine the value of a:

(1) If you examine the coordinates of some of the points on the graph, you will notice that the y-coordinates are twice as large as the x-coordinates, and different in sign. From this you can conclude that the equation is $y = -2x$.

(2) Choose a point on the line, substitute its coordinates into the equation $y = ax$ and solve for a. Any point on the line will do, as long as you are certain of the coordinates. We will use $(1, -2)$:
$$y = ax$$
$$-2 = a(1)$$
$$a = -2$$

We can then conclude that the equation is $y = -2x$.

81a. The amount that the customer pays will be ten times the number of CD's bought.

2 CD's will cost $2(10) = 20$ dollars
3 CD's will cost $3(10) = 30$ dollars
4 CD's will cost $4(10) = 40$ dollars
5 CD's will cost $5(10) = 50$ dollars

81b. If we let x represent the number of CD's bought, and y represent the cost of them, then we have a set of points:

(2,20), (3,30), (4,40), (5,50)

This is just another way of representing the relationship between the number of CD's bought and the amount paid.

We notice that in each case, the y-coordinate is ten times the x-coordinate; that is, $y = 10x$.

81c.   See answer section in the text.

Problem Set 9.3

1. $m = \dfrac{1-2}{3-4} = \dfrac{-1}{-1} = 1$

5. $m = \dfrac{-11-(-3)}{-6-(-1)} = \dfrac{-8}{-5} = \dfrac{8}{5}$

9. $m = \dfrac{7-(-7)}{-3-(-3)} = \dfrac{14}{0}$ *undefined*

13. $m = \dfrac{-6-0}{4-0} = \dfrac{-6}{4} = -\dfrac{3}{2}$

17. $m = \dfrac{0.38-0.56}{0.49-0.37} = \dfrac{-0.18}{0.12} = -1.5$

21.   A vertical line has undefined slope.

25.   Using the points $(-4,0)$ and $(0,6)$, the slope is $m = \dfrac{6-0}{0-(-4)} = \dfrac{6}{4} = \dfrac{3}{2}$.

29. $-2x + 7y = 28$

   intercepts: $(0,4)$ and $(-14,0)$

   $m = \dfrac{0-4}{-14-0} = \dfrac{2}{7}$

33. $x = 7$ is a vertical line, and so the slope is undefined.

37. $x + y = -11$
   intercepts: $(0,-11)$, $(-11, 0)$

   $m = \dfrac{0-(-11)}{-11-0} = \dfrac{11}{-11} = -1$

41. $(-3, 5)$, $(-6, 3)$ $m = \dfrac{3-5}{-6-(-3)} = \dfrac{2}{3}$

   $(3, -3)$, $(6, -1)$ $m = \dfrac{-1-(-3)}{6-3} = \dfrac{2}{3}$

   The lines are parallel.

45. $(0,6)$, $(8,0)$

   $m = \dfrac{0-6}{8-0} = -\dfrac{3}{4}$

   $(-5,0)$, $(0, -7)$

   $m = \dfrac{-7-0}{0-(-5)} = -\dfrac{7}{5}$

   The lines are neither parallel nor perpendicular.

49. $(5,3)$, $(5, -3)$

   $m = \dfrac{-3-3}{5-5} = \dfrac{-6}{0}$

   undefined slope (vertical line)

   $(7, 6)$, $(7, -6)$

   $m = \dfrac{-6-6}{7-7} = \dfrac{-12}{0}$

   undefined slope (vertical line)
   The lines are parallel.

53. $y = 7$ and $y = -7$ are both horizontal lines, and therefore are parallel.

57. $2x - 4y = 9$
intercepts: $\left(0, -\frac{9}{4}\right)$, $\left(\frac{9}{2}, 0\right)$

$$m = \frac{0 - \left(-\frac{9}{4}\right)}{\frac{9}{2} - 0} = \frac{9}{4} \cdot \frac{2}{9} = \frac{1}{2}$$

$9x + 2y = 4$
intercepts: $(0, 2)$, $\left(\frac{4}{9}, 0\right)$

$$m = \frac{0 - 2}{\frac{4}{9} - 0} = -2 \cdot \frac{9}{4} = -\frac{9}{2}$$

The lines are neither parallel nor perpendicular.

61. Starting from the y-intercept and moving towards the x-intercept, the graph rises approximately 4 units and runs approximately 2 units. The slope should be approximately $\frac{4}{2} = 2$.
Answer: E

65. Let x be the change in elevation.

$$0.05 = \frac{\text{rise}}{\text{run}} = \frac{x}{1800}$$

$(0.05)(1800) = x$
$x = 90$ Answer: 90 meters

69. $(168, 413), (211, 761)$

$$m = \frac{761 - 413}{211 - 168}$$

$$m = \frac{348}{43} \approx 8.09$$

73. The x-axis has zero slope because between any two points the change in y is zero, no matter what the change is x is. This makes the slope $= \dfrac{0}{\text{change in x}} = 0$.

77. Mentally visualize a line passing through quadrants II, III and IV (but not I) It will fall from left to right and so will have a negative slope. Or, sketch the line through $(-4,1)$ and $(1, -4)$. It passes through quadrants II, III and IV, but not I. The line falls, and the slope is -1.

Problem Set 9.4

Note: Graphs required in the solutions may be found in the answer section of the text.

1. $y = 2x + 5$ $m = 2$, y-intercept is $(0,5)$
An equation of the form $y = mx + b$ represents a line with slope m and y-intercept $(0,b)$.

5. $y = 6x + 0$ $m = 6$, y-intercept is $(0,0)$
An equation of the form $y = mx + b$ represents a line with slope m and y-intercept $(0,b)$.

9. $x = -7$
This is a vertical line through $(-7,0)$, and so has undefined slope. There is no y-intercept, as it does not intersect the y-axis.

13. $y - 4 = 5(x - 3)$ The slope is $m = 5$ and a point on the line is $(3, 4)$.
An equation of the form $y - y_1 = m(x - x_1)$ represents a line with slope m and point (x_1, y_1).

17. $y = \frac{1}{2}(x + 6)$

$y - 0 = \frac{1}{2}(x - (-6))$ The slope is $m = \frac{1}{2}$ and a point on the line is $(-6,0)$.

An equation of the form $y - y_1 = m(x - x_1)$ represents a line with slope m and point (x_1, y_1).

21.　$m = -\frac{2}{11}$,　(0,5)　　Since you are given the slope and y-intercept, begin with the slope-intercept equation of the line: $y = mx + b$.

$y = -\frac{2}{11}x + 5$　　Now you can rewrite it to general form. Multiply both sides by 11.

$11y = -2x + 55$　　Put all terms on the left to obtain the form $Ax + By + C = 0$.

$2x + 11y - 55 = 0$　　This is general form.

　　　 __COMMENT__　*You may have come up with $-2x - 11y + 55 = 0$. If you multiply both sides by -1, you will obtain $2x + 11y - 55 = 0$. It is often preferred to have a positive first coefficient.*

25.　　m undefined, (0,0)　　When a line has undefined slope, the slope-intercept equation and the point-slope equation *do not apply*. However, a line with undefined slope is a vertical line, which has equation $x = h$. Since this is a vertical line going through the origin, this line is precisely the y-axis, and has equation $x = 0$.

29.　　$(-1, 7)$, $m = \frac{2}{3}$　　Since you are given a point and the slope, begin with the point-slope equation of the line:　$y - y_1 = m(x - x_1)$.

$y - 7 = \frac{2}{3}(x + 1)$　　Now you can rewrite it to general form. Multiply both sides by 3.

$3(y - 7) = 3\left[\frac{2}{3}(x + 1)\right]$

$3y - 21 = 2(x + 1)$
$3y - 21 = 2x + 2$
$2x - 3y + 23 = 0$

33.　　(0,6), (−3,0)　　Calculate slope first.

$m = \frac{0 - 6}{-3 - 0} = 2$

　　　　　　　　　　Notice that the y-intercept (0,6) is given, so it is convenient to set
$y = 2x + 6$　　up the slope-intercept equation.
$2x - y + 6 = 0$　　This is general form.

　　　 __COMMENT__　*Instead of using slope-intercept form, you can set up point-slope form with $m = 2$ and either one of the points. You will obtain the same results at the end.*

37.　　$(-2, 8)$, $(-5, 8)$　　Notice that the y-coordinates are equal; this means that these points lie on horizontal line with equation $y = 8$ or $y - 8 = 0$.

41.　　See answer section in the text for the graph.
　　　The idea here is to plot the given point (1,2) and use the slope to plot a second point.

$m = \frac{3}{4} = \frac{\text{change in y}}{\text{change in x}}$, so move 3 units up from (1,2) and then 4 units to the right.

The second point will be (5, 5). You may repeat this process starting at (5,5) to obtain a third point for a check; you will obtain (9,8). Connect these points with a straight edge and you have the graph of the line.

45. $(-6, 0)$ $m = 2$ Begin at the point $(-6,0)$.

$m = \frac{2}{1} = \frac{\text{change in y}}{\text{change in x}}$.   Move up 2 units and right 1 unit to obtain new points.

49. $(3, 4)$, $m = 0$ A line with slope of zero is a horizontal line. All you need to do is draw a horizontal line through (3,4).

53. Notice that both lines are in slope-intercept form, so you can "see" what the slope is.

$y = \frac{2}{3}x + 7$ $(m = \frac{2}{3})$

$y = -\frac{3}{2}x - 2$ $(m = -\frac{3}{2})$ These lines are perpendicular, since the product of their slopes is -1.

57. Notice that both of these equations are in point-slope form.

$y - 8 = 4(x + 5)$ $(m = 4)$ The lines are neither parallel nor perpendicular.
$y + 4 = 5(x - 3)$ $(m = 5)$

61. A line parallel to the x-axis is a horizontal line. Since it goes through (2,3), every point has y-coordinate 3, therefore the equation is $y = 3$ or $y - 3 = 0$.

65. $(-5, 8)$, $m = \frac{2}{3}$ Begin with the point-slope equation.

$y - 8 = \frac{2}{3}(x + 5)$ Now you may rewrite to general form. Multiply both sides by 3.

$3(y - 8) = 3\left[\frac{2}{3}(x + 5)\right]$

$3y - 24 = 2(x + 5)$ Carry out multiplication on the right side to remove parentheses.
$3y - 24 = 2x + 10$ Collect all terms on the left side.
$2x - 3y + 34 = 0$

69. x-intercept $(3,0)$ with $m = 7$ Begin with the point-slope equation.
$y - 0 = 7(x - 3)$ Now rewrite to general form.
$y = 7x - 21$
$7x - y - 21 = 0$

73. $(-7, -2)$, $m = 0$ A line with slope of zero is a horizontal line. Since it passes through the point $(-7, -2)$, every point has y-coordinate -2, so the equation is $y = -2$ or $y + 2 = 0$.

section   9.4

77. There is more than one way to do this.

(Option I) If the y-intercept and slope can be determined from the graph of a line, this may be the simplest way to proceed. The y-intercept is observed to be (0,1). The slope is a negative number, since the line falls from left to right. If you will measure the rise and run between two points, you will find the slope to be -2. The equation of the line is $y = -2x + 1$. (This is slope-intercept form, which was easiest to construct from the y-intercept and slope. It is also the form you would use when graphing on a graphics calculator. However, this equation could be written in general form if desired.)

(Option II) Choose two points on the line so that they have integer coefficients (no fractions) and calculate the slope. Use the slope and y-intercept (which is easily observed from the graph) and write the slope-intercept form of the line.

81. The line $y - 1 = \frac{1}{2}(x + 3)$ is in point-slope form $y - y_1 = m(x - x_1)$ and so the slope is $\frac{1}{2}$. Any line perpendicular to it will have slope -2.

Using $m = -2$ and the point $(3,5)$, we can set up the point-slope equation:

$y - 5 = -2(x - 3)$
$y - 5 = -2x + 6$
$2x + y - 11 = 0$ This is general form.

Problem Set 9.5

Note: Graphs required in the solutions may be found in the answer section of the text.

1. You need to determine if the point makes the inequality a true statement. Substitute in $x = 0$ and $y = 0$: $0 - 4(0) = 0 < 8$ (true)
Therefore $(0,0)$ is a solution to the given inequality.

5. $12 - 4(1) = 8$
$8 < 8$? (false) $(12,1)$ is not a solution.

9. $-3(2) + 4(4)$
$= -6 + 16 = 10$
$10 \geq 24$? (false) $(2,4)$ is not a solution.

13. A point is in the solution set of the linear inequality if it lies either in the shaded region or on the line (if the line is a solid line). All you need to do is plot the point; $(0,0)$ lies in the shaded region and so is a solution.

17. The point $(0, -4)$ lies in the shaded region, so is a solution.

21. The point $(1, 8)$ lies in the shaded region, so is a solution.

25. Since the line is solid, only A or C are possibilities. Choose a point from the shaded half-plane and determine which inequality it satisfies, A or C.

$(0, -4)$ $\qquad$ $0 - 3(-4) = 12 \geq 3$ $\qquad$ The graph corresponds to C.

29. $x + 4y + 4 \geq 0$ $\qquad$ Choose a test point, say $(0,0)$.
$0 + 4(0) + 4 \geq 0$
$4 \geq 0$ $\qquad$ The test point satisfies the inequality. Shade in the half-plane above the boundary line.

33. $y \leq 3$ $\qquad$ This inequality represents all points whose y-coordinates are less than or equal to 3, and all such points lie on or below the line $y = 3$. Shade in the region below the boundary line.

37. $x + y \geq 4$ $\qquad$ Graph the *solid* line $x + y = 4$. (Intercepts: $(0,4)$, $(4,0)$)
$0 + 0 \geq 4$ (false) $\qquad$ Test $(0,0)$. It does *not* satisfy the inequality, so shade in the half-plane that *does not contain (0,0)*.

41. $-2x + 5y - 10 > 0$ $\qquad$ Graph the *dashed* line $-2x + 5y - 10 = 0$. If you write $-2x + 5y = 10$, it is a little easier to find the intercepts.

$-2(0) + 5(0) - 10 > 0$

$-10 > 0$ (false) $\qquad$ The test point $(0,0)$ does not satisfy the inequality, so shade in the half-plane that does not contain $(0,0)$.

45. $x > -4$ $\qquad$ Graph the *dashed* line $x = -4$, a vertical line through $(-4,0)$. Notice that the inequality represents all points whose x-coordinate is greater than -4; therefore we have all points to the right of the dashed line.

49. $x + y < 0$ $\qquad$ Graph the *dashed* line $x + y = 0$.
$1 + 1 < 0$ $\qquad$ The test point $(1,1)$ was chosen because $(0,0)$ lies on the line.
$2 < 0$ (false) $\qquad$ Shade in the region that does not contain the point $(1,1)$.

53. $y > \frac{2}{5}x - 4$ $\qquad$ Graph the *dashed* line $y = \frac{2}{5}x - 4$. Notice that the slope is $\frac{2}{5}$ and the y-intercept is $(0,-4)$.

$0 > \frac{2}{5}(0) - 4$

$0 > -4$ (true) $\qquad$ Shade in the half-plane containing $(0,0)$.

page 205

57.   2 more than the y-coordinate: y + 2
       "at least" means "no less than" , which corresponds to ≥

         x ≥ y + 2   Graph the solid line x = y + 2.   (Suggestion: find the intercepts.)
         0 ≥ 0 + 2   Test the point (0,0).
         0 ≥ 2 (false)   Shade in the half-plane that does not contain (0,0).

61.   "at least" means "no less than" which corresponds to ≥
       2 more than three times the y-coordinate: 3y + 2
         x ≥ 3y + 2   Graph the solid line x = 3y + 2.
         0 ≥ 3(0) + 2   Test the point (0,0).
         0 ≥ 2 (false)   Shade in the half-plane that does not contain (0,0).

65.   The line passes through (0,0) and (1,2) and so has equation y = 2x. Since the shaded region
       is above the line, and the line is solid, the inequality is y ≥ 2x.

69.   Graph the solid line x + y = 150 . Determine the correct half-plane, but shade in only that
       portion that falls in quadrant I.

73.   x ≥ 1 and x ≤ 3

       The graph of x ≥ 1 consists of all points on or to the right of the vertical line x = 1.
       The graph of x ≤ 3 consists of all points on or to the left of the vertical line x = 3.
       The solution is the *intersection* of the two regions; that is, where they overlap. Suggestion:
       use different colors to shade in each region to make it easier to identify the intersection.

Solutions to All Review Exercises — Chapter Nine

Note: Graphs for the odd-numbered exercises may be found in the answer section of the text.

1.   See answer section in the text.   2.   (5,1) I
3.   See answer section in the text.   4.   (2, −3) IV
5.   See answer section in the text.   6.   (0, −5) y-axis
7.   Plot the points given. They do not even approximately lie in a straight line. See text for graph.

8.

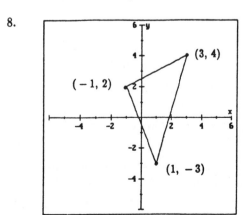

9. $4(10) - 5(8) + 80$
    $= 40 - 40 + 80$
    $= 80$
    $\neq 0$
    $(10, 8)$ is not a solution.

10. $4(8) - 5(-10) + 80$
    $= 32 + 50 + 80$
    $\neq 0$ (Notice we do not have to finish the
    calculation to know it will not be zero.)
    $(8, -10)$ is not a solution.

11. $4(6) - 3(4) = 24 - 12 = 12 \not> 12$
    $(6,4)$ is not a solution.

12. $4(0) - 3(-5) = 0 + 15 = 15 > 12$
    $(0, -5)$ is a solution.

In #13 – #18, a point is a solution to the inequality if it lies on the line or in the shaded region. Just locate the point in the plane to test it.

13. The point $(0,0)$ lies in the shaded region, so it is a solution to the inequality.

14. $(-6,0)$ is not a solution.

15. $(0, -6)$ is a solution.

16. $(4, -1)$ is a solution.

17. $(1, 4)$ is not a solution.

18. $(0,2)$ is a solution.

19. $4x + 7y + 28 = 0$ To find the x-intercept, set $y = 0$ and solve for x.
    $4x + 28 = 0$
    $4x = -28$
    $x = -7$ The x-intercept is the point $(-7, 0)$.

    $4x + 7y + 28 = 0$ To find the y-intercept, set $x = 0$ and solve for y.
    $7y + 28 = 0$
    $y = -4$ The y-intercept is the point $(0, -4)$.

20. $3x - 11y = 66$ To find the x-intercept, set $y = 0$ and solve for x.
    $3x = 66$
    $x = 22$ The x-intercept is the point $(22,0)$.

    $3x - 11y = 66$ To find the y-intercept, set $x = 0$ and solve for y.
    $-11y = 66$
    $y = -6$ The y-intercept is the point $(0, -6)$.

21. $7x = 9y$ To find the x-intercept, set $y = 0$ and solve for x.
    $7x = 0$
    $x = 0$ Since the x-intercept is $(0,0)$, the y-intercept is also $(0,0)$.

22. The equation $x = 5$ represents a vertical line. All points on this line have x-coordinate 5. The line intersects the x-axis at $(5,0)$ and so this is the x-intercept. Since this line is parallel to the y-axis, there can be no y-intercept.

Chapter Nine Review

23. The equation $y = 9$ represents a horizontal line, parallel to the x-axis. There can be no x-intercept. It intersects the y-axis at (0,9).

24. $y = 5x + 15$

Notice that this equation is given in slope-intercept form $y = mx = b$, and so we may simply read off the y-intercept: (0,15).

$0 = 5x + 15$ To find the x-intercept, set $y = 0$ and solve for x.

$x = -3$ The x-intercept is the point $(-3, 0)$.

25. $y = \frac{3}{4}x - 12$

This equation is in slope-intercept form $y = mx + b$. Therefore, the y-intercept is (0,−12).

$0 = \frac{3}{4}x - 12$ To find the x-intercept, set $y = 0$ and solve for x.

$4(0) = 4\left(\frac{3}{4}x - 12\right)$ Multiply both sides by 4 to eliminate the fractions.

$0 = 3x - 48$

$3x = 48$

$x = 16$ The x-intercept is (16,0).

26. $y - .1 = 5(x - 2)$ To find the x-intercept, set $y = 0$ and solve for x.

$-1 = 5(x - 2)$

$-1 = 5x - 10$

$9 = 5x$

$x = \frac{9}{5}$ The x-intercept is the point $(\frac{9}{5}, 0)$.

$y - 1 = 5(x - 2)$ To find the y-intercept, set $x = 0$ and solve for y.

$y - 1 = 5(-2)$

$y - 1 = -10$

$y = -9$ The y-intercept is the point $(0, -9)$.

27. (5, 8) and (3, 14)   Use the formula for slope: $m = \frac{y_2 - y_1}{x_2 - x_1}$

$m = \frac{14 - 8}{3 - 5} \quad \frac{6}{-2} = -3$

28. $(-3, -2)$ and $(12, -8)$ Use the formula for slope: $m = \frac{y_2 - y_1}{x_2 - x_1}$

$m = \frac{-8 - (-2)}{12 - (-3)} = \frac{-6}{15} = -\frac{2}{5}$

29. Every vertical line has undefined slope. There is no need for any calculation.

page 208

30. Every horizontal line has a slope with value zero. There is no need for any calculation.

> SUGGESTION *It is worth your while to notice how problems that involve horizontal or vertical lines are handled. Questions about intercepts and slope are easily and quickly answered because of the very nature of these special lines, and we do not need any calculations. This can be crucial when taking a test, because it saves time.*

31. $y = \frac{5}{7}x + 8$ You are expected to notice that this line is given in slope-intercept form $y = mx + b$ and so we can read off the slope: $\frac{5}{7}$

32. $y - 4 = -\frac{7}{9}(x + 4)$ You are expected to notice that this line is given in point-slope form $y - y_1 = m(x - x_1)$, and we can read off the slope: $-\frac{7}{9}$

33. An equation of the form $y = k$ always represents a horizontal line, and a horizontal line has slope $m = 0$.

34. An equation of the form $x = h$ always represents a vertical line, and a vertical does not have slope, it is undefined.

35. $2x - 11y + 66 = 0$ There is more than one way to do this, but perhaps the easiest way is to rewrite this equation in slope-intercept form.

$11y = 2x + 66$

$y = \frac{2}{11}x + 6$ The slope is $\frac{2}{11}$.

36. $\frac{x}{2} - \frac{y}{5} = 1$ There is more than one way to do this, but perhaps the easiest way is to rewrite this equation in to slope-intercept form.

$\frac{y}{5} = \frac{x}{2} - 1$

$y = \frac{5}{2}x - 5$ The slope is $\frac{5}{2}$.

37. $y = \frac{2}{3}x + 8$ is in slope-intercept form $y = mx + b$, and so the slope is $\frac{2}{3}$. Since parallel lines have the same slope, any line parallel to $y = \frac{2}{3}x + 8$ will have slope $\frac{2}{3}$.

38. $y = \frac{8}{5}x - 7$ is in slope-intercept form $y = mx + b$ and so the slope of this line is $\frac{8}{5}$. The line we are interested in is perpendicular to this line, and therefore the product of their slopes is -1. This means the slope of the line we are interested in is the *negative reciprocal* of $\frac{8}{5}$, and that number is $-\frac{5}{8}$.

Chapter Nine Review

39. (5, 8) and (4,6) $(-1, 3)$ and $(1, 2)$

$m = \dfrac{6 - 8}{4 - 5} = \dfrac{-2}{-1} = 2$ $m = \dfrac{2 - 3}{1 - (-1)} = \dfrac{-1}{2} = -\dfrac{1}{2}$

The product of the slopes is -1, so the lines are perpendicular.

40. $(-1, 0)$ and (4,7) $(2, -3)$ and $(-3, 4)$

$m = \dfrac{7 - 0}{4 - (-1)} = \dfrac{7}{5}$ $m = \dfrac{4 - (-3)}{-3 - 2} = \dfrac{7}{-5} = -\dfrac{7}{5}$

The lines are neither parallel nor perpendicular.

41. $(-1,1)$, $(2, -4)$ $(3, -5)$, $(-3, 5)$

$m = \dfrac{-4 - 1}{2 - (-1)} = \dfrac{-5}{3} = -\dfrac{5}{3}$ $m = \dfrac{5 - (-5)}{-3 - 3} = \dfrac{10}{-6} = -\dfrac{5}{3}$

The lines have the same slope, so are parallel.

42. (3,4) and (3,5) $(-2, 8)$ and (6,8)

The easiest way to go here is to notice   Notice that the y-coordinates are the same.
that these points have the same   This is a horizontal line.
x-coordinate, so the line is vertical.

A vertical line and a horizontal line are always perpendicular.

43. Plot the points and use a straight edge to draw a line through them. (See answer section in the
     text for the graph.)

SUGGESTION   *Graph paper makes the best graphs, and saves you time. If you must make
your sketches without it, use a straight edge to make your axes, make sure that your axes are
perpendicular, and take care that the tick marks on the axes are equally spaced. Failure to do
these things can cause you trouble when you graph your lines.*

*Remember also, one of the reasons for making graphs in the first place is to provide a visual
display of information, often for other people. If a graph is too small, sloppy, poorly labeled, or
inaccurate, then it is useless.*

Chapter Nine Review

44. Graph for #44:

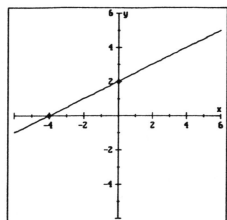

45. See answer section in the text.

Use the slope to locate a second point.

$m = \frac{4}{3} = \frac{\text{change in y}}{\text{change in x}}$, which means from $(-3, -5)$ you can move 4 units up and then 3 units right to locate a new point.

46. Use the slope to locate a second point.

$m = -\frac{3}{4} = \frac{\text{change in y}}{\text{change in x}}$, which means from

$(0, 5)$ you can move 3 units down and then 4 units to the right to locate a new point; or, you can move 3 units up and 4 units to the left.
The important thing is that the new point is located so that the resulting line *falls* from left to right.
Remember, the slope is negative.

Graph for #46:

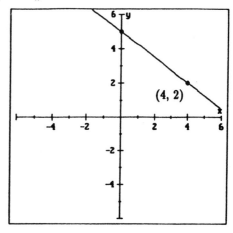

47. $y - 3 = \frac{2}{5}(x + 2)$

There is more than one way to proceed, but you need two points to plot in any case. Use $(-2, 3)$ as the first one, and use the slope to locate a second one. See #45 or #46 for a review on how this is done. See answer section in the text for the graph.

See textbook for graphs for
odd-numbered exercises.

48. Graph for #48:

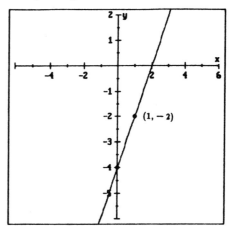

49. $2x = -8$
    $x = -4$
    This is a vertical line passing through $(-4, 0)$. See answer section in the text for the graph.

    ┌─────────────────────────────┐
    │ See textbook for graphs for │
    │ odd-numbered exercises.   │
    └─────────────────────────────┘

50. $y + 5 = 0$
    $y = -5$

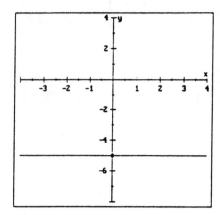

51. A line with undefined slope is a vertical line. Draw a vertical line through $(-2, 3)$. See answer section in the text.

52. A line parallel to the x-axis is horizontal.

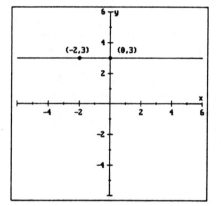

53. See answer section in the text. Any line perpendicular to the x-axis must be *vertical.* This particular vertical line passes through $(2,4)$, so start by plotting that point.

54. Rewrite as $2x - 5y = 10$ and plot the intercepts.

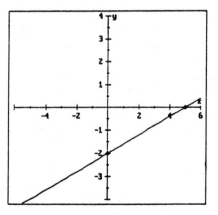

55. $x \geq 4$
Start by graphing the solid *vertical*
line $x = 4$. Shade in the region to the
right of the line.
See answer section in the text for the graph.

56. $y < -3$

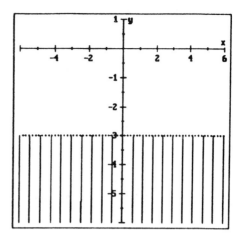

┌─────────────────────────────────┐
│ See textbook for graphs for   │
│   │
│ odd-numbered exercises.   │
└─────────────────────────────────┘

57. See answer section in the text for the graph.   Some suggestions:

$x + y \leq 0$ Start by graphing the solid line $x + y = 0$.
$1 + 1 \leq 0$? $(1,1)$ is used for a test point, since $(0,0)$ is on the boundary line.
$2 \leq 0$ (false)   Shade in the half-plane that does not contain $(1,1)$.

58. $y > 3x - 5$ Start by graphing the dashed line $y = 3x - 5$.
$0 > 3(0) - 5$?   Does $(0,0)$ satisfy the inequality? Yes, so shade in the
$0 > -5$ (true)   half-plane that contains $(0,0)$.

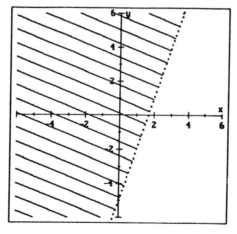

59. $x > 5y$ You may want to rewrite this:

$y < \frac{1}{5} x$ Now graph the dashed line $y = \frac{1}{5} x$.

$1 < \frac{1}{5}(1)$? $(1,1)$ is used as a test point, since $(0,0)$ is on the boundary line.
                                    $(1,1)$ does <u>not</u> satisfy the inequality, so shade in the half-plane that
                                    does not contain $(1,1)$.   See answer section in the text for the graph.

60. $4x - 5y - 20 \geq 0$ Start by graphing the solid line $4x - 5y = 20$.
$4(0) - 5(0) - 20 \geq 0$?   Does $(0,0)$ satisfy the inequality?
$-20 \geq 0$ (false)   Shade in the half-plane that does <u>not</u> contain $(0,0)$.

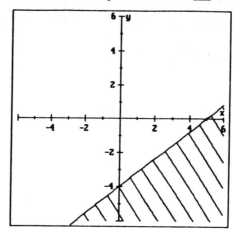

61. $(4, 5)$, $m = \frac{1}{2}$ Since you are given a point and the slope, begin with the point-slope
equation $y - y_1 = m(x - x_1)$.

$y - 5 = \frac{1}{2}(x - 4)$ Now you can rewrite to general form. Multiply both sides by 2.

$2(y - 5) = 2\left[\frac{1}{2}(x - 4)\right]$

$2y - 10 = x - 4$
$0 = x - 2y + 6$ The general form is $x - 2y + 6 = 0$.

62. $(-2, 4)$ and $(0, -2)$

$m = \dfrac{-2 - 4}{0 - (-2)} = -3$ Find the slope : $m = \dfrac{y_2 - y_1}{x_2 - x_1}$

$y = -3x - 2$ We are given the y-intercept $(0, -2)$, so set
up slope-intercept form $y = mx + b$.

This can be left as is, or rewritten in general form depending on your instructor's preference.
To write in general form, bring all terms to the left, and arrange so that the variables are in
alphabetical order, and the constant term is last:
$y + 3x + 2 = 0$
$3x + y + 2 = 0$ This is general form.

63.   A vertical line is one of the simplest to deal with. All vertical lines have an equation of the
form x = h, where h is the x-coordinate of each and every point. Since our line goes through
$(5, -7)$, *every* point has x-coordinate 5. The equation is x = 5 or x − 5 = 0.

64.   A horizontal line is one of the simplest to deal with. All horizontal lines have an equation of the
form y = k, where k is the y-coordinate of each and every point. Since our line goes through
$(5, -7)$, *every* point has y-coordinate −7. The equation is y = −7 or y + 7 = 0.

65. y-intercept $(0, -2)$, $m = \frac{3}{8}$

$y = \frac{3}{8}x - 2$ Use the slope-intercept equation, $y = mx + b$. Multiply both sides by 8.

$8y = 8\left(\frac{3}{8}x - 2\right)$
$8y = 3x - 16$
$3x - 8y - 16 = 0$ This is general form.

66. x-intercept $(5,0)$ and $m = -\frac{3}{7}$ Use point-slope form. *Slope-intercept form requires the y-intercept, which we don't have.*

$y - 0 = -\frac{3}{7}(x - 5)$ Simplify.

$y = -\frac{3}{7}x + \frac{15}{7}$ This is point-slope form. If you want general form, keep reading:

$7y = -3x + 15$

$3x + 7y - 15 = 0$ This is general form.

67. through $(-4, 5)$, slope undefined

Since the slope is undefined, we know it is a vertical line. A vertical line is one of the simplest to deal with. All vertical lines have an equation of the form $x = h$, where h is the x-coordinate of each and every point. Since our line goes through $(-4, 5)$, *every* point has x-coordinate -4. The equation is $x = -4$ or $x + 4 = 0$.

68. Since the slope is zero, it is a horizontal line. A horizontal line is one of the simplest to deal with. All horizontal lines have an equation of the form $y = k$, where k is the y-coordinate of each and every point. Since our line goes through $(-4, 5)$, *every* point has y-coordinate 5. The equation is $y = 5$ or $y - 5 = 0$.

69. Any line that is parallel to the y-axis is a vertical line. Since it passes through $(-9, 11)$, the equation of the line is $x = -9$ or $x + 9 = 0$. See #67 for a related problem.

70. A line that is perpendicular to the y-axis is a horizontal line. Since it passes through $(-9, 11)$, all y-coordinates are 11, and so the equation of the line is $y = 11$ or $y - 11 = 0$. See #68 for a related problem.

| COMMENT | *Remember that inequalities correspond to shaded graphs and equations correspond to lines only. Keep this in mind when working #71 – 78.*

71. This must be a vertical line, so it is D.

72. This must be a horizontal line, so it is C.

Chapter Nine Review

73. This is a line that rises from left to right, so it must be **F.**

74. This is a line that falls from left to right, so it must be **A.**

75. The shaded graphs go with the inequalities. The graph will require a dashed line with y-intercept (0,1), so it must be **E.**

76. It requires a solid line, so can be only either G or H. It can't be G, because $3x + 5y + 15 = 0$ is not the equation of a vertical line. The correct graph must be **H.**

77. This corresponds to a solid vertical line, so it must be **G.**

78. The dashed line will pass through (0,0), so it must be **B.**

79. We notice that the slope of the roof is the same whether we compute it from the edge of the roof to the brace, or from the edge of the roof to the peak. Let ℓ be the length of the brace.

The slope from the edge to the peak: $\dfrac{5}{24}$

The slope from the edge to the brace : $\dfrac{\ell}{18}$

These are equal, so $\dfrac{5}{24} = \dfrac{\ell}{18}$. Cross-multiplying, you get:

$5(18) = 24\ell$

$\ell = \dfrac{90}{24} = \dfrac{15}{4}$ or $3\dfrac{3}{4}$ The length of the brace is $3\dfrac{3}{4}$ ft.

(This problem can also be worked using similar triangles.)

80. $x + y \leq 47$ Graph:
Graph the solid line $x + y = 47$.
The intercepts are easy to use.
Intercepts: (0, 47), (47, 0).
Using (0,0) as a test point, we have $0 + 0 < 47$,
a true statement. Shade in the half-plane that
contains (0,0), but only that portion in the first
quadrant.

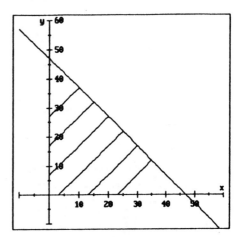

Chapter Nine Review

1. See answer section in the text.

2. yes; graph:

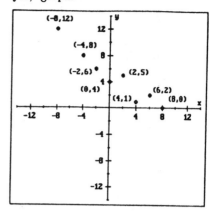

3a. (0,0)   3b. x-axis   3c. IV   3d. II

4a. (0,4), (18,0)   4b. (0,5), (−7,0)   4c. $(0,-2), (\frac{1}{2}, 0)$ 4d. (0,−7), (2,0)

5. See answer section in the text for graphs for odd-numbered exercises..

6a. x = −4

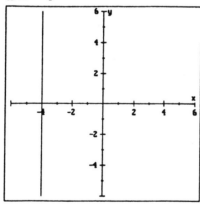

6b. 2x − 6 = 0   which is x = 3

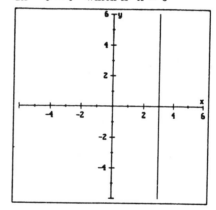

6c. y = 2

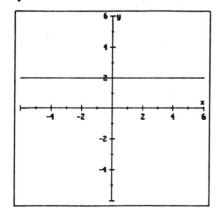

6d. 3y + 6 = 0   which is y = −2

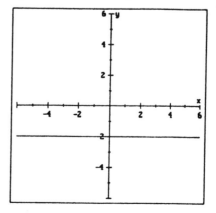

Chapter Nine Mastery Test

7a. $m = \dfrac{0-5}{7-2} = \dfrac{-5}{5} = -1$

7b. $m = \dfrac{4-(-4)}{-6-(-2)} = \dfrac{8}{-4} = -2$

7c. $m = \dfrac{7-5}{2-2} = \dfrac{2}{0}$ undefined

7d. $m = \dfrac{-4-(-4)}{2-(-2)} = \dfrac{0}{4} = 0$

8a.   The first line has slope -1; the second line has slope 1. They are perpendicular.

8b.   Both lines have slope -3. They are parallel.

8c.   These are slope-intercept equations. The first line has slope $\dfrac{2}{3}$, the second line has slope $\dfrac{7}{3}$. They are neither parallel nor perpendicular.

8d.   The first line is a vertical line, and the second line is a horizontal line. They are perpendicular.

9a. $y-(-1) = 2(x-4)$ (point-slope form)

$y + 1 = 2x - 8$

$2x - y - 9 = 0$ (general form)

9b. $m = \dfrac{9-4}{0-(-2)} = \dfrac{5}{2}$

$y = \dfrac{5}{2}x + 9$ (slope-intercept form)

$5x - 2y + 18 = 0$ (general form)

9c.   Begin with $y = mx + b$,
using $b = 0$ and $m = -3$.
$y = -3x$
$3x + y = 0$ (general form)

9d.   This is a horizontal line passing through the origin. The only line with these properties is the x-axis, which has equation $y = 0$.

10a.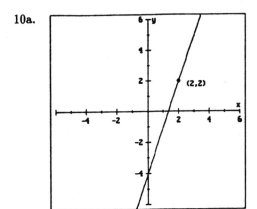

10b.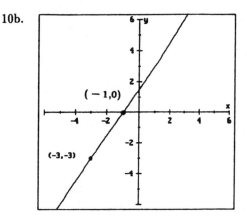

10c.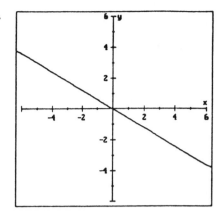

10d.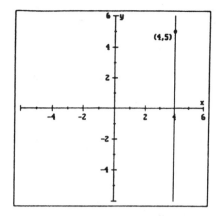

11a. $m = \frac{4}{3}$; y-intercept $= (0,2)$

11b. $m = -\frac{5}{11}$; y-intercept $= (0,9)$

11c. $6x - 7y + 14 = 0$

$7y = 6x + 14$

$y = \frac{6}{7}x + 2$

$m = \frac{6}{7}$; y-intercept $= (0,2)$

11d. $y = 8$

This is a horizontal line, so has slope 0.

It intersects the y-axis at $(0,8)$.

12a.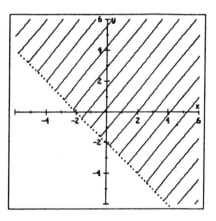

12b.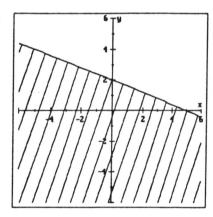

12c.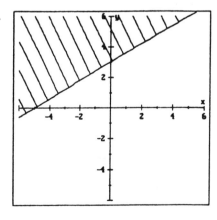

12d.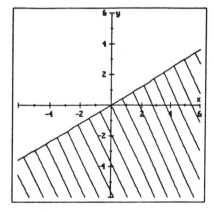

Chapter Nine Mastery Test

1. See answer section in the text. When factoring, always look first for a greatest common factor.

2. See answer section in the text. This factors by trial and error.

3. See answer section in the text. This factors by trial and error.

4. See answer section in the text. This factors by trial and error.

5. $7x^2 + 7x - 42$ First factor out the GCF.
$= 7(x^2 + x - 6)$ Now apply trial and error.
$= 7(x + 3)(x - 2)$

6. $x^2 - 10x + 25$ Notice that $25 = 5^2$ and $10 = 2(5)$. This is a perfect square trinomial.
$= (x - 5)^2$ Use the form $(a - b)^2 = a^2 - 2ab + b^2$.

7. $x^2 - 25$ Use the form $a^2 - b^2 = (a + b)(a - b)$, with x for a and 5 for b.
$= (x + 5)(x - 5)$

8. $4x^2 - 28x + 49$
$= (2x)^2 - 28x + 7^2$ Use the form $(a - b)^2 = a^2 - 2ab + b^2$.
$= (2x - 7)^2$

9. $2x^2 - 98$ Always start by looking for a common factor. Factor out 2.
$= 2(x^2 - 49)$ This is now more recognizable; use $a^2 - b^2 = (a + b)(a - b)$.
$= 2(x + 7)(x - 7)$ Don't drop that factor of 2! It is part of the answer.

10. $3x^2 + 27$ Factor out 3, but don't remove it.
$= 3(x^2 + 9)$ $x^2 + 9$ is a sum of squares, so we can go no further.

11. $ax - 5a + bx - 5b$ When you have four or more terms, try grouping.
$= a(x - 5) + b(x - 5)$ We now have a common expression $(x - 5)$ to factor out.
$= (x - 5)(a + b)$

12. $x^2 - y^2 + 3x + 3y$ When you have four or more terms, try grouping.
$= (x + y)(x - y) + 3(x + y)$
     Now factor out the common expression $(x + y)$.
$= (x + y)(x - y + 3)$

Cumulative Review

13. $\dfrac{30x^2y - 35xy^2}{5xy}$

Because the denominator is a monomial, there is more than one way to proceed on this one.

(Option I) Break up $\dfrac{30x^2y - 35xy^2}{5xy}$ into the difference of two fractions:

$\dfrac{30x^2y - 35xy^2}{5xy} = \dfrac{30x^2y}{5xy} - \dfrac{35xy^2}{5xy}$ Now reduce each.

$\qquad\qquad = 6x - 7y$

(Option II) Factor the numerator of $\dfrac{30x^2y - 35xy^2}{5xy}$ and then reduce.

$\dfrac{30x^2y - 35xy^2}{5xy} = \dfrac{5xy(6x - 7y)}{5xy} = 6x - 7y$

14. $\dfrac{m^2 - m - 2}{m^2 - 4m + 4}$ Factor the numerator and denominator.

$= \dfrac{(m - 2)(m + 1)}{(m - 2)^2}$ Now reduce by dividing out factors common to the numerator and denominator.

$= \dfrac{m + 1}{m - 2}$

15. $\dfrac{5m - m^2 + 7}{m^2 - 5m - 7}$ Rewrite the numerator in standard form.

$= \dfrac{- m^2 + 5m + 7}{m^2 - 5m - 7}$ Factor -1 out of the numerator.

$= \dfrac{-(m^2 - 5m - 7)}{m^2 - 5m - 7}$ Notice that $m^2 - 5m - 7$ can be canceled out.

$= -1$

16. $\dfrac{11a - 7}{5a - 2} - \dfrac{a - 3}{5a - 2}$

$= \dfrac{11a - 7 - (a - 3)}{5a - 2}$

$= \dfrac{10a - 4}{5a - 2}$

$= \dfrac{2(5a - 2)}{5a - 2}$

$= 2$

17. $\left(\dfrac{6a + 6}{3a - 6} \right)\left(\dfrac{5a - 10}{4a + 4} \right)$

$= \dfrac{6(a + 1)5(a - 2)}{3(a - 2)4(a + 1)}$

$= \dfrac{6(5)}{3(4)}$

$= \dfrac{5}{2}$

Cumulative Review

18. $\left(\dfrac{x^2 - 9}{x^2 - 4}\right)\left(\dfrac{x^2 - 3x + 2}{x^2 + 2x - 3}\right)$

$= \dfrac{(x + 3)(x - 3)(x - 2)(x - 1)}{(x + 2)(x - 2)(x + 3)(x - 1)}$

$= \dfrac{x - 3}{x + 2}$

19. $\dfrac{5m^2 + 15m}{m + 5} \div \dfrac{m^2 - 9}{2m + 10}$

$= \dfrac{(5m^2 + 15m)(2m + 10)}{(m + 5)(m^2 - 9)}$

$= \dfrac{5m(m + 3)(2)(m + 5)}{(m + 5)(m + 3)(m - 3)}$

$= \dfrac{10m}{m - 3}$

20. $\dfrac{5}{x - 2} + \dfrac{2}{x + 3}$

$= \dfrac{5(x + 3) + 2(x - 2)}{(x - 2)(x + 3)}$

$= \dfrac{5x + 15 + 2x - 4}{(x - 2)(x + 3)}$

$= \dfrac{7x + 11}{(x - 2)(x + 3)}$

21. $\dfrac{\dfrac{16}{a^2} - \dfrac{1}{9}}{\dfrac{4}{a} - \dfrac{1}{3}}$ Multiply numerator and denominator by $9a^2$, which is the LCD.

$= \dfrac{\left(\dfrac{16}{a^2} - \dfrac{1}{9}\right)9a^2}{\left(\dfrac{4}{a} - \dfrac{1}{3}\right)9a^2}$

$= \dfrac{144 - a^2}{36a - 3a^2}$

$= \dfrac{(12 + a)(12 - a)}{3a(12 - a)}$

$= \dfrac{12 + a}{3a}$ or $\dfrac{a + 12}{3a}$

22. $\dfrac{1 - \dfrac{1}{x} - \dfrac{20}{x^2}}{\dfrac{1}{x} - \dfrac{5}{x^2}}$

$= \dfrac{\left(1 - \dfrac{1}{x} - \dfrac{20}{x^2}\right)x^2}{\left(\dfrac{1}{x} - \dfrac{5}{x^2}\right)x^2}$

$= \dfrac{x^2 - x - 20}{x - 5}$

$= \dfrac{(x - 5)(x + 4)}{x - 5}$

$= x + 4$

23. $\dfrac{3}{a + 3} + \dfrac{a + 1}{a - 2} - \dfrac{4a - 3}{(a + 3)(a - 2)}$

$= \dfrac{3(a - 2) + (a + 1)(a + 3) - (4a - 3)}{(a + 3)(a - 2)}$

$= \dfrac{3a - 6 + a^2 + 4a + 3 - 4a + 3}{(a + 3)(a - 2)}$

$= \dfrac{a^2 + 3a}{(a + 3)(a - 2)}$

$= \dfrac{a(a + 3)}{(a + 3)(a - 2)}$

$= \dfrac{a}{a - 2}$

Cumulative Review

24. $\left(1 + \frac{1}{m}\right) \div \left(1 - \frac{1}{m^2}\right)$ Rewrite this in fraction form:

$= \dfrac{1 + \frac{1}{m}}{1 - \frac{1}{m^2}}$ Now multiply numerator and denominator by m^2.

$= \dfrac{\left(1 + \frac{1}{m}\right) m^2}{\left(1 - \frac{1}{m^2}\right) m^2}$

$= \dfrac{m^2 + m}{m^2 - 1}$

$= \dfrac{m(m + 1)}{(m + 1)(m - 1)}$

$= \dfrac{m}{m - 1}$

25. $(2x - 3)(3x + 2) = 0$

$2x - 3 = 0 \qquad 3x + 2 = 0$

$x = \dfrac{3}{2} \qquad\qquad x = -\dfrac{2}{3}$

26.   You will need to rewrite this one first. Expand the left side, and then collect all terms to the left side.
$(x - 3)(x - 2) = 6$
$x^2 - 5x + 6 = 6$
$x^2 - 5x = 0$ Now factor the left side and set each factor equal to zero.
$x(x - 5) = 0$
$x = 0 \ , \ x = 5$

27.   You will need to rewrite this one. Multiply out both sides, and then collect all terms to the left side.
$(3x - 5)(x - 5) = (x - 4)(x + 5)$
$3x^2 - 20x + 25 = x^2 + x - 20$
$2x^2 - 21x + 45 = 0$
$(2x - 15)(x - 3) = 0$

$x = \dfrac{15}{2} \ , \ x = 3$

28. $\dfrac{12}{2x - 1} = 4$ You may cross-multiply. (Allowed when you have fraction = fraction.)
Consider $4 = \frac{4}{1}$.

$12 = 4(2x - 1)$

page 223

$$12 = 8x - 4$$
$$16 = 8x$$
$$x = 2$$

29. $\dfrac{-3x}{4x-3} = 2$ You may cross-multiply. (Allowed when you have fraction = fraction.) Consider $2 = \dfrac{2}{1}$.

$$-3x = 2(4x - 3)$$
$$-3x = 8x - 6$$
$$-11x = -6$$

$$x = \dfrac{6}{11}$$

30. $\dfrac{y}{2} = \dfrac{3}{x}$ We need to "release" x from the denominator. Cross-multiply so as to rewrite this equation. (Allowed when you have fraction = fraction.)

$yx = 6$ Now divide by y.

$$x = \dfrac{6}{y}$$

31. $x^2 - ax - 3x + 3a = 0$ The left side needs factoring. When there are four or more terms, try grouping.

$x(x-a) - 3(x-a) = 0$ Factor out the common expression $(x - a)$.

$(x - a)(x - 3) = 0$ Now set each factor equal to zero and solve <u>for x.</u>

$$x - a = 0 \; ; \; x - 3 = 0$$
$$x = a \; ; \; x = 3$$

32. $(2x + 1)(x + 2)(x - 5) = 0$ Set each factor equal to zero and solve for x.

$2x + 1 = 0 \quad x + 2 = 0 \quad x - 5 = 0$

$x = -\dfrac{1}{2} \; ; \quad x = -2; \quad x = 5$

33. See answer section in the text. It is convenient here to find the intercepts and plot them. Notice that 12 is divisible by both 3 and 4.

34.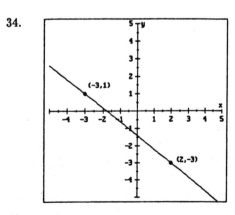

Cumulative Review

35. See answer section in the text.
    All points will have y-coordinate -3.

36. All points will have x-coordinate 2.

37. Use the formula for slope: $m = \dfrac{y_2 - y_1}{x_2 - x_1}$ where $(x_1, y_1) = (3, -4)$ and $(x_2, y_2) = (4, 3)$.

$$m = \frac{3 - (-4)}{4 - 3}$$

$$m = 7$$

38. One way to do this is to put the line in slope-intercept form:
    $3x + 2y = 6$

    $2y = -3x + 6$

    $y = -\dfrac{3}{2}x + 3$ This is in the form $y = mx + b$, so the slope is $-\dfrac{3}{2}$.

39. To write the equation of a line, we need a point on the line and the slope. We are given a point on the line. Since the line is parallel to $y = 3x - 4$, these lines have the same slopes. The line $y = 3x - 4$ has slope 3 (it is written in slope-intercept form $y = mx + b$) and so 3 is also the slope of the line we are seeking.

    Using point-slope form: $y - 2 = 3(x - 1)$

    This is certainly an equation for the line. Your instructor may want another form, so you should be able to take one form and rewrite it to obtain another.
    slope-intercept form: $y = 3x - 1$
    general form: $3x - y - 1 = 0$

    $\boxed{COMMENT}$ *Of all lines, equations of horizontal or vertical lines are the easiest to find.*
    horizontal line: $y = k$ where k is the y-coordinate of any and all points on the line
    vertical line: $x = h$ where h is the x-coordinate of any and all points on the line

40. This problem is similar to #39. We are given a point on the line, but we need the slope. Since the line is *perpendicular* to $y = 3x - 4$, which has slope 3, the line we seek has slope $-\dfrac{1}{3}$. Remember, perpendicular lines have slopes that are *negative reciprocals* of each other.

$\boxed{\text{Cumulative Review}}$

Using point-slope form: $y - 2 = -\frac{1}{3}(x - 1)$

Rewriting to slope-intercept form will give: $y = -\frac{1}{3}x + \frac{7}{3}$

Rewriting to general form will give: $x + 3y - 7 = 0$
While all of these are valid equations for the line, your instructor may have a preference.

41.   Negative. This is the same question as #77, section 9.3. See the solution in this manual.

42.   Remember that $\geq$ or $\leq$ means that the line is included.

Graph:

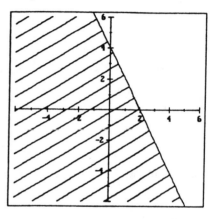

43.   See answer section in the text. Remember that $>$ or $<$ means that the line is not included, and so should be drawn in using dashes or dots. If you don't recall what to do next, see steps given for #42 above.

44.   Remember that $<$ or $>$ means that the line is not included, and should be drawn in with dashes or dots.
Notice that in this problem, you can't use $(0,0)$ as a test point because it lies on the boundary line. Choose another test point, but make things easy on yourself - perhaps $(1,1)$.
Since $1 < 2(1)$, the point $(1,1)$ satisfies the inequality and so the region containing $(1,1)$ should be shaded.

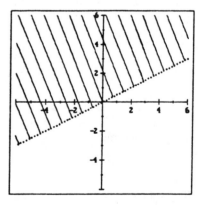

45.   This one is simpler than others because it contains only one variable. First simplify by dividing both sides by 4:
$y \geq 2$
This is the set of all points whose y-coordinate is 2 or greater. The solution set will be the line $y = 2$, and all points above it.

Cumulative Review

46. Let w be the width of the window; then 2w + 1 is the length. Since the area of a rectangle is the product of the width and length, we have:

$w(2w + 1) = 36$ Multiply out the left side, and collect all terms to the left side.

$2w^2 + w = 36$ You need zero on one side.

$2w^2 + w - 36 = 0$ Now factor the left side.

$(2w + 9)(w - 4) = 0$ Set each factor equal to zero and solve.

$w = -\frac{9}{2}$; $w = 4$ We reject the negative value because w represents width.

The width is 4 ft, and the length is $2(4) + 1 = 9$ ft. As a check, we see that $4(9) = 36$.

47. Let w be the width. Then the length is w + 3, and the diagonal brace is $3 + (w + 3) = w + 6$. We use the Pythagorean Theorem:

$$w^2 + (w + 3)^2 = (w + 6)^2$$
$$w^2 + w^2 + 6w + 9 = w^2 + 12w + 36$$
$$w^2 - 6w - 27 = 0$$
$$(w - 9)(w + 3) = 0$$
$$w = 9, \quad w = -3$$ We reject $w = -3$ because width is nonnegative.

The width is 9 ft, but let's be sure to answer the question asked. The length of the diagonal brace is w + 6, which is $9 + 6 = 15$ ft.

Note: You could have started this problem with letting d be the length of the brace. The legs of the triangle would then be $d - 3$ for the longer one and $(d - 3) - 3 = d - 6$ for the shorter leg. Proceed with the Pythagorean Theorem as before.

48. If y varies inversely as x, then there is a constant k such that $y = \frac{k}{x}$.

Find k: $10 = \dfrac{k}{\frac{4}{5}}$; $k = 10\left(\frac{4}{5}\right) = 8$

Now set up the complete equation: $y = \frac{8}{x}$ and use this to find y when x is $\frac{7}{8}$.

$y = \dfrac{8}{\frac{7}{8}} = 8\left(\frac{8}{7}\right) = \frac{64}{7}$

49. Let n be the first integer and n + 2 be the next odd integer.

Their reciprocals: $\frac{1}{n}$ and $\dfrac{1}{n + 2}$

Cumulative Review

The sum of their reciprocals is $\frac{8}{15}$, so we write:

$\frac{1}{n} + \frac{1}{n+2} = \frac{8}{15}$ The LCD is $15n(n+2)$. Multiply both sides by this LCD.

$15n(n+2)\left(\frac{1}{n} + \frac{1}{n+2}\right) = 15n(n+2)\frac{8}{15}$

$15(n+2) + 15n = 8n(n+2)$

$15n + 30 + 15n = 8n^2 + 16n$

$8n^2 - 14n - 30 = 0$ Divide both sides by 2.

$4n^2 - 7n - 15 = 0$

$(4n+5)(n-3) = 0$

$n = -\frac{5}{4}$ $n = 3$ We reject $-\frac{5}{4}$; it is not an integer.

The two consecutive odd integers are 3 and 5.

50.   Let r be the speed of the boat in still water.   (You may wish to set up a table to organize your information.)

Upstream: rate $= r - 4$ distance $= 8$
Downstream: rate $= r + 4$ distance $= 40$

The time to go upstream is the same as the time to go downstream. Since $R \cdot T = D$, we divide distance by rate to get expressions for the time for each:

Time upstream: $\frac{8}{r-4}$ Time downstream: $\frac{40}{r+4}$

And now set these times equal to each other:

$\frac{8}{r-4} = \frac{40}{r+4}$ We may cross-multiply. (Allowed when you have fraction = fraction.)

$8(r+4) = 40(r-4)$
$8r + 32 = 40r - 160$
$32r = 192$
$r = 6$ The speed of the boat in still water is 6 km/h.

Cumulative Review

Problem Set 10.1

1. (2, 5)

    $x + y = 7$ $x - y = -3$
    $2 + 5$ $2 - 5$
    $7 = 7$ $-3 = -3$

    (2,5) is a solution of the system.

5. (−2, −3)

    $2x + 5y = 11$
    $2(-2) + 5(-3)$
    $-4 \ - 15 \ = - 19$

    (−2, −3) does not satisfy the first equation, so it is not a solution. We need not check the second equation.

9. (3, −7)

    $y = 2x - 1$
    $-7 = 2(3) - 1$
    $-7 = 6 - 1$
    $-7 = 5$ (false)

    (3, −7) is not a solution.

13. (3, 5)

    The vertical line x = 3 contains all points with x-coordinate 3, so it contains (3,5). The horizontal line y = 5 contains all points with y-coordinate 5, so it contains (3,5).

    (3,5) is a solution.

17. (0.1, −0.2)

    $4x - 3y = 1$ $2x + \ y = 0$
    $4(0.1) - 3(-0.2)$ $2(0.1) + (-0.2)$
    $= 0.4 + 0.6$ $= 0.2 - 0.2$
    $= 1$ $= 0$ (0.1, −0.2) is a solution.

21. $x + 3y = 6$ intercepts: (0,2) , (6,0)
    $x + 2y = 2$ intercepts: (0,1) , (2,0)

    The lines intersect at (−6, 4).

25. $2x + y = 10$ intercepts: (0,10) , (5, 0)
    $x - 6 = 0$ vertical line x = 6

    The lines intersect at (6, −2).

29. $5x - 4y = 20$ intercepts: (0,−5) , (4, 0)
    $2x + y = 8$ intercepts: (0, 8) , (4, 0)

    The lines intersect at (4, 0).

33. $x - 5y = 0$ intercept: (0,0)   second point: (5, 1)   (there are many choices)

     $4x + y = 0$ intercept: (0,0)   second point: (−1, 4)

The lines intersect at (0,0).

37. $y = -\frac{2}{3}x - 1$ y-intercept: (0, −1)   second point: (3, −3)

     $y = \frac{3}{2}x + 12$ y-intercept: (0, 12)   second point: (2, 15)

The lines intersect at (−6, 3).

41. $-4x + 8y = 12$ intercepts: $(0, \frac{3}{2})$, (−3, 0)

     $5x - 10y = 15$ intercepts: $(0, -\frac{3}{2})$, (3, 0)

When sketched, the lines appear to be parallel. To confirm this, determine the slopes of the lines by writing them in slope-intercept form:

$-4x + 8y = 12$ $5x - 10y = 15$

$8y = 4x + 12$ $-10y = -5x + 15$

$y = \frac{4x}{8} + \frac{12}{8}$ $y = \frac{-5x}{-10} + \frac{15}{-10}$

$y = \frac{1}{2}x + \frac{3}{2}$ $(m = \frac{1}{2})$ $y = \frac{1}{2}x - \frac{3}{2}$ $(m = \frac{1}{2})$

The lines have equal slopes, so are parallel. There is no point of intersection, so no solution.

45. $y = \frac{3}{7}x - 17$ $m = \frac{3}{7}$ $y = -\frac{7}{8}x + 8$ $m = -\frac{7}{8}$

(a) The lines have different slopes so must intersect at a single point.
(b) One point of intersection means that there is one solution to the system.
(c) If a system has one solution, we call it independent and consistent.

49. $6x + 2y = 4$ $9x = 6 - 3y$

     $2y = -6x + 4$ $3y = -9x + 6$

     $y = -3x + 2$ $m = -3$ $y = -3x + 2$ $m = -3$

(a) The lines are the same. We say that the lines coincide, or are coincident.
(b) There are infinitely many solutions. (Each point on the line is a solution.)
(c) The system is dependent.

53. $\quad \frac{x}{6} + \frac{y}{4} = \frac{1}{3} \quad$ LCD $= 12$ $\qquad\qquad \frac{x}{3} + \frac{y}{2} = \frac{1}{4} \quad$ LCD $= 12$

$\quad 12\left(\frac{x}{6} + \frac{y}{4}\right) = 12\left(\frac{1}{3}\right) \qquad\qquad 12\left(\frac{x}{3} + \frac{y}{2}\right) = 12\left(\frac{1}{4}\right)$

$\quad 2x + 3y = 4 \qquad\qquad\qquad\qquad 4x + 6y = 3$

$\quad 3y = -2x + 4 \qquad\qquad\qquad\quad 6y = -4x + 3$

$\quad y = -\frac{2}{3}x + \frac{4}{3} \qquad\qquad\qquad\quad y = -\frac{2}{3}x + \frac{1}{2}$

(a) Both lines have slope $m = -\frac{2}{3}$ but different y-intercepts, so the lines are parallel.
(b) There are no points of intersection, so there can be no solution.
(c) The system is inconsistent.

57. See answer section in the text. The solution to the system is the point of intersection of the two graphs. It appears to be $(2,2)$, but we must check, by substituting $(2,2)$ into the equations:

$2 = 3(2) - 4 \qquad$ true
$2 = -2(2) + 6 \qquad$ true $\qquad\qquad$ We have confirmed that $(2,2)$ is the solution.

61. The intercepts are easy to find, and so are good points for plotting.

$x + y = 4 \qquad\qquad$ intercepts: $(0,4), (4, 0)$
$x - y = 1 \qquad\qquad$ intercepts: $(0,-1)$, $(1, 0)$

From the graph, the point of intersection is estimated to be $(2.5, 1.5)$.
See answer section in the text for the graph.

65. We like to have convenient points for graphing, that is, those with integer coordinates. Unfortunately, such points are not always easy to find, and this is the case with the first equation. We suggest that you go with the intercepts: $(0, -3\frac{3}{10})$, $(\frac{11}{5}, 0)$. Expressed as decimals, we have $(0, -3.3)$ and $(2.2, 0)$. Place these carefully when you plot them.

In the second equation, first notice that $(0,0)$ satisfies the equation. For a second point, use $x = 4$, then $\frac{4}{4} + \frac{y}{5} = 0$ becomes $\frac{y}{5} = -1$ and so $y = -5$. This gives the point $(4, -5)$. From the graph (see answer section in the text) we estimate the point of intersection to be $(1.2, -1.5)$.

Understanding the graphical significance of the solution to a system of equations is important and can be helpful in many ways in problem solving. However, having to estimate from a graph, even when using a graphing utility (such as a graphics calculator or computer software), still presents some drawbacks. In the next few sections we will be showing you more reliable and quicker methods to solving systems of equations.

Problem Set 10.2

1. $y = 2x - 3$
   $x + y = -9$

   Substitute $2x - 3$ for y in the second equation:
   $x + 2x - 3 = -9$
   $3x = -6$
   $x = -2$

   Using $y = 2x - 3$ with $x = -2$, solve for y:
   $y = 2(-2) - 3$
   $y = -4 - 3$
   $y = -7$ Solution: $(-2, -7)$

5. $x = 6 - 5y$
   $2x + 9y = 4$

   Substitute $6 - 5y$ for x in the second equation:
   $2(6 - 5y) + 9y = 4$
   $12 - 10y + 9y = 4$
   $y = 8$

   Using $x = 6 - 5y$ with $y = 8$, solve for x:
   $x = 6 - 5(8)$
   $x = 6 - 40$
   $x = -34$ Solution: $(-34, 8)$

9. $2x - y - 6 = 0$
   $x - y - 6 = 0$

   Solve for y in the first equation: $y = 2x - 6$
   Substitute $y = 2x - 6$ into the second equation:
   $x - (2x - 6) - 6 = 0$

   $-x + 6 - 6 = 0$

   $x = 0$

   Using $y = 2x - 6$ with $x = 0$, solve for y: $y = 2(0) - 6$
   $y = -6$
   Solution: $(0, -6)$

13. $2x - 5y = 9$
    $x - 3 = 0$

    From the second equation, $x = 3$.
    Substituting $x = 3$ into the first equation gives $2(3) - 5y = 9$.
    $6 - 5y = 9$
    $-5y = 3$

    $y = -\frac{3}{5}$

    Solution: $(3, -\frac{3}{5})$

17. $y = -2x$
    $5x - y = -7$

    Substituting $y = -2x$ into the second equation gives $5x - (-2x) = -7$.
    $5x + 2x = -7$
    $7x = -7$
    $x = -1$

    Using $y = -2x$ with $x = -1$ gives
    $y = -2(-1) = 2$.

    Solution: $(-1, 2)$

21. $x + y = 21$
    $x - y = 3$

    Solving for y in the first equation gives $y = 21 - x$; substituting this into the second equation gives $x - (21-x) = 3$.
    $x - 21 + x = 3$
    $2x = 24$
    $x = 12$

    Using $y = 21 - x$ with $x = 12$ gives
    $y = 21 - 12 = 9$.

    Solution: $(12, 9)$

25. $5x - 2y = 11$

$3x + 3y = 15$

Begin with $3x + 3y = 15$ and divide both

sides by 3: $x + y = 5$

This gives $x = 5 - y$, which we can
substitute into the first equation:

$5(5 - y) - 2y = 11$
$25 - 5y - 2y = 11$
$25 - 7y = 11$
$-7y = -14$
$y = 2$

Using $x = 5 - y$ with $y = 2$, we have
$x = 5 - 2 = 3$

Solution: $(3, 2)$

29. $y = x + 3$

$\frac{x}{2} - \frac{y}{5} = 3$

Substitute $y = x + 3$ into the second

equation: $\frac{x}{2} - \frac{x + 3}{5} = 3$

$10\left(\frac{x}{2} - \frac{x + 3}{5}\right) = 10(3)$

$5x - 2(x + 3) = 30$
$5x - 2x - 6 = 30$
$3x = 36$
$x = 12$

Using $y = x + 3$ with $x = 12$, we have
$y = 12 + 3 = 15$

Solution: $(12, 15)$

33. See answer section in the text. Think "identical" when using the word "identity".

37. $y = 5x - 3$ Substituting $y = 5x - 3$ into the second equation gives:
$10x - 2y = 6$ $10x - 2(5x - 3) = 6$
                              $10x - 10x + 6 = 6$
                              $6 = 6$ (an identity)

Answer: The lines are the same (coincident) so there are an infinite number of solutions.

41. $x = 2y - 5$ Substituting $x = 2y - 5$ into the second
$\frac{x}{6} + \frac{y}{8} = 1$ equation gives $\frac{2y - 5}{6} + \frac{y}{8} = 1$.

$24\left(\frac{2y - 5}{6} + \frac{y}{8}\right) = 24(1)$

$4(2y - 5) + 3y = 24$
$8y - 20 + 3y = 24$
$11y = 44$
$y = 4$

Using $x = 2y - 5$ with $y = 4$ gives $x = 2(4) - 5 = 8 - 5 = 3$.   Solution: $(3,4)$

45.

$$\boxed{1} \quad \frac{x}{4} - \frac{y}{2} = \frac{7}{24}$$

$$\boxed{2} \quad \frac{x}{3} + \frac{y}{2} = 0$$

It will be convenient to eliminate the fractions in both equations.

$$\boxed{1} \quad 24\left(\frac{x}{4} - \frac{y}{2}\right) = 24\left(\frac{7}{24}\right) \qquad\qquad \boxed{2} \quad 6\left(\frac{x}{3} + \frac{y}{2}\right) = 6(0)$$

$$6x - 12y = 7 \qquad\qquad\qquad\qquad\qquad 2x + 3y = 0$$

We have an equivalent system that is free of fractions:

$6x - 12y = 7$

$2x + 3y = 0$

Solving for x in the second equation gives $x = -\frac{3}{2}y$. Substituting this into the first equation gives:

$$6\left(-\frac{3}{2}y\right) - 12y = 7$$

$$-9y - 12y = 7$$

$$-21y = 7$$

$$y = -\frac{1}{3}$$

Using $x = -\frac{3}{2}y$ with $y = -\frac{1}{3}$ gives $x = -\frac{3}{2}\left(-\frac{1}{3}\right) = \frac{1}{2}$.   Solution: $\left(\frac{1}{2}, -\frac{1}{3}\right)$

49. We will show you the setup here, and leave solving the system to you.

Let n be the number of nickels, and d be the number of dimes.

5n   the amount of money in nickels, expressed in cents

10d   the amount of money in dimes, expressed in cents

The total amount of money is \$1.10, which is 110 <u>cents</u>. This gives us:

$5n + 10d = 110$ Notice that by expressing these quantities in terms of cents instead of dollars, we avoid decimals.

The total number of coins is 18, so $n + d = 18$.   This gives the system:

$5n + 10d = 110$
$n + d = 18$ $\Big\}$ We leave it up to you to solve this system.

Answer: There are 14 nickels and 4 dimes.

53. $x = 4.91y - 5.899$
$2.1x - 9.9y = -12.84$

Substituting $x = 4.91y - 5.899$ into the second equation gives:
$2.1(4.91y - 5.899) - 9.9y = -12.84$
$10.311y - 12.3879 - 9.9y = -12.84$
$0.411y = -0.4521$
$y = -1.1$

Using $x = 4.91y - 5.899$ with $y = -1.1$ gives
$x = 4.91(-1.1) - 5.899$
$x = -5.401 - 5.899$
$x = -11.3$

Solution: $(-11.3, -1.1)$

57. $y = 2x - 1$
$y = 3x - 2$

These are already in the form $y = mx + b$. See answer section in the text. Solution: $(1,1)$

Problem Set 10.3

1. $x + 2y = 6$ Add.
    $\underline{-x + 3y = 4}$
       $5y = 10$
        $y = 2$

$x + 2(2) = 6$
$x + 4 = 6$

$x = 2$
Solution: $(2,2)$

5. $6x + 2y = -1$
    $12x - y = 3$ Multiply by 2.

    $6x + 2y = -1$ Add the equations.
    $\underline{24x - 2y = 6}$
    $30x \quad\quad = 5$
    $x \quad\quad = \frac{1}{6}$

$12\left(\frac{1}{6}\right) - y = 3$

$2 - y = 3$
$y = -1$

Solution: $\left(\frac{1}{6}, -1\right)$

9. $2x + 3y = -9$ multiply by 2

$-4x + 5y = -37$

$4x + 6y = -18$

$\underline{-4x + 5y = -37}$

$11y = -55$

$y = -5$

$2x + 3(-5) = -9$
$2x - 15 = -9$
$2x = 6$
$x = 3$

Solution: $(3, -5)$

13. $2x + 5y = -3$ multiply by 3

$3x + 8y = -5$ multiply by -2

$6x + 15y = -9$

$\underline{-6x - 16y = 10}$

$-y = 1$

$y = -1$

$2x + 5(-1) = -3$
$2x - 5 = -3$
$2x = 2$
$x = 1$

Solution: $(1, -1)$

17. $2x - 13y = 38$ multiply by 5

$5x + 27y = 95$ multiply by -2

$10x - 65y = 190$

$\underline{-10x - 54y = -190}$

$-119y = 0$

$y = 0$

$2x - 13(0) = 38$
$2x = 38$
$x = 19$

Solution: $(19, 0)$

21. $2x + 6 = 0$ put constant on right

$3x + 2y = 1$

$2x = -6$ multiply by 3

$3x + 2y = 1$ multiply by -2

$6x = -18$

$\underline{-6x - 4y = -2}$

$-4y = -20$

$y = 5$

The first equation gives us $x = -3$.
Solution: $(-3, 5)$

25. Infinitely many solutions. The equations represent the same line, and the solution set is the set of *all* points lying on the line.

29. $6x = 9 - 3y$ Start by writing each equation in the form $Ax + By = C$.

$4y = 12 - 8x$

$6x + 3y = 9$ Divide both sides by 3.

$8x + 4y = 12$ Divide both sides by 4.

page 236

section   10.3

$$2x + y = 3$$
$$2x + y = 3$$
$$0 = 0$$

Subtraction of the two equations yields the identity $0 = 0$. This means that the equations represent the same line. There are an infinite number of solutions and we say the system is dependent.

33. $y = 2x - 1$
   $7x - 4y = -1$

Since we are given $y = 2x - 1$ (y is solved for in terms of x) it makes sense to use the substitution method.

$7x - 4(2x - 1) = -1$
$7x - 8x + 4 = -1$
$-x = -5$
$x = 5$

Using $y = 2x - 1$ with $x = 5$, solve for y:
$y = 2(5) - 1 = 9$.

Solution: (5,9)

37. $11x + 9y = 1$
   $5x - 7y = 6$

Rather than solve for x or y and substitute, it will be easier to use the addition-subtraction method.

$55x + 45y = 5$
$-55x + 77y = -66$
$122y = -61$
$y = -\frac{1}{2}$

Using the first equation with $y = -\frac{1}{2}$, solve for x: $11x + 9(-\frac{1}{2}) = 1$
$11x - \frac{9}{2} = 1$

$11x = \frac{11}{2}$; $x = \frac{1}{2}$

Solution: $\left(\frac{1}{2}, -\frac{1}{2}\right)$

41. ① $\frac{x}{4} - \frac{y}{3} = \frac{5}{12}$

   ② $\frac{x}{2} - \frac{2y}{3} = 1$

It will be easier to solve the problem if we first remove the fractions.

① $12\left(\frac{x}{4} - \frac{y}{3}\right) = 12\left(\frac{5}{12}\right)$

$3x - 4y = 5$

② $6\left(\frac{x}{2} - \frac{2y}{3}\right) = 6(1)$

$3x - 4y = 6$

We have the equivalent system $3x - 4y = 5$
                                $3x - 4y = 6$

Subtracting gives: $0 = -1$ (contradiction)

A contradiction implies that there is no point (x,y) that satisfies both equations; therefore there is no solution to the system. We say it is inconsistent.

45. The sum of twice m and three times n is 34: $2m + 3n = 34$
    The sum of three times m and twice n is 31: $3m + 2n = 31$

We will multiply the first equation by 3, and the second by -2:

$$
\begin{array}{rcl}
6m + 9n &=& 102 \\
-6m - 4n &=& -62 \\
\hline
5n &=& 40 \\
n &=& 8
\end{array}
$$

Now substitute into the original first equation to find m.

$2m + 3(8) = 34$
$2m = 10$
$m = 5$

The solution: $m = 5$, $n = 8$

49. We will show one way to analyze the problem in order to set up the system. (You may find other ways.) We leave the solving of the system to you.

Let y be the total number of units of Y used. Since there are 12 on hand, these have a value of $(12)(6) = \$72$. This means that we will be using 5280 $(5208 + 72)$ worth of materials.
Let x be the number of units of X used.

Cost for x units of X: $24x$ $\Big\}$ The money for materials is $\$5280$, so $24x + 6y = 5280$.
Cost for y units of Y: $6y$

Now carefully reread the problem, and see if you agree with the following:
1 unit of X will produce 4 units of P
1 unit of Y will produce 10 units of P

From this, we have:
x units of X will produce 4x units of P $\Big\}$ $4x = 10y$
y units of Y will produce 10y units of P

The system is: $24x + 6y = 5280$
                $4x = 10y$

We suggest that you simplify each equation before solving the system. Once it is solved, you will have $x = 200$ and $y = 80$. Remember, x units of X will produce 4x units of P, so 200 units of X will produce $4(200) = 800$ units of P.

53. $2x - 4y = 6$ $x + 3y = 3$

     $4y = 2x - 6$ $3y = -x + 3$

     $y = \frac{1}{2}x - \frac{3}{2}$ $y = -\frac{1}{3}x + 1$

See answer section in the text for the graph.   Solution: $(3,0)$

Problem Set 10.4

1. Let the two numbers be x and y.

    $x + y = 100$ sum is 100

    $\underline{x - y = 66}$ difference is 66

    $2x = 166$

    $x = 83$

    $83 + y = 100$

    $y = 17$

    The two numbers are 83 and 17.

5. Let the two numbers be x and y.

    $\dfrac{x + y}{2} = 82$ average is 82

    $x + y = 164$ (fraction cleared)

    $\underline{x - y = 6}$ range is 6

    $2x\ \ \ = 170$

    $x = 85$

    $x + y = 164$

    $85 + y = 164$

    $y = 79$

    The two numbers are 85 and 79.

9. $b = 12 + 2a$ The measure of angle b is 12° more than twice the measure of angle a.

    $a + b = 180$ The sum of their measures is 180°.

    $2a - b = -12$ Noticing the presence of a single b in both equations, we rewrite the

    $\underline{a + b = 180}$ first equation to take advantage of the addition-subtraction method.

    $3a = 168$ The equations were added.

    $a = 56$

    Using $a + b = 180$ with $a = 56$,

    $56 + b = 180$

    $b = 124$ The two angles are 56° and 124°.

13. The total value of the dimes is 0.10(d), and the total value of the coins is $21.40.

	value per coin	·	number of coins	=	value
nickels	0.05	·	n	=	0.05n
dimes	0.10	·	d	=	0.10d

$\left(\text{value of nickels}\right) + \left(\text{value of dimes}\right) = 21.40$

$0.05n + 0.10(d) = 21.40$ Multiply both sides by 100 to remove decimals.

$5n + 10d = 2140$

$\left(\text{number of nickels}\right) + \left(\text{number of dimes}\right) = 268$

$n \quad + \quad d \quad = 268$

The system is: $5n + 10d = 2140$
                   $n + d = 268$ Multiply by -5.

                   $5n + 10d = 2140$
                   $\underline{-5n - 5d = -1340}$
                   $5d = 800$
                   $d = 160$ There are 160 dimes.

                   $n + d = 268$
                   $n + 160 = 268$
                   $n = 108$ There are 108 nickels.

17.

p_1 = principal invested at a rate of 7% $\left.\begin{array}{c}\\ \\ \end{array}\right\}$ $p_1 + p_2 = \$12,000$
p_2 = principal invested at at rate of 9%

$0.07p_1$ = the dollar amount of interest earned on the investment of p_1 dollars
$0.09p_2$ = the dollar amount of interest earned on the investment of p_2 dollars

	principal	rate	time	interest
principal at 7%	p_1	0.07	1	$0.07p_1$
principal at 9%	p_2	0.09	1	$0.09p_2$

$0.07p_1 + 0.09p_2 = 890$ Multiply both sides by 100 to remove decimals.

$7p_1 + 9p_2 = 89000$

The system is: $p_1 + p_2 = 12000$ Multiply both sides by -7.
                   $7p_1 + 9p_2 = 89000$

                   $-7p_1 - 7p_2 = -84000$
                   $\underline{7p_1 + 9p_2 = 89000}$
                   $2p_2 = 5000$
                   $p_2 = 2500$ \$2,500 was invested at 9%.

                   $p_1 + p_2 = 12000$
                   $p_1 + 2500 = 12000$
                   $p_1 = 9500$ \$9,500 was invested at 7%.

21. s_1 = number of liters of the 33% strong solution of disinfectant $\quad\quad$ } $s_1 + s_2 = 80$ liters
     s_2 = number of liters of the 5% strong solution of disinfectant

The table given reminds you how to calculate the actual amount of disinfectant in each quantity of solution:

$0.33s_1$ = amount of pure disinfectant in the 33% solution
$0.05s_2$ = amount of pure disinfectant in the 5% solution

$0.12(80) = 9.6$ = amount of pure disinfectant in the 12% solution (the mixture)

	percent per liter	·	liters	=	amt of disinfectant
33% soln	0.33	·	s_1	=	$0.33s_1$
5% soln	0.05	·	s_2	=	$0.05s_2$
12% mix	0.12	·	80	=	$0.12(80)$ or 9.6

The amount of pure disinfectant in the mixture is the combined amounts of pure disinfectant in the two solutions to be mixed together, so:

$0.33s_1 + 0.05s_2 = 9.6$ $\quad\quad\quad\quad$ Multiply both sides by 100 to remove the decimals.

$33s_1 + 5s_2 = 960$

The system is: $\quad\quad\quad$ $s_1 + s_2 = 80$ $\quad\quad\quad$ Multiply both sides by -5.

$\quad\quad\quad\quad\quad\quad\quad\quad$ $33s_1 + 5s_2 = 960$

$\quad\quad\quad\quad\quad\quad\quad\quad$ $-5s_1 - 5s_2 = -400$

$\quad\quad\quad\quad\quad\quad\quad\quad$ $\underline{33s_1 + 5s_2 = 960}$

$\quad\quad\quad\quad\quad\quad\quad\quad$ $28s_1 = 560$

$\quad\quad\quad\quad\quad\quad\quad\quad$ $s_1 = 20$ $\quad\quad\quad\quad\quad$ Use 20 liters of the 33% solution.

$\quad\quad\quad\quad\quad\quad\quad\quad$ $s_1 + s_2 = 80$

$\quad\quad\quad\quad\quad\quad\quad\quad$ $20 + s_2 = 80$

$\quad\quad\quad\quad\quad\quad\quad\quad$ $s_2 = 60$ $\quad\quad\quad\quad\quad$ Use 60 liters of the 5% solution.

25.

x = amount of 80% solution
y = amount of water (0% solution)
$\left.\right\}$ $x + y = 1000$

0.80x = amount of insecticide in the x liters of 80% solution
0   = amount of insecticide in the y liters of water
0.04(1000) = 40 = amount of insecticide in the final product
$\left.\right\}$ $0.80x + 0 = 40$

The system is:
$$0.80x \quad\quad = 40$$
$$x \; + y \quad = 1000$$

Dividing the first equation by 0.80 gives $x = 50$. Substituting that into the second equation gives $y = 950$. The nurseryman will use 50 liters of the 80% solution of insecticide with 950 liters of water.

29.

	time used for smaller doors	+	time used for larger doors	= time available
Machine A	2 S	+	3 L	= 410
Machine B	1 S	+	2 L	= 230

The system is:
$2S + 3L = 410$
$S + 2L = 230$ We will multiply the second equation by -2.

$$\begin{array}{r} 2S + 3L = 410 \\ -2S - 4L = -460 \\ \hline -L = -50 \\ L = 50 \end{array}$$

Substitute into $S + 2L = 230$ and solve for S:
$S + 2(50) = 230; \; S = 130$ Answer: 50 large doors, 130 small doors

33.   Let r_1 be the rate of the slower train, and r_2 be the rate of the faster train.

	Rate	·	Time	=	Distance
slower train	r_1	·	0.5	=	$0.5\,r_1$
faster train	r_2	·	0.5	=	$0.5\,r_2$

$0.5\,r_1 + 0.5r_2 = 89$ The sum of their distances is 89 km after $\frac{1}{2}$ hour.
$r_2 = r_1 + 10$ The faster train's rate is 10 km/h more than the slower train's rate.

Since the second equation is solved for r_2, substitution would be reasonable to use:
$0.5r_1 + 0.5(r_1 + 10) = 89$
$0.5\,r_1 + 0.5\,r_1 + 5 = 89$
$r_1 + 5 = 89$
$r_1 = 84$ The slower train averages 84 km/h; the faster train averages 94 km/h.

section   10.4

37. Let r_1 be the rate of the slower train, and r_2 be the rate of the faster train. Notice from the times mentioned we can decide how long each train traveled.

	Rate	$\cdot$	Time	$=$	Distance
slower train	r_1	$\cdot$	1	$=$	r_1
faster train	r_2	$\cdot$	0.5	$=$	$0.5\, r_2$

$r_1 + 0.5\, r_2 = 135$ The sum of their distances is 135 km.
$r_2 = r_1 + 15$ The faster train's rate is 15 km/h more than the slower train's rate.

Since r_2 is solved for in the second equation, substitution would be reasonable to use. We leave the solving of the system to you.

$r_1 = 85$ The slower train averages 85 km/h.
$r_2 = r_1 + 15 = 100$ The faster train averages 100 km/h.

Problem Set 10.5

Note: Graphs required in the solutions may be found in the answer section of your text.

1.   (2,1) must satisfy *both* inequalities to be a solution of the system.

     $x - 2y > -2$ $x + y \geq 2$
     $2 - 2(1) > -2$ $2 + 1 \geq 2$ (2,1) is a solution.
     $2 - 2 > -2$ $3 \geq 2$ (true)
     $0 > -2$ (true)

5. $(5, -3)$

     $x - 2y > -2$ $x + y \geq 2$
     $5 - 2(-3) > -2$ $5 - 3 \geq 2$ $(5, -3)$ is a solution.
     $5 + 6 > -2$ $2 \geq 2$ (true)
     $11 > -2$ (true)

9. $(-4, 0)$

     $x > -3$
     $-4 > -3$ (false)

Since $(-4, 0)$ does not satisfy the first inequality, $(-4, 0)$ is not a solution to the system. You need not check the second inequality; whether $(-4, 0)$ satisfies it or not is of no consequence once we know that the point does not satisfy the first inequality.

13. (2,7) lies in the shaded region, so it is a solution. If you have trouble seeing this, it can help to extend the lines.

17. $(-2, 7)$ lies in the shaded region, so it is a solution.

21. The point $(0, -1)$ lies on the boundary of the shaded region. This is a solid line, meaning that $(0, -1)$ would be included in the solution set.

25. First graph the solution set for each inequality.

$x - y \geq 0$ The line $x - y = 0$ (or $y = x$) is a line through $(0,0)$ with slope $m = 1$. Using the test point $(2,1)$:
$$2 - 1 \geq 0$$
$$1 \geq 0 \quad \text{(true)}$$

The correct half-plane is the one below the line $x - y = 0$.

$x + y \geq 0$ The line $x + y = 0$ (or $y = -x$) is line through $(0,0)$ with slope $m = -1$. Using the test point $(2, 1)$,
$$2 + 1 \geq 0$$
$$3 \geq 0 \quad \text{(true)}$$

The correct half-plane is the one above the line $x + y = 0$.

The intersection of the two solution sets represents the solution set for the system. Look to see where the regions overlap. Shading with two colors can help.

29. $x \geq 0$
$y \geq 0$ } These represent the first quadrant (with the axes.)

$3x + 4y \leq 12$ The correct half-plane for $3x + 4y \leq 12$ is the one below the boundary line. (Use $(0,0)$ as a test point to check.)

See answer section in the text.

33. $5x + 3y \leq 15$ Graph the solid line $5x + 3y = 15$. Intercepts: $(0,5)$, $(3, 0)$
$5(0) + 3(0) \leq 15$ Test point: $(0,0)$
$0 \leq 15$ (true)   The correct half-plane is the one that contains $(0,0)$.

$x - 2y > 4$ Graph the dashed line $x - 2y = 4$. Intercepts: $(0,-2)$, $(4, 0)$
$0 - 2(0) > 4$ Test point: $(0,0)$
$0 > 4$ (false)   The correct half-plane does not contain $(0,0)$.

The solution to the system is represented by the intersection of the two solution sets. It is a good idea to mark your two half-planes with different types of markings or colors so it is easy to identify where they overlap. See answer section in the text.

page 244

section   10.5

37.

$y - 1 \leq \frac{2}{3}(x - 2)$ Graph the solid line $y - 1 = \frac{2}{3}(x - 2)$. Slope: $\frac{2}{3}$ Point: $(2,1)$

$0 - 1 \leq \frac{2}{3}(0 - 2)$ Test point: $(0,0)$

$-1 \leq -\frac{4}{3}$ (false)   The correct half-plane does not contain $(0,0)$.

$y + 1 \geq \frac{3}{2}(x - 1)$ Graph the solid line $y + 1 = \frac{3}{2}(x - 1)$. Slope: $\frac{3}{2}$ Point: $(1, -1)$

$0 + 1 \geq \frac{3}{2}(0 - 1)$ Test point: $(0,0)$

$1 \geq -\frac{3}{2}$ (true)   The correct half-plane contains $(0,0)$.

The solution to the system is represented by the intersection of the two solution sets.

41. $x \leq 3$ Graph the solid vertical line $x = 3$. You don't need a test point. This inequality represents all points with an x-coordinate less than or equal to 3, so the correct half-plane is to the left of $x = 3$.

   $x \geq -2$ Graph the solid vertical line $x = -2$. The correct half-plane is to the right of $x = -2$.

The solution to the system is represented by the intersection of these two solution sets. Use different markings or colors for the two sets to make it easier to see where they overlap. See answer section in the text.

45. $\left.\begin{matrix} x \geq 0 \\ y \geq 0 \end{matrix}\right\}$ These inequalities represent the first quadrant (with the axes.)

$x + y \leq 4$ Graph the solid line $x + y = 4$. Intercepts: $(0,4)$, $(4, 0)$

$0 + 0 \leq 4$ Test point: $(0,0)$

$0 \leq 4$ (true)   The correct half-plane is the one below the boundary line.

See answer section in the text.

49. The correct response is B.
The shaded region contains points on or to the left of the line $x = 2$; that is, points with x-coordinate less than or equal to 2. Such points satisfy $x \leq 2$.
It also contains points on or above the line $y = 2$; that is, points with y-coordinate greater than or equal to 2. Such points satisfy $y \geq 2$.

53. $x + 5y > 5$ Graph the dashed line $x + 5y = 5$. Intercepts: $(0,1)$, $(5,0)$
The correct half-plane is the one that *does not* contain $(0,0)$.

$x - 2y > -2$ Graph the dashed line $x - 2y = -2$. Intercepts: $(0,1)$, $(-2,0)$
The correct half-plane is the one that contains $(0,0)$.

$x < 3$ Graph the dashed vertical line $x = 3$. The correct half-plane is
left of the line, since all the x-coordinates are *less than* 3.

The solution set is the intersection of the solutions to each inequality.
It is a good idea to shade in the different half-planes with different types of markings or colors,
to make it easier to see where they overlap. See answer section in the text.

57. $3x + y \geq 3$ Graph the solid line $3x + y = 3$. Intercepts: $(0,3)$, $(1,0)$
The correct half-plane is the one that *does not* contain $(0,0)$.

$3x + y \leq 6$ Graph the solid line $3x + y = 6$. Intercepts: $(0,6)$, $(2,0)$
The correct half-plane is the one that contains $(0,0)$.

$x > 0 , y > 0$ These indicate that we are concerned with only the first quadrant,
including boundaries.

The solution set is the intersection of the solutions to each inequality.
It is a good idea to shade in the different half-planes with different types of markings or colors,
to make it easier to see where they overlap. See answer section in the text.

61.   First determine the equations for the boundary lines. There is more than one way to do this,
but we suggest:

(1) Determine the y-intercept.

(2) Does the line rise from left to right? Then the slope is positive.
Does the line fall from left to right? Then the slope is negative.

(3) Step from one point (perhaps the y-intercept) to another point on the graph, measuring
rise and run. Calculate the slope. Be sure to assign the correct sign.

(4) Set up the slope-intercept form for the lines: $y = mx + b$

Finally, when using the form $y = mx + b$:
shaded region above the line $y \geq mx + b$
shaded region below the line $y \leq mx + b$

See answer section in the text.

65. $2x + y \geq 6000$ Graph the solid line $2x + y = 6000$. Intercepts: $(0,6000)$, $(3000,0)$
                            The correct region is the one that does not contain $(0,0)$.

     $x \geq 0$ Since x and y are amounts of items produced, they must be
     $y \geq 0$ nonnegative. Geometrically, these inequalities simply refer to the first
                          quadrant, including boundaries.

Essentially, the final region is that portion of the solution to $2x + y \geq 6000$ that falls in
the first quadrant, including all boundaries.

Solutions to All Review Exercises — Chapter Ten

1. $(-2, 3)$
     $5x + 3y = -1$ $3x + 2y = 0$
     $5(-2) + 3(3)$ $3(-2) + 2(3)$
     $-10 + 9$ $-6 + 6$
     $-1 = -1$ $0 = 0$

The point $(-2, 3)$ satisfies both equations, so it is a solution.

2. $(5, -4)$
                          $2x - 3y = 22$
     $6x + 7y = 2$ $2(5) - 3(-4) = 22$
     $6(5) + 7(-4) = 2$ $10 + 12 = 22$
     $30 - 28 = 2$ $22 = 22$
          $2 = 2$

$(5, -4)$ is a solution.

3. $(8,5)$
         $5x - 8y = 0$ $8x + 5y = 36$
     $5(8) - 8(5) = 0$ $8(8) + 5(5) = 36$
         $40 - 40 = 0$ $64 + 25 = 36$
              $0 = 0$ (true) $89 = 36$ (false)

$(8,5)$ is not a solution.

4. $(-1, -3)$
     $4x - 2y > 1$ $3x - 5y \geq 12$
     $4(-1) - 2(-3) > 1$ $3(-1) - 5(-3) \geq 12$
     $-4 + 6 > 1$ $-3 + 15 \geq 12$
          $2 > 1$ (true) $12 \geq 12$ (true)

$(-1, -3)$ is a solution.

5. (4, −3)

$2x + 3y < 0$ $\qquad\qquad\qquad$ $3x - 5y \geq 27$

$2(4) + 3(-3) < 0$ $\qquad\qquad$ $3(4) - 5(-3) \geq 27$

$8 - 9 < 0$ $\qquad\qquad\qquad$ $12 + 15 \geq 27$

$-1 < 0$ (true) $\qquad\qquad$ $27 \geq 27$ (true)

The point (4, −3) satisfies both inequalities, so it is a solution.

6. (2,7)

$3x - y > -1$

$3(2) - 7 > -1$

$6 - 7 > -1$

$-1 > -1$ (False)

There is no need to check the second inequality. (2,7) is not a solution.

7. See answer section in the text.

To graph these, first find the intercepts:

$x + 2y = -4$ ` $(-4,0)$ $\qquad\qquad$ $(0, -2)$

$3x + y = 3$ $\qquad\qquad$ $(1,0)$ $\qquad\qquad$ $(0,3)$

8. $x - y = 5$

$2x + y = 4$

These are easy to graph if you find
and plot the intercepts.

$x - y = 5$ $\qquad$ $(5,0)$ $(0, -5)$

$2x + y = 4$ $\qquad$ $(2,0)$ $(0,4)$

The point of intersection is (3, −2).

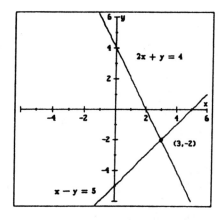

9. $3x - 2y = 6$ $\qquad$ Intercepts: $(0,-3)$, $(2, 0)$

$x = 4$ $\qquad\qquad$ This is a vertical line passing through (4, 0).

$\qquad\qquad\qquad$ The lines intersect at (4, 3).

Chapter Ten Review

10. $y = \frac{2}{3}x + 1$ The point of intersection is (6,5).

$y = \frac{1}{2}x + 2$

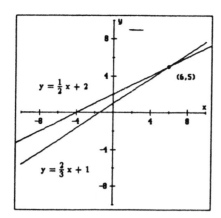

11. $y = -2x + 3$
$2x + y = 4$

Rewrite the second equation as $y = -2x + 4$. The lines have the same slope, -2.
They are not the same line, because they have different y-intercepts; so they must be parallel.
There is no solution, because there are no points of intersection.

12. $y = -\frac{x}{2} + 1$
$3x + 6y = 6$

Rewrite the second equation to slope-intercept form:
$6y = -3x + 6$
$y = -\frac{1}{2}x + 1$

These lines are identical, and so there are infinitely many solutions.

13. $x = 2y - 9$ Use substitution.
$5x + 6y = 3$

$5(2y - 9) + 6y = 3$
$10y - 45 + 6y = 3$
$16y = 48$
$y = 3$ Now solve for x.

$x = 2(3) - 9$
$x = -3$

solution: $(-3, 3)$

Chapter Ten Review

14. $y = 7x + 11$ Use substitution.
       $5x - 2y = -22$

       $5x - 2(7x + 11) = -22$
       $5x - 14x - 22 = -22$
       $5x - 14x = 0$
       $\quad\quad -9x = 0$
       $\quad\quad\quad x = 0$ Now substitute into either of the original equations.
       $y = 7(0) + 11$
       $y = 11$ Solution: $(0,11)$

15. $7x - 3y = -14$
       $4x + y = -8$ Solve for y.

       $y = -8 - 4x$ Substitute into first equation.

       $7x - 3(-8 - 4x) = -14$
       $7x + 24 + 12x = -14$
       $19x = -38$
       $\quad x = -2$ Now find y. The easiest equation for this is $y = -8 - 4x$.

       $y = -8 - 4(-2)$
       $y = 0$ Solution: $(-2,0)$

16. $7x + y = 0$ Solve for y.
       $20x + 3y = -1$

       $y = -7x$ Substitute into the second equation.

       $20x + 3(-7x) = -1$
       $\quad 20x - 21x = -1$
       $\quad\quad\quad -x = -1$
       $\quad\quad\quad\quad x = 1$ Now find y.

       $y = -7(1) = -7$ Solution: $(1,-7)$

17. $y = \frac{3}{7}x + 4$

       $3x - 7y = 0$ Rewrite this as $y = \frac{3}{7}x$.

The lines have the same slopes, but are not the same line. There is no solution.

18. $y = -\frac{2}{5}x + 1$ Substitute into the second equation.

       $6x + 15y = 15$

       $6x + 15\left(-\frac{2}{5}x + 1\right) = 15$

       $6x - 6x + 15 = 15$

              $15 = 15$

    or $0 = 0$ This is an identity.

                        There are an infinite number of solutions.

19. $7x + 11y = -1$ Add.

       $\underline{9x - 11y = 49}$

       $16x \qquad = 48$

       $x = 3$ Now find y. Use either equation.

       $7(3) + 11y = -1$

         $21 + 11y = -1$

              $11y = -22$

                $y = -2$ Solution: $(3, -2)$

20. $12x + 5y = 49$ Subtract.

     $\underline{-(12x - 7y = -11)}$

            $12y = 60$

              $y = 5$ Now find x. Use either equation.

       $12x + 5(5) = 49$

       $12x = 24$

          $x = 2$ Solution: $(2,5)$

21. $5x - 7y = 4$ Multiply both sides by -3.

    $15x + 8y = -46$

    $-15x + 21y = -12$ Add.

    $\underline{15x + \quad 8y = -46}$

           $29y = -58$

       $y = -2$

    $5x - 7y = 4$

    $5x - 7(-2) = 4$

    $5x + 14 = 4$

    $5x = -10$

    $x = -2$ solution: $(-2, -2)$

Chapter Ten Review

22.	$2x + 23y = 45$		Multiply by 3.

	$3x - 4y = -8$		Multiply by -2.

	$6x + 69y = 135$		Add.

	$\underline{-6x + 8y = 16}$

	$\quad\quad\quad 77y = 151$

	$\quad\quad\quad\quad y = \frac{151}{77}$		Now find x.

	$3x - 4\left(\frac{151}{77}\right) = -8$

	$3x = -\frac{12}{77}$

	$x = \frac{-4}{77}$		Solution: $\left(-\frac{4}{77}, \frac{151}{77}\right)$

23.	$6x - 15y = 9$

	$4x - 10y = 6$

	Notice that we can first simplify each equation.

	1️⃣ $6x - 15y = 9$	Divide both sides by 3.

	$\quad 2x - 5y = 3$

	2️⃣ $4x - 10y = 6$	Divide both sides by 2.

	$\quad 2x - 5y = 3$

	These equations represent the same line. There are infinitely many solutions.

24.	$6x - 4y = 8$

	$9x - 6y = 9$

	We will simplify each equation first.

	1️⃣ $6x - 4y = 8$

	$\quad 3x - 2y = 4$

	2️⃣ $9x - 6y = 9$

	$\quad 3x - 2y = 3$

	Now subtract the simplified equations.

	$3x - 2y = 4$

	$\underline{-(3x - 2y = 3)}$

	$\quad\quad\quad 0 = 1$		This is a contradiction. No solution.

25. $y = 17x - 33$
    $5x - 4y = -57$

    We will use substitution.

$5x - 4(17x - 33) = -57$
$5x - 68x + 132 = -57$
$-63x = -189$
$x = 3$

$y = 17(3) - 33$
$y = 51 - 33$
$y = 18$

Solution: $(3, 18)$

26. $14x - 13y = 10$
    $7x + 6y = 130$

Since it is not convenient to solve for y in either equation, we will use addition-subtraction. Multiply the second equation by -2, and add.

$$14x - 13y = 10$$
$$\underline{-14x - 12y = -260}$$
$$-25y = -250$$
$$y = 10$$

Now find x.

$7x + 6(10) = 130$
$7x = 70$
$x = 10$

Solution: $(10, 10)$

27. $y = \frac{2}{11}x - 7$
    $2x + 19y = -13$

Since y is solved for in the first equation, we will use substitution.

$2x + 19\left(\frac{2}{11}x - 7\right) = -13$

$2x + \frac{38}{11}x - 133 = -13$

$2x + \frac{38}{11}x = 120$

Multiply both sides by 11.

$22x + 38x = 1320$

$60x = 1320$
$x = 22$

Now find y.

$y = \frac{2}{11}(22) - 7$

$y = 2(2) - 7 = -3$

Solution: $(22, -3)$

28. $6x + 9y = 6$
$6x - 3y = 2$

Since we have the expression 6x in both equations, it will be easiest to use addition-subtraction. Subtract the equations:

$$6x + 9y = 6$$
$$-\,(6x - 3y = 2)$$
$$12y = 4$$
$$y = \tfrac{1}{3} \qquad \text{Now find x.}$$

$6x - 3(\tfrac{1}{3}) = 2$

$6x - 1 = 2$

$x = \tfrac{1}{2}$ \qquad Solution: $(\tfrac{1}{2}, \tfrac{1}{3})$

29.   ① $\tfrac{x}{2} + \tfrac{y}{3} = 3$ \qquad It will be more convenient to rewrite these equations in a form without fractions.

② $\tfrac{x}{4} - \tfrac{y}{5} = 7$

① $\quad 6\left(\tfrac{x}{2} + \tfrac{y}{3}\right) = 6(3)$ \qquad\qquad ② $\quad 20\left(\tfrac{x}{4} - \tfrac{y}{5}\right) = 20(7)$

$\qquad 3x + 2y = 18$ \qquad\qquad\qquad\qquad $\quad 5x - 4y = 140$

$3x + 2y = 18$ \qquad Multiply both sides by 2.
$5x - 4y = 140$

With $x = 16$, solve for y:

$6x + 4y = 36$ \qquad\qquad\qquad $3x + 2y = 18$
$\underline{5x - 4y = 140}$ \qquad\qquad\qquad $3(16) + 2y = 18$
$11x = 176$ \qquad\qquad\qquad\qquad $48 + 2y = 18$
$x = 16$ \qquad\qquad\qquad\qquad\qquad $2y = -30$
\qquad\qquad\qquad\qquad\qquad\qquad\qquad $y = -15$

Solution: $(16, -15)$

30. $y + 11 = 7x - 1$
$19x + 15y = 68$

It is not difficult to solve for y in the first equation, so we will do that and use substitution.

$y = 7x - 12$
$19x + 15(7x - 12) = 68$
$19x + 105x - 180 = 68$
$124x = 248$

$x = 2$ \qquad\qquad\qquad Now find y. It will be simplest to use $y = 7x - 12$.
$y = 7(2) - 12 = 2$ \qquad Solution: $(2,2)$

Chapter Ten Review

31. $y = 7x$
       $x = 7y$

       Substitute $y = 7x$ into the second equation.

       $x = 7(7x)$
       $x = 49x$
       $-48x = 0$
       $x = 0$ Now find y. Use $y = 7x$.

       $y = 7(0) = 0$ Solution: (0,0)

32. $5x - 2y = 25$
       $11x + 9y = 189$

       It is not convenient to solve for y or x in either equation, so we will not use substitution. We
       will use addition-subtraction. First multiply each equation by a suitable factor.
       $5x - 2y = 25$ Multiply by 11. $11x + 9y = 189$ Multiply by 5.
       $55x - 22y = 275$ $55x + 45y = 945$

       Now subtract the new equations:

       $55x - 22y = 275$
       $-(55x + 45y = 945)$
       _____
       $-67y = -670$
       $y = 10$ Now find x.

       $5x - 2(10) = 25$
       $5x = 45$
       $x = 9$ Solution: (9,10)

33.   First rewrite each equation.

       $2(x - 1) = 3(y + 4)$ $3(x + 1) = 6(y + 5)$
       $2x - 2 = 3y + 12$ $3x + 3 = 6y + 30$
       $2x - 3y = 14$ $3x - 6y = 27$ Divide both sides by 3.
                                                $x - 2y = 9$

       We have the equivalent system:
       $2x - 3y = 14$
       $x - 2y = 9$ Multiply both sides by -2.

       $2x - 3y = 14$ Find x: $x - 2y = 9$
       $-2x + 4y = -18$ $x - 2(-4) = 9$
       _____ $x = 1$
       $y = -4$

                                Solution: $(1, -4)$

34. $x = 7$
    $5(y - 9) = 6(x + 3)$

Notice that we already know the value of x, it is seven. To find y, substitute $x = 7$ into the second equation. We leave the details to you on this one.   Solution: $(7, 21)$

35. $3x + 4y = 2x - y + 5$
    $4x - 5y = 2x + 7y - 12$

First we simplify each equation by combining like terms. This will result in a new, but equivalent system.

$x + 5y = 5$
$2x - 12y = -12$ This equation can be divided on both sides by 2.

$\quad x + 5y = 5$ We are now ready to subtract these equations.
$\underline{-(x - 6y = -6)}$
$\qquad\quad 11y = 11$

$\qquad\quad\quad y = 1$ Now find x. The equation $x + 5y = 5$ is probably simplest.

$x + 5(1) = 5$
$x = 0$ Solution: $(0,1)$

36. $\boxed{1}$ $\dfrac{x}{3} - \dfrac{y}{5} = 16$

    $\boxed{2}$ $\dfrac{x}{5} + \dfrac{y}{2} = -9$

It will help considerably to rewrite each equation in a form without fractions.

$\boxed{1}$ $\dfrac{x}{3} - \dfrac{y}{5} = 16$ The LCD is 15.   Multiply both sides by 15.

$\quad 15\left(\dfrac{x}{3} - \dfrac{y}{5}\right) = 15(16)$

$\quad\quad 5x - 3y = 240$

$\boxed{2}$ $\dfrac{x}{5} + \dfrac{y}{2} = -9$ The LCD is 10, so multiply both sides by 10.

$\quad 10\left(\dfrac{x}{5} + \dfrac{y}{2}\right) = 10(-9)$

$\quad\quad 2x + 5y = -90$

We begin again with our new, but equivalent, system:

$5x - 3y = 240$ We will multiply the first by 2 and the second by 5.
$2x + 5y = -90$

$\quad 10x - 6y = 480$ Subtract.
$\underline{-(10x + 25y = -450)}$
$\qquad\quad -31y = 930$

$$y = -30$$

$5x - 3(-30) = 240$
$5x = 150$
$x = 30$ Solution: $(30, -30)$

$\boxed{COMMENT}$ *On #37– #42, start by finding the slope of each line. If the slopes are different, the system is independent and consistent (and has one solution); if the slopes are the same then the lines might be parallel (inconsistent) or the same line (dependent) — you will have to dig a little deeper to find out which.*

37.   Both lines are in slope-intercept form, $y = mx + b$. The slope of the first line is $\frac{5}{3}$; the slope of the second line is $\frac{3}{5}$. Since the slopes are different, the lines intersect at one point only. The system is independent and consistent.

38.   These lines are in slope-intercept form, $y = mx + b$.
The slopes are the same, but the y-intercepts are different. The lines are parallel, so there are no points of intersection, and so no solution to the system. It is inconsistent.

39.   If you rewrite the second equation in slope-intercept form, you get $y = \frac{2}{7}x - 3$.
This is the same as the first equation, telling us that the two original equations represent the same line. The system is dependent.

40.   If you rewrite the first equation in slope-intercept form, you get $y = -\frac{3}{4}x + 2$. This line has slope $-\frac{3}{4}$.
The second equation need not be rewritten, because although it is not in slope-intercept form it *is* in point-slope form, and we can read off the slope, which is $\frac{4}{3}$.
What we have found here is that the lines are actually perpendicular, which means there is exactly one point of intersection. The system is independent and consistent.

41.   Put each equation into slope-intercept form to make it easier to find the slope.

$2x = 3y - 4$	$3x = 2y - 4$
$3y = 2x + 4$	$2y = 3x + 4$
$y = \frac{2}{3}x + \frac{4}{3}$	$y = \frac{3}{2}x + 2$
$m = \frac{2}{3}$	$m = \frac{3}{2}$

The lines have different slopes, so they intersect in one point. The system is independent and consistent.

42.   Rewrite the first equation as $y = \frac{2}{9}x$; the second as $y = \frac{9}{2}x$. These lines have different slopes and so the system is independent and consistent.

$\boxed{\text{Chapter Ten Review}}$

43. Let the first number be x and the second number y. From the information given, we have:

$$x + y = 1$$

$$x = \frac{3}{5}y + 9$$

We will leave the solving to you. Suggestion: since x is solved for in the second equation, substitution would be sensible to use. The solution is $(6, -5)$.

44. Let a be the smaller angle and b be the larger angle. The system would be:

$$a + b = 180$$
$$b = 42 + 2a$$

We will leave the solving to you. Suggestion: since b is solved for in the second equation, substitution would be sensible to use. Answer: 46° and 134°

45. In previous chapters, we often set up charts to help organize the information. You may wish to do so (see Chapter 4 for a review.)
Here is another way to organize the information:

n = number of nickels   0.05n = value of the nickels in dollars
d = number of dimes   0.10d = value of the dimes in dollars

(number of nickels) + (number of dimes) = total number of coins
n + d = 680

(value of the nickels) + (value of the dimes) = sum of all the money
0.05n + 0.10d = 60.00   Multiply both sides by 100 to remove decimals.
5n + 10d = 6000

We have the system:
n + d = 680
5n + 10d = 6000

We leave the solving to you. Either substitution or addition-subtraction works well.
Answer: There are 520 dimes and 160 nickels.

46. Let b be the amount invested at bonds, and s be the amount invested in stocks.

	principal	rate	time	interest(dollars)
7% (bonds)	b	0.07	1	0.07b
10% (stocks)	s	− 0.10	1	− 0.10s

Notice that we use a negative rate for the stocks, since they lost value.

Chapter Ten Review

The system is:

b + s = 8000
0.07b − 0.10s = 220

Either substitution or addition-subtraction works fine here. However, we suggest that you start by multiplying both sides of the second equation by 100 to remove the decimals. This gives:

b + s = 8000
7b − 10s = 22000

We leave the details of solving this to you. Answer: There was $6000 invested in bonds.

47. Let a be the amount of chemicals worth $6 per liter, and b be the amount of chemicals worth $9 per liter.

	cost per liter	liters	cost in dollars
chemicals costing less	6	a	6a
chemicals costing more	9	b	9b
mixture	7.65	100	765

The system is:

a + b = 100
6a + 9b = 765

Either substitution or addition-subtraction works fine here. We suggest that you start by dividing both sides by 3 (6, 9 and 765 are all divisible by 3) to simplify the equation.
We leave the details of solving the system to you.
Answer: 45 liters at $6 per liter, and 55 liters at $9 per liter.

48. Let b be the length of the base and a be the length of each of the other two sides.

2a + b = 37   The perimeter is 37 cm.
b = a − 5   The base is five centimeters shorter than the other two sides.

This is easily solved by substitution, since b is already solved for in the second equation. Just substitute it into the first equation and solve for a. We leave the details of solving to you.
Answer: The base is 9 cm. (Each of the other two sides are 14 cm.)

49.   Let x = speed of the slower train and y = speed of the faster train.

Slower Train: rate = x km/h

time = $\frac{1}{2}$ hour

distance = rate × time = $x\left(\frac{1}{2}\right)$ or $\frac{x}{2}$ km

Faster Train: rate = y km/h

time = $\frac{1}{2}$ hour

distance = rate × time = $y\left(\frac{1}{2}\right)$ or $\frac{y}{2}$ km

Setting up the system:

$y = x + 8$ The rate of the faster train is 8 km/h more than the slower train.

$\frac{x}{2} + \frac{y}{2} = 100$ Since they are traveling in opposite directions, the sum of their individual

distances equals the number of kilometers that they are apart.

We leave the details of solving the system to you, but suggest that you multiply the second equation by 2 to eliminate the fractions. Also, since the first equation is solved for y, substitution is quite reasonable to use.
Answer: The slower train's speed is 96 km/h and the faster train's speed is 104 km/h.

50.   Let r be the rate of the current, and b be the rate of the boat.

	rate	time	distance
downstream	$b + r$	1.5	48
upstream	$b - r$	2	36

To set up the system, we apply the formula rate × time = distance to each trip the boat makes:

$(b + r)1.5 = 48$
$(b - r)2 = 36$

Suggestion: Divide the first equation by 1.5 and the second by 2 to simplify each before doing anything else. You will then find that addition-subtraction is quite easy to use. We will leave the details to you. The rate of the boat is 25 km/h, and the rate of the current is 7 km/h.

    (a) Graph the first line.   If it is easy to find the intercepts, use them.
        Remember that > and < corresponds to a dashed line, while ≤ and ≥ gives a solid line.
    (b) Use (0,0) as a test point to see what side of the line to shade. (Remember: if (0,0) is on
        the boundary line, pick another test point.)
    (c) Repeat this procedure for the second inequality. When shading, use a different color or
        different markings than you did for the first line.
    (d) The solution set is where the two regions overlap.
    (e) The inequalities $x \geq 0$ and $y \geq 0$ simply refer to the first quadrant.

Graphs for the odd-numbered exercises are in your book. Graphs for the even-numbered exercises are here. A few suggestions are made to help you out, and #53 is worked out in detail.

51.   Suggestions: Use the intercepts to graph each dashed line.   See section 10.5 if you need review. See answer section in the text for the graph.

52.   The intercepts are easy to find. Use them to make your lines. Notice that both lines are solid.

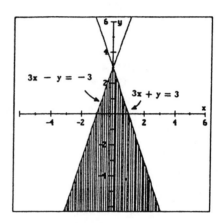

53. $y > \frac{1}{2}x - 3$ Graph the dashed line $y = \frac{1}{2}x - 3$. Slope: $\frac{1}{2}$ y-intercept: $(0,-3)$

     $0 > \frac{1}{2}(0) - 3$ Test point: $(0,0)$

     $0 > -3$ (true)   The correct half-plane contains $(0,0)$.

     $y \leq -\frac{1}{2}x + 2$ Graph the solid line $y = -\frac{1}{2}x + 2$. Slope: $-\frac{1}{2}$ y-intercept: $(0,2)$

     $0 \leq -\frac{1}{2}(0) + 2$ Test point: $(0,0)$

     $0 \leq 2$ (true)   The correct half-plane contains $(0,0)$.

The solution to the system is represented by the intersection of the solutions to the individual inequalities. See answer section in the text for the graph.

Chapter Ten Review

54.   Make sure to note that one line is
       dashed, the other solid.

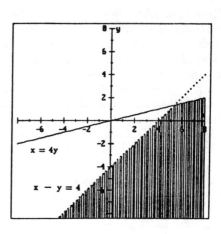

55.   Remember, $x \geq 0$ and $y \geq 0$ simply refer to the first quadrant - you don't have to do
       anything special with these equations, just remember that the final region cannot go outside
       of the first quadrant.
       Graph $x + y = 3$ by using the intercepts, and then check the test point $(0,0)$.
       Then shade in the portion of the correct half-plane *that lies in the first quadrant.*
       See answer section in the text for the graph.

56.   Remember, $x \geq 0$ and $y \geq 0$ simply refer to the first quadrant - you don't have to do
       anything special with these equations, just remember that the final region cannot go outside
       of the first quadrant.
       Graph $3x + 4y = 12$ by using the intercepts, and then check the test point $(0,0)$.
       Then shade in the portion of the correct half-plane *that lies in the first quadrant.*

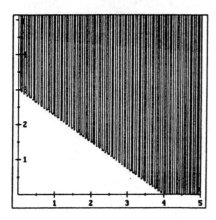

Solutions To All Mastery Test Problems − Chapter Ten

1a. $2x - y = 8$
       $2x - 5y = 0$ Soln: $(5,2)$
       See answer section in the text.

1b. $3x + y = 0$
       $x + y = 2$ Soln: $(-1,3)$
       See answer section in the text.

Chapter Ten Mastery Test

1c. $x + y = 3$
     $x + y = -2$
     Parallel lines. No solution.
     See answer section in the text.

1d. $-2x + 4y = -6$
     $3x - 6y = 9$
     Coincident lines. Infinitely many solutions.
     See answer section in the text.

2a. The lines have equal slopes, but different y-intercepts. They are parallel and the system is inconsistent.

2b. The slopes are $-\frac{3}{4}$ and $\frac{3}{4}$. Since the lines have different slopes, there will be one point of intersection (and one solution.) The system is consistent, with independent equations.

2c. Notice that the second equation can be obtained from the first by multiplying both sides by 2. These equations are equivalent, and represent the same line. The system is consistent, with dependent equations.

2d. The slopes are $\frac{3}{4}$ and $\frac{4}{9}$. Since the lines have different slopes, there will be one point of intersection (and one solution.) The system is consistent, with independent equations.

3a. $y = 3x - 4$
     $4x - 5y = 42$

     $4x - 5(3x - 4) = 42$
     $4x - 15x + 20 = 42$
     $-11x = 22$
     $x = -2$

     $y = 3(-2) - 4$
     $y = -10$ Solution: $(-2, -10)$

3b. $x + 7y = -9$
     $6x - 11y = 52$

     $x = -9 - 7y$
     $6(-9 - 7y) - 11y = 52$
     $-53y = 106$
     $y = -2$

     $x = -9 - 7(-2)$
     $x = 5$ Solution: $(5, -2)$

3c. $y = x - 8$

     $x - y = 1$

     $x - (x - 8) = 1$

     $8 = 1$

     Contradiction; no solution.

3d. $y = \frac{2}{3}x + \frac{1}{3}$

     $2x - 3y + 1 = 0$

     $2x - 3(\frac{2}{3}x + \frac{1}{3}) + 1 = 0$

     $2x - 2x - 1 + 1 = 0$

     $0 = 0$

     Infinitely many solutions.

4a. $3x - 7y = -13$

     $\underline{2x + 7y = 3}$ Add.

     $5x = -10$

     $x = -2$

     Now find y by substituting -2 for

     x in the second equation.

4b. $3x + 19y = 16$ Multiply by -2.

     $6x - 11y = -17$

     $-6x - 38y = -32$

     $\underline{6x - 11y = -17}$ Add.

     $-49y = -49$

Chapter Ten Mastery Test

$$2(-2) + 7y = 3$$
$$-4 + 7y = 3$$
$$7y = 7$$
$$y = 1 \qquad \text{Solution: } (-2,1)$$

$$y = 1$$

$$3x + 19(1) = 16$$

$$x = -1 \qquad \text{Solution: } (-1,1)$$

4c.
$$3x - 2y = 6$$
$$-6x + 4y = -12$$

$$6x - 4y = 12$$
$$\underline{-6x + 4y = -12} \quad \text{add}$$
$$0 = 0$$

Infinitely many solutions.

4d.
$$y = \frac{3}{5}x + 2$$
$$9x - 15y = 20$$

Rewrite the first equation as $3x - 5y = -10$ and then multiply both sides by -3:

$$-9x + 15y = 30$$
$$\underline{9x - 15y = 20} \quad \text{Add.}$$
$$0 = 50 \qquad \text{Contradiction; no solution.}$$

5a. Let the numbers be x and y, such that y is five less than two-thirds of x.

$$x + y = 25 \qquad \text{The sum of the two numbers is 25.}$$
$$y = \frac{2}{3}x - 5 \qquad \text{y is five less than two-thirds of x.}$$

We leave the solving to you. Substitution is reasonable to use.

Answer: 7 and 18

5b. Let d be the number of dimes and q be the number of quarters.

	(value per coin)	• (number of coins) =	total value
dimes	0.10	d	0.10d
quarters	0.25	q	0.25q

$$0.10d + 0.25q = 100$$
$$d + q = 760$$

$$10d + 25q = 10000$$
$$\underline{10d + 10q = 7600}$$
$$15q = 2400$$
$$q = 160$$

$$d + 160 = 760; \quad d = 600$$

There are 600 dimes and 160 quarters.

5c. Let x be the number of milliliters of the 30% solution and y be the number of milliliters of the 80% solution.

	(% of medicine) •	(number of milliliters)	= amount of medicine
30% solution	0.30	x	0.30x
80% solution	0.80	y	0.80y
50% mixture	0.50	25	0.50(25) or 12.5

$0.30x + 0.80y = 12.5$ 　　　　　 $3x + 8y = 125$

　$x + y = 25$ 　　　　　　　　　 $\underline{3x + 3y = 75}$

　　　　　　　　　　　　　　　　 $5y = 50$

　　　　　　　　　　　　　　　　　 $y = 10$

$x + 10 = 25$

$x = 15$

Answer: Mix 15 ml of the 30% solution with 10 ml of the 80% solution.

5d.

Tanker 　　　　　　　　 Jet

speed: x mi/hr 　　　　 speed: y mi/h

time: $\frac{1}{2}$ hr 　　　　　　 time: $\frac{1}{2}$ hr 　　　　 $\boxed{\text{Distance} = \text{Rate} \times \text{Time}}$

distance: $\frac{x}{2}$ miles 　　 distance: $\frac{y}{2}$ miles

$\frac{x}{2} + \frac{y}{2} = 350$ 　　　 The sum of their distances is 350.

$y = x + 250$ 　　　　 The speed of the jet is 250 mi/h more than the speed of the tanker.

Multiply both sides of the first equation by 2: 　　　 $x + y = 700$

Substitute $y = x + 250$ into the new first equation: $x + (x + 250) = 700$

　　　　　　　　　　　　　　　　　　　　　 $2x = 450$

　　　　　　　　　　　　　　　　　　　 $x = 225$ 　　　 $225 + 250 = 475$

Answer: The speed of the tanker is 225 mi/h and the speed of the jet is 475 mi/h.

6a. $3x + y \geq -3$
   $x + 3y \geq -3$

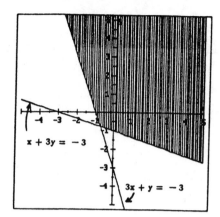

b. $x - y < 1$
   $x + y > 2$

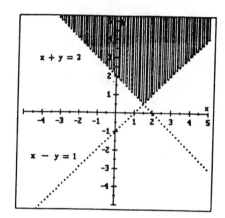

6c. $x \geq 0$
   $y \geq 0$
   $x + 2y < 4$

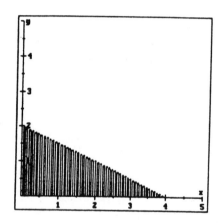

6d. $x \geq 0$
   $y \geq 0$
   $x \leq 4$
   $y \leq 2$

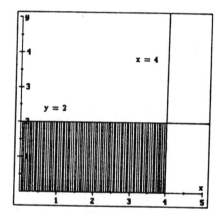

Chapter Ten Mastery Test

Problem Set 11.1

1a. $\sqrt{16} = 4$ because $4^2 = 16$.   1b. $\sqrt{169} = 13$ because $13^2 = 169$.
We know that 16 has two square roots, 4 and −4. However, the notation $\sqrt{x}$ means ONLY the positive square root of x (or zero, if x = 0), NOT BOTH SQUARE ROOTS. Since this notation will come up in later courses, it is worth your while to remember this.

1c. $\sqrt{19^2} = 19$ 1d. $\sqrt{(-17)^2} = |-17| = 17$

5a. $\sqrt{100} = \sqrt{10^2} = 10$ 5b. $\sqrt{1{,}000{,}000} = \sqrt{10^6} = \sqrt{(10^3)^2}$
$$= 10^3 = 1000$$

5c. $\sqrt{0.01} = \sqrt{(0.1)^2} = 0.1$ 5d. $\sqrt{2500} = \sqrt{(50)^2} = 50$

9a. $\sqrt{\dfrac{1}{49}} = \sqrt{\left(\dfrac{1}{7}\right)^2} = \dfrac{1}{7}$ 9b. $\sqrt{\dfrac{25}{64}} = \sqrt{\left(\dfrac{5}{8}\right)^2} = \dfrac{5}{8}$

9c. $\sqrt{\dfrac{9}{100}} = \sqrt{\left(\dfrac{3}{10}\right)^2} = \dfrac{3}{10}$

13a. $\sqrt[4]{3^4} = 3$ 13b. $\sqrt[4]{(-3)^4} = |-3| = 3$

13c. There is more than one correct way to proceed. It is helpful to start by finding a factor of 160,000 that is a power of 4. Since $16 = 2^4$, we start there:

$$\sqrt[4]{160{,}000} = \sqrt[4]{16(10000)} = \sqrt[4]{16(10^4)} = \sqrt[4]{2^4(10^4)} = \sqrt[4]{(20)^4} = 20$$

13d. There is more than one correct way to proceed. Remember to look for fourth powers. Since $3^4 = 81$, it seems reasonable to check the value of $(0.3)^4$. Since $(0.3)^4 = 0.0081$, we have:

$$\sqrt[4]{0.0081} = \sqrt[4]{(0.3)^4} = 0.3$$

Another approach: $\sqrt[4]{0.0081} = \sqrt[4]{\dfrac{81}{10000}} = \sqrt[4]{\dfrac{3^4}{10^4}} = \sqrt[4]{\left(\dfrac{3}{10}\right)^4} = \dfrac{3}{10} = 0.3$

17. See answer section in the text.   21.   See answer section in the text.

25. $5 + \sqrt{8} \approx 5 + \sqrt{9} = 5 + 3 = 8$ 29. $\sqrt[3]{28} \approx \sqrt[3]{27} = 3$
The best choice is D.   The best choice is A.

33. If a and b are nonnegative, then $\sqrt{a^2} = a$, and $\sqrt{b^2} = b$.
This gives $\sqrt{a^2} + \sqrt{b^2} = a + b$.

37. $\sqrt{v^2 + 6v + 9} = \sqrt{(v + 3)^2} = v + 3$

41. $\sqrt{a^2 b^4} = \sqrt{(ab^2)^2} = ab^2$

45. $c = \sqrt{7^2} + \sqrt{24^2} = 7 + 24 = 31$

49. $c = 15 + \sqrt{(10)(4.9)} = 15 + \sqrt{49} = 15 + \sqrt{7^2} = 15 + 7 = 22$

53. A calculator that displays eight digits would display an approximation of: 299.10032
You will have to do the rounding yourself. Rounded to the nearest thousandth, we have 299.100.

The key sequence on many scientific calculators is:

8 9 4 6 1 $\boxed{\sqrt{}}$ $\boxed{=}$

The key sequence on many graphics calculators is:

$\boxed{\text{2nd}}$ $\boxed{x^2}$ 8 9 4 6 1 $\boxed{\text{Enter}}$ (the square root is located above x^2)

$\boxed{SUGGESTION}$ *Calculators do not all operate in the same way. Ask your instructor or perhaps another student if you need help with yours.*

57. A calculator the displays eight digits would display an approximation of: 0.1264911
You will need to do the rounding yourself. Rounded to the nearest thousandth, we have 0.126.
See calculator steps above given in #53.

61. How you do this depends on your calculator; however, on scientific and graphics calculators, π is built into the calculator. You should see the symbol π on a key or above a key. Be sure to take advantage of this, and use this built-in value. In general you will get better approximations than if you type in your own approximation of π.
Answer: 1.33

$\boxed{COMMENTS\ ON\ USING\ PI}$

(1) Since $\pi = 3.141592654...$ is an irrational number, the digits never end nor repeat. How many are stored in your calculator depends on what you have, but there will certainly be more than you will want to remember. Take advantage of your calculator's ability to remember!

(2) In doing exercise #61, you might notice that you get the same answer (rounded to the

nearest hundredth) whether you use the built-in π or just 3.14. So why do we suggest you use the built-in value?
Calculate $\sqrt[4]{3.14}$. A calculator that displays eight digits will read 1.3311666.
Now calculate $\sqrt[4]{\pi}$, using the built-in approximation of π. You will get 1.3313354.

They are __not__ the same after the thousandths position, and the calculation using the built-in π is more accurate. The error in using 3.14 can be small (as in this problem) or large, depending on the calculations involved. If your calculator has π built-in, use it!

65. $\sqrt{v^2} = |\,v\,|$ The radical notation indicates only the principal square root, which is nonnegative.

         Example: $\sqrt{(-7)^2} = \sqrt{49} = 7$, *which is the absolute value of −7.*

69. $c = \sqrt{10^2 + 24^2} = \sqrt{100 + 576} = \sqrt{676} = 26$

<u>Problem Set 11.2</u>

1a. $\sqrt{5}\sqrt{7} = \sqrt{5(7)} = \sqrt{35}$ 1b. $\sqrt{10}\sqrt{10} = \sqrt{10^2} = 10$

1c. $\sqrt{7v}\sqrt{7v} = \sqrt{(7v)^2} = 7v$ 1d. $\sqrt{5y^3}\sqrt{5y^3} = \sqrt{(5y^3)^2} = 5y^3$

5. $\sqrt{98} = \sqrt{2(49)} = \sqrt{2}\sqrt{49} = (\sqrt{2})\,(7) = 7\sqrt{2}$

9. $\sqrt{600} = \sqrt{100(6)} = \sqrt{100}\sqrt{6} = 10\sqrt{6}$

13. $\sqrt{18a^2b^2} = \sqrt{9a^2b^2(2)} = \sqrt{9a^2b^2}\,\sqrt{2} = 3ab\sqrt{2}$

17. $\sqrt{7}\sqrt{35} = \sqrt{7(35)} = \sqrt{7(7)(5)} = \sqrt{7^2(5)} = \sqrt{7^2}\sqrt{5} = 7\sqrt{5}$

21. $\sqrt{18}\sqrt{12} = \sqrt{18(12)} = \sqrt{6(3)(6)(2)} = \sqrt{6^2(6)} = \sqrt{6^2}\sqrt{6} = 6\sqrt{6}$

25. $\sqrt{26w}\sqrt{39w} = \sqrt{26w(39w)} = \sqrt{13(2w)(13)(3w)} = \sqrt{(13w)^2 6} = \sqrt{13w^2}\sqrt{6} = 13w\sqrt{6}$

29. $\sqrt{\dfrac{36}{49}} = \sqrt{\left(\dfrac{6}{7}\right)^2} = \dfrac{6}{7}$

33. $\sqrt[3]{24} = \sqrt[3]{8(3)} = \sqrt[3]{8} \cdot \sqrt[3]{3} = 2\,\sqrt[3]{3}$

37. $\sqrt[4]{1250} = \sqrt[4]{10(125)} = \sqrt[4]{2(5)(5)(5)(5)} = \sqrt[4]{2(5^4)} = \sqrt[4]{5^4} \cdot \sqrt[4]{2} = 5\sqrt[4]{2}$

41. $\sqrt{\frac{1}{5}} \cdot \sqrt{\frac{9}{5}} = \sqrt{\left(\frac{1}{5}\right)\left(\frac{9}{5}\right)} = \sqrt{\frac{9}{5^2}} = \sqrt{\left(\frac{3}{5}\right)^2} = \frac{3}{5}$

45. $\sqrt{v^6 w^8} = \sqrt{(v^3)^2 (w^4)^2} = \sqrt{(v^3)^2} \sqrt{(w^4)^2} = v^3 w^4$

49. $\sqrt[4]{w^8} = \sqrt[4]{(w^2)^4} = w^2$

53. $s = \sqrt[3]{999} \approx \sqrt[3]{1000} = 10$ 　　　　　The best response is B.

57. This is equivalent to $\sqrt{20+80} = \sqrt{100} = 10$.

61. $\sqrt[3]{\frac{24a^3 b}{3b^4}} = \sqrt[3]{\frac{8a^3}{b^3}} = \frac{2}{b}$

65. $c = \sqrt{8^2 + 15^2} = \sqrt{64 + 225} = \sqrt{289} = 17$

The hypotenuse is 17 centimeters.

Problem Set 11.3

1. Like. The radicands of both square roots are 17.

5. Unlike. The radicands, 2 and 5, are different.

9. $-4\sqrt{10} + 3\sqrt{10}$
$= (-4 + 3)\sqrt{10}$
$= (-1)\sqrt{10} = -\sqrt{10}$

13. $6\sqrt{7} - 3\sqrt{7} + \sqrt{7}$
$= (6 - 3 + 1)\sqrt{7}$
$= 4\sqrt{7}$

17. $2\sqrt{3} + 3\sqrt{2} + \sqrt{3} - \sqrt{2}$
$= 2\sqrt{3} + \sqrt{3} + 3\sqrt{2} - \sqrt{2}$
$= (2 + 1)\sqrt{3} + (3 - 1)\sqrt{2}$
$= 3\sqrt{3} + 2\sqrt{2}$

21. $\sqrt{125} - \sqrt{5}$
$= \sqrt{25 \cdot 5} - \sqrt{5}$
$= \sqrt{25}\sqrt{5} - \sqrt{5}$
$= 5\sqrt{5} - \sqrt{5} = 4\sqrt{5}$

25. $3\sqrt{32} - 2\sqrt{50}$

$= 3\sqrt{16 \cdot 2} - 2\sqrt{25 \cdot 2}$

$= 3\sqrt{16}\sqrt{2} - 2\sqrt{25}\sqrt{2}$

$= 3 \cdot 4\sqrt{2} - 2 \cdot 5\sqrt{2}$

$= 12\sqrt{2} - 10\sqrt{2}$

$= 2\sqrt{2}$

29. $2\sqrt{72} - 3\sqrt{12} - \sqrt{48}$

$= 2\sqrt{36 \cdot 2} - 3\sqrt{4 \cdot 3} - \sqrt{16 \cdot 3}$

$= 2\sqrt{36}\sqrt{2} - 3\sqrt{4}\sqrt{3} - \sqrt{16}\sqrt{3}$

$= 2 \cdot 6\sqrt{2} - 3 \cdot 2\sqrt{3} - 4\sqrt{3}$

$= 12\sqrt{2} - 6\sqrt{3} - 4\sqrt{3}$

$= 12\sqrt{2} - 10\sqrt{3}$

33. $\sqrt[3]{24} - \sqrt[3]{375}$

$= \sqrt[3]{8 \cdot 3} - \sqrt[3]{125 \cdot 3}$

$= \sqrt[3]{8}\,\sqrt[3]{3} - \sqrt[3]{125}\,\sqrt[3]{3}$

$= 2\sqrt[3]{3} - 5\sqrt[3]{3}$

$= (2 - 5)\sqrt[3]{3} = -3\sqrt[3]{3}$

37. $\sqrt[3]{28} + \sqrt[3]{124}$

$\approx \sqrt[3]{27} + \sqrt[3]{125}$

$\approx 3 + 5$

$= 8$

The best response is A.

41. $\sqrt{8a} + \sqrt{18a}$

$= \sqrt{4 \cdot 2a} + \sqrt{9 \cdot 2a}$

$= 2\sqrt{2a} + 3\sqrt{2a}$

$= (2 + 3)\sqrt{2a}$

$= 5\sqrt{2a}$

45. $4x\sqrt{3xy} - 8\sqrt{75x^3 y}$

$= 4x\sqrt{3xy} - 8\sqrt{25x^2 3xy}$

$= 4x\sqrt{3xy} - 8\sqrt{25x^2}\sqrt{3xy}$

$= 4x\sqrt{3xy} - 8 \cdot 5x\sqrt{3xy}$

$= 4x\sqrt{3xy} - 40x\sqrt{3xy}$

$= (4x - 40x)\sqrt{3xy}$

$= -36x\sqrt{3xy}$

49. $\sqrt{6(10)(15)} - (10 - 15)\sqrt{4}$

$= \sqrt{900} - (-5)(2)$

$= 30 + 10$

$= 40$

Notice also:

$\sqrt{6(10)(15)} = \sqrt{2(3)(2)(5)(3)(5)}$

$= \sqrt{2^2 3^2 5^2} = 2 \cdot 3 \cdot 5 = 30$

53. $\sqrt[5]{100{,}000} - \sqrt[3]{27{,}000}$

$= \sqrt[5]{10^5} - \sqrt[3]{27(1000)}$

$= 10 - \sqrt[3]{3^3 10^3}$

$= 10 - \sqrt[3]{(30)^3}$

$= 10 - 30 = -20$

57. The area of a rectangle is the product of the length and width.

Area $= (8\sqrt{2})(5\sqrt{3})$

$= 40\sqrt{6} \ \text{cm}^2$

$\approx 98.0 \ \text{cm}^2$

61. For the moment look only at one of the small squares. Let s be the length of a side of a small square. The formula for the area of one of them is $A = s^2$.

Each small square has area 5 cm^2, so we will solve $5 = s^2$.

$5 = s^2$
$s = \sqrt{5} \ \text{cm}$ (We reject $-\sqrt{5}$, since s is a length.)

Now look again at the large rectangular grid, which is composed of the small squares, each with side of length s.
The large rectangular grid is 5s in length, or $5\sqrt{5}$ cm; its width is 3s, which is $3\sqrt{5}$ cm.

Problem Set 11.4

1. $\sqrt{2}\left(\sqrt{2} - \sqrt{3}\right)$

$= \sqrt{2}\sqrt{2} - \sqrt{2}\sqrt{3}$

$= 2 - \sqrt{6}$

5. $3\sqrt{5}\left(\sqrt{10} - 4\sqrt{15}\right)$

$= 3\sqrt{5}\sqrt{10} - 3\sqrt{5}\left(4\sqrt{15}\right)$

$= 3\sqrt{5}\sqrt{5}\sqrt{2} - 12\sqrt{5}\sqrt{5}\sqrt{3}$

$= 3 \cdot 5\sqrt{2} - 12 \cdot 5\sqrt{3}$

$= 15\sqrt{2} - 60\sqrt{3}$

9. Use FOIL to multiply this out:

$\left(7 + 2\sqrt{11}\right)\left(2 - 3\sqrt{11}\right)$

$= 7(2) - 7\left(3\sqrt{11}\right) + 2\left(2\sqrt{11}\right) - \left(2\sqrt{11}\right)\left(3\sqrt{11}\right)$

$= 14 - 21\sqrt{11} + 4\sqrt{11} - 6(11)$

$= 14 - 17\sqrt{11} - 66$

$= -52 - 17\sqrt{11}$

13. Use FOIL to multiply this out:

$$\left(3\sqrt{5} - \sqrt{7}\right)\left(2\sqrt{5} + 3\sqrt{7}\right)$$

$$= \left(3\sqrt{5}\right)\left(2\sqrt{5}\right) + \left(3\sqrt{5}\right)\left(3\sqrt{7}\right) - \left(\sqrt{7}\right)\left(2\sqrt{5}\right) - \left(\sqrt{7}\right)\left(3\sqrt{7}\right)$$

$$= 6(5) + 9\sqrt{35} - 2\sqrt{35} - 3(7)$$

$$= 30 + 7\sqrt{35} - 21$$

$$= 9 + 7\sqrt{35}$$

17. $\left(2\sqrt{5} + \sqrt{7}\right)\left(2\sqrt{5} - \sqrt{7}\right)$ Use the form $a^2 - b^2 = (a + b)(a - b)$.

$$= \left(2\sqrt{5}\right)^2 - \left(\sqrt{7}\right)^2$$

$$= 2^2\left(\sqrt{5}\right)^2 - 7$$

$$= 4(5) - 7$$

$$= 20 - 7 = 13$$

21. $\dfrac{18}{\sqrt{3}} \cdot \dfrac{\sqrt{3}}{\sqrt{3}} = \dfrac{18\sqrt{3}}{3} = 6\sqrt{3}$

25. $\sqrt{\dfrac{2}{3}} = \dfrac{\sqrt{2}}{\sqrt{3}} \cdot \dfrac{\sqrt{3}}{\sqrt{3}} = \dfrac{\sqrt{6}}{3}$

29. $\dfrac{10\sqrt{5}}{3\sqrt{2}} \cdot \dfrac{\sqrt{2}}{\sqrt{2}} = \dfrac{10\sqrt{10}}{3(2)} = \dfrac{5\sqrt{10}}{3}$

33. $\dfrac{6}{\sqrt{7} + \sqrt{5}} \cdot \dfrac{\sqrt{7} - \sqrt{5}}{\sqrt{7} - \sqrt{5}}$

$$= \dfrac{6\left(\sqrt{7} - \sqrt{5}\right)}{\left(\sqrt{7} + \sqrt{5}\right)\left(\sqrt{7} - \sqrt{5}\right)}$$

$$= \dfrac{6\left(\sqrt{7} - \sqrt{5}\right)}{7 - 5} = 3\left(\sqrt{7} - \sqrt{5}\right) \quad \text{or} \quad 3\sqrt{7} - 3\sqrt{5}$$

37. $\dfrac{\sqrt{10} + \sqrt{15}}{\sqrt{5}} \cdot \dfrac{\sqrt{5}}{\sqrt{5}}$

$$= \dfrac{\left(\sqrt{10} + \sqrt{15}\right)\sqrt{5}}{5}$$

$$= \dfrac{\sqrt{10}\sqrt{5} + \sqrt{15}\sqrt{5}}{5}$$

$$= \dfrac{5\sqrt{2} + 5\sqrt{3}}{5}$$

$$= \sqrt{2} + \sqrt{3}$$

41. See answer section in the text.

45. $\sqrt{7}\left(\sqrt{5} - \sqrt{7}\right)$

$$= \sqrt{35} - \sqrt{49}$$

$$\approx \sqrt{36} - \sqrt{49}$$

$$= 6 - 7$$

$$= -1 \qquad \text{The best response is C.}$$

49. $6\sqrt{12} - 7\sqrt{75}$

$= 6\sqrt{4 \cdot 3} - 7\sqrt{25 \cdot 3}$

$= 6\sqrt{4}\sqrt{3} - 7\sqrt{25}\sqrt{3}$

$= 6(2)\sqrt{3} - 7(5)\sqrt{3}$

$= 12\sqrt{3} - 35\sqrt{3}$

$= -23\sqrt{3}$

53. $\left(\sqrt{y} + \sqrt{3}\right)\left(\sqrt{y} + 4\sqrt{3}\right)$ Use FOIL.

$= \left(\sqrt{y}\right)^2 + \left(\sqrt{y}\right)\left(4\sqrt{3}\right)$

$\quad + \left(\sqrt{3}\right)\left(\sqrt{y}\right) + 4\left(\sqrt{3}\right)^2$

$= y + 4\sqrt{3y} + \sqrt{3y} + 4(3)$

$= y + 5\sqrt{3y} + 12$

57. $\dfrac{x - y}{\sqrt{x} + \sqrt{y}} \cdot \dfrac{\sqrt{x} - \sqrt{y}}{\sqrt{x} - \sqrt{y}}$

$= \dfrac{(x - y)\left(\sqrt{x} - \sqrt{y}\right)}{\left(\sqrt{x} + \sqrt{y}\right)\left(\sqrt{x} - \sqrt{y}\right)}$

$= \dfrac{(x - y)\left(\sqrt{x} - \sqrt{y}\right)}{\left(\sqrt{x}\right)^2 - \left(\sqrt{y}\right)^2}$

$= \dfrac{(x - y)\left(\sqrt{x} - \sqrt{y}\right)}{x - y} = \sqrt{x} - \sqrt{y}$

61. $\sqrt{5 + 4} + \sqrt{3(5) + 1}$

$= \sqrt{9} + \sqrt{16}$

$= 3 + 4 = 7$

$x = 5$ is a solution.

65. Using $r = \sqrt[3]{\dfrac{3V}{4\pi}}$, we have $r = \sqrt[3]{\dfrac{3(15,000)}{4\pi}} \approx 15.3$ ft

If your calculator has a built-in approximation for π, you should use it rather than type in your own approximation.   It is easier, and more accurate.

If you have trouble using your calculator, ask your instructor or perhaps another student for help. They do not all work in the same way, so it is not practical here to try to tell you what keys to press.

<u>Problem Set 11.5</u>

1. $\left(\sqrt{x}\right)^2 = 5^2$

$x = 25$

check: $\sqrt{25} = 5$

5. $\left(\sqrt{-3w}\right)^2 = 9^2$

$-3w = 81 \; ; \; w = -27$

check: $\sqrt{-3(-27)} = \sqrt{81} = 9 \; \checkmark$

| *Remember! When solving radical equations, you MUST CHECK for extraneous solutions!* |

9. $\left(\sqrt{m+3}\right)^2 = 5^2$

$m + 3 = 25; \quad m = 22$

check: $\sqrt{22 + 3} = \sqrt{25} = 5$

13. $\left(\sqrt{3z + 28}\right)^2 = 5^2$

$3z + 28 = 25; \; 3z = -3; \; z = -1$

check: $\sqrt{3(-1) + 28} = \sqrt{25} = 5$

17. $\sqrt{3 - 2x} - 3 = 2$

$\sqrt{3 - 2x} = 5$

$\left(\sqrt{3 - 2x}\right)^2 = 5^2$

$3 - 2x = 25$

$-2x = 22$

$x = -11$

Before you square both sides, the radical must isolated on one side. Add 3 to both sides. Now square both sides.

check: $\sqrt{3 - 2(-11)} - 3 = 2$?

$\sqrt{25} - 3 = 5 - 3 = 2$ ✓

21. $\left(\sqrt{y + 2}\right)^2 = \left(\sqrt{2y - 5}\right)^2$

$y + 2 = 2y - 5$

$y = 7$

check: $\sqrt{7 + 2} = \sqrt{2(7) - 5}$?

$\sqrt{9} = \sqrt{14 - 5}$

$3 = 3$ ✓

25. $\left(2\sqrt{m + 1}\right)^2 = \left(\sqrt{5m + 1}\right)^2$

$4(m + 1) = 5m + 1$
$4m + 4 = 5m + 1$
$m = 3$

check: $2\sqrt{3 + 1} = \sqrt{5(3) + 1}$?

$2\sqrt{4} = \sqrt{16}$
$2(2) = 4$
$4 = 4$ ✓

__COMMENT__ *When checking, be sure that you substitute back into the original equation, not into a subsequent equation where the radicals have been removed. It is possible to have a value that "checks" in an equation obtained after a radical has been removed, but does __not__ check in the original equation. It would be extraneous.*

29. $\left(\sqrt{4t-3}\right)^2 = t^2$

Square both sides.

$4t - 3 = t^2$

$t^2 - 4t + 3 = 0$

$(t - 3)(t - 1) = 0$

$t - 3 = 0$	$t - 1 = 0$
$t = 3$	$t = 1$

check: t = 3

$\sqrt{4(3) - 3} = 3$?

$\sqrt{12 - 3} = 3$; $\sqrt{9} = 3$; $3 = 3$ $\checkmark$

check: t = 1

$\sqrt{4(1) - 3} = 1$?

$\sqrt{4 - 3} = 1$; $\sqrt{1} = 1$; $1 = 1$ $\checkmark$

33. $\sqrt{x^2 - 5x} - 6 = 0$

$\sqrt{x^2 - 5x} = 6$

$\left(\sqrt{x^2 - 5x}\right)^2 = 6^2$

$x^2 - 5x = 36$

$x^2 - 5x - 36 = 0$

$(x - 9)(x + 4) = 0$

$x - 9 = 0$	$x + 4 = 0$
$x = 9$	$x = -4$

Before you can square both sides, the radical must be isolated on one side. Add 6 to both sides. Now square both sides.

check: x = 9 (yes)

check: x = − 4 (yes)

37. $\left(\sqrt{x - 1}\right)^2 = (x - 7)^2$

$x - 1 = x^2 - 14x + 49$

$x^2 - 15x + 50 = 0$

$(x - 10)(x - 5) = 0$

$x - 10 = 0$	$x - 5 = 0$
$x = 10$	$x = 5$

check: x = 10

$\sqrt{10 - 1} = 10 - 7$?

$\sqrt{9} = 3$

$3 = 3$ $\checkmark$

check: x = 5

$\sqrt{5 - 1} = 5 - 7$?

$\sqrt{4} = -2$

$2 = -2$ (false)

Notice that x = 5 does not check.   The only valid solution to the problem is x = 10.

41. $\left(\sqrt{3y^2 + 5y + 6}\right)^2 = (y + 3)^2$ check: $y = \frac{3}{2}$

$3y^2 + 5y + 6 = y^2 + 6y + 9$

$2y^2 - y - 3 = 0$

$(2y - 3)(y + 1) = 0$

$2y - 3 = 0 \qquad y + 1 = 0$

$y = \frac{3}{2} \qquad\qquad y = -1$

$\sqrt{3\left(\frac{3}{2}\right)^2 + 5\left(\frac{3}{2}\right) + 6} = \frac{3}{2} + 3$?

$\sqrt{3\left(\frac{9}{4}\right) + \frac{15}{2} + 6} = \frac{9}{2}$

$\sqrt{\frac{27}{4} + \frac{30}{4} + \frac{24}{4}} = \frac{9}{2}$

$\sqrt{\frac{81}{4}} = \frac{9}{2}$

$\frac{9}{2} = \frac{9}{2}$ ✓

check: $y = -1$

$\sqrt{3(-1)^2 + 5(-1) + 6} = -1 \overset{?}{+} 3$

$\sqrt{3 - 5 + 6} = 2$

$\sqrt{4} = 2$

$2 = 2$ ✓

45. $\left(\sqrt[3]{x + 1}\right)^3 = (-2)^3$ check: $\sqrt[3]{-9 + 1} = -2$?

$x + 1 = -8$ $\sqrt[3]{-8} = -2$

$x = -9$ $-2 = -2$ ✓

49. $\sqrt{3.7x + 31.69} = \sqrt{7.8x + 9.14}$ Square both sides.

$\left(\sqrt{3.7x + 31.69}\right)^2 = \left(\sqrt{7.8x + 9.14}\right)^2$

$3.7x + 31.69 = 7.8x + 9.14$

$-4.1x = -22.55$

$x = 5.5$

Problem Set 11.6

1. Let the number be x.

seven more than a number: $x + 7$

the square root of that quantity: $\sqrt{x + 7}$

$$\sqrt{x + 7} = 3 \qquad \text{Square both sides.}$$

check : x = 2

$$\left(\sqrt{x + 7}\right)^2 = 3^2$$
$$x + 7 = 9$$
$$x = 2$$

$$\sqrt{2 + 7} = 3 \quad ?$$
$$\sqrt{9} = 3$$
$$3 = 3 \quad \checkmark$$

5. Let the number be x; its square root is $\sqrt{x}$.

$$2\sqrt{x} = 18$$
$$\sqrt{x} = 9$$
$$\left(\sqrt{x}\right)^2 = 9^2$$
$$x = 81$$

check: x = 81

$$2\sqrt{81} = 18 \quad ?$$
$$2(9) = 18$$
$$18 = 18 \quad \checkmark$$

9. Let the number be x.
one more than the number: x + 1
the square root of one more than the number: $\sqrt{x + 1}$
five times that quantity: $5\sqrt{x+1}$
7 more than the number: x + 7

$$5\sqrt{x+1} = x + 7$$

check: x = 3

$$5\sqrt{3 + 1} = 3 + 7 \quad ?$$
$$5(2) = 10$$
$$10 = 10 \quad \checkmark$$

$$\left(5\sqrt{x+1}\right)^2 = (x + 7)^2$$
$$25(x + 1) = x^2 + 14x + 49$$
$$25x + 25 = x^2 + 14x + 49$$
$$x^2 - 11x + 24 = 0$$

check: x = 8

$$(x - 3)(x - 8) = 0$$

$$5\sqrt{8 + 1} = 8 + 7 \quad ?$$
$$5(3) = 15$$

$$x - 3 = 0 \qquad x - 8 = 0$$
$$x = 3 \qquad\quad x = 8$$

$$15 = 15 \quad \checkmark$$

13.
$$c^2 = a^2 + b^2$$
$$c = \sqrt{a^2 + b^2}$$

$$c = \sqrt{5^2 + 7^2}$$
$$c = \sqrt{25 + 49}$$
$$c = \sqrt{74}$$
$$c \approx 8.60 \text{ cm}$$

17.
$$c^2 = a^2 + b^2$$
$$25^2 = 24^2 + b^2$$
$$625 = 576 + b^2$$
$$b^2 = 49$$
$$b = 7 \text{ cm}$$

21. $d = \sqrt{(5-2)^2 + (11-7)^2}$

$= \sqrt{3^2 + 4^2}$

$= \sqrt{9 + 16}$

$= \sqrt{25} = 5$

25. $d = \sqrt{(0-(-1))^2 + (1-0)^2}$

$= \sqrt{1^2 + 1^2}$

$= \sqrt{2} \approx 1.41$

29. $d = \sqrt{(-1-5)^2 + (-1-(-2))^2}$

$= \sqrt{(-6)^2 + (1)^2}$

$= \sqrt{36 + 1}$

$= \sqrt{37} \approx 6.08$

33. Let w be the length of the wire.

$w^2 = 90^2 + 20^2$
$w^2 = 8100 + 400$
$w^2 = 8500$
$w = 10\sqrt{85} = \sqrt{8500} \approx 92.20$

The wire is about 92.2 meters in length.

37. $C = 30\sqrt{n + 45}$

(a) Use n = 0 and calculate C:

$C = 30\sqrt{0 + 45} = 30\sqrt{45}$
$\approx 30(6.7082039) \approx \$201.25$

(b) Use n = 20 and calculate C:

$C = 30\sqrt{20 + 45} = 30\sqrt{65}$
$\approx 30(8.0622577) \approx \$241.87$

(c) Use C = 360 and calculate n:

$360 = 30\sqrt{n + 45}$
$12 = \sqrt{n + 45}$
$12^2 = \left(\sqrt{n + 45}\right)^2$
$144 = n + 45$
$n = 99$ (99 stereo speaker boxes)

(d) Use C = 600 and calculate n:

$600 = 30\sqrt{n + 45}$
$20 = \sqrt{n + 45}$
$20^2 = \left(\sqrt{n + 45}\right)^2$
$400 = n + 45$
$n = 355n$ (355 stereo speaker boxes)

41. $p = 3\sqrt{\dfrac{L}{9.8}}$ Divide both sides by 3.

$\dfrac{p}{3} = \sqrt{\dfrac{L}{9.8}}$ Square both sides.

$\left(\dfrac{p}{3}\right)^2 = \left(\sqrt{\dfrac{L}{9.8}}\right)^2$

$\dfrac{p^2}{9} = \dfrac{L}{9.8}$ Multiply both sides by 9.8.

$9.8\left(\dfrac{p^2}{9}\right) = L$

$L = \dfrac{9.8p^2}{9}$

Problem Set 11.7

1a. $7^{\frac{1}{2}} = \sqrt{7}$ The exponent $\frac{1}{n}$ indicates the principal n^{th} root.

1b. $v^{\frac{1}{6}} = \sqrt[6]{v}$

1c. $x^{\frac{3}{4}} = \left(x^3\right)^{\frac{1}{4}} = \sqrt[4]{x^3}$ 1d. $w^{\frac{2}{5}} = \left(w^2\right)^{\frac{1}{5}} = \sqrt[5]{w^2}$

5. $4^{\frac{1}{2}} = \sqrt{4} = 2$ 9. $0^{\frac{1}{4}} = \sqrt[4]{0} = 0$

13.   When you have to evaluate a number raised to a fractional power, it may be easier to evaluate the root first.

$$64^{\frac{2}{3}} = \left((64)^{\frac{1}{3}}\right)^2 = \left(\sqrt[3]{64}\right)^2 = (4)^2 = 16$$

17. $144^{\frac{3}{2}} = \left((144)^{\frac{1}{2}}\right)^3 = \left(\sqrt{144}\right)^3 = 12^3 = 1728$

21. $(0.01)^{\frac{3}{2}} = \left((0.01)^{\frac{1}{2}}\right)^3 = \left(\sqrt{0.01}\right)^3 = \left(\sqrt{\dfrac{1}{100}}\right)^3 = \left(\dfrac{1}{10}\right)^3 = \dfrac{1}{1000} = 0.001$

25. $\left(\dfrac{8}{27}\right)^{\frac{2}{3}} = \left(\left(\dfrac{8}{27}\right)^{\frac{1}{3}}\right)^2 = \left(\dfrac{2}{3}\right)^2 = \dfrac{4}{9}$

29. $8^{-\frac{1}{3}} = \dfrac{1}{8^{\frac{1}{3}}} = \dfrac{1}{\sqrt[3]{8}} = \dfrac{1}{2}$

33a. $9^{\frac{1}{2}} + 16^{\frac{1}{2}} = \sqrt{9} + \sqrt{16} = 3 + 4 = 7$

33b. $\left(9 + 16\right)^{\frac{1}{2}} = \sqrt{9 + 16} = \sqrt{25} = 5$

37. $\dfrac{11^{\frac{5}{3}}}{11^{\frac{2}{3}}} = 11^{\frac{5}{3} - \frac{2}{3}} = 11^{\frac{3}{3}} = 11^{1} = 11$

41. $48^{\frac{1}{2}} \approx 49^{\frac{1}{2}} = \sqrt{49} = 7$ 　　　　The best response is D.

45. $x^{\frac{1}{3}}x^{\frac{2}{5}} = x^{\frac{1}{3} + \frac{2}{5}} = x^{\frac{5 + 6}{15}} = x^{\frac{11}{15}}$

49. $\left(x^{\frac{1}{3}}y^{\frac{2}{3}}\right)^{\frac{3}{4}} = \left(x^{\frac{1}{3}}\right)^{\frac{3}{4}}\left(y^{\frac{2}{3}}\right)^{\frac{3}{4}} = x^{\frac{1}{4}}y^{\frac{1}{2}}$

53. $\dfrac{x^{\frac{2}{5}} y^{\frac{3}{4}}}{x^{-\frac{1}{5}} y^{-\frac{1}{4}}} = x^{\frac{2}{5} + \frac{1}{5}} y^{\frac{3}{4} + \frac{1}{4}} = x^{\frac{3}{5}} y^{\frac{4}{4}} = x^{\frac{3}{5}} y$

57. $97^{\frac{2}{3}} \approx 21.11$

This can be done by several methods. If you have trouble with this, we suggest you ask your instructor or perhaps another student for help. Not all calculators work the same way and we do not want to confuse you with steps that do not work on your model.

| SUGGESTION | It will be very much worth your time to learn to use the features on your calculator.
You might want to form a calculator group with other students in the class who have the same model as yours. You can learn a lot from each other.

| SUGGESTION | If there is more than one way to work a problem, try it both ways. This has the advantage of checking against errors, and you will begin to learn which methods are best for which problems. You will also gain some flexibility in your approach to problem-solving, which is probably one of the most important skills you can develop.

1. $\sqrt{64} = \sqrt{8^2} = 8$

2. $\sqrt[3]{64} = \sqrt[3]{4^3} = 4$

3. $\sqrt[6]{64} = \sqrt[6]{2^6} = 2$

4. $\sqrt{0} = 0$

5. $\sqrt{196} = \sqrt{14^2} = 14$

6. $\sqrt{(-47)^2} = |-47| = 47$

7. $\sqrt{160,000} = \sqrt{400^2} = 400$

8. $\sqrt{0.0049} = \sqrt{(0.07)^2} = 0.07$

9. $\sqrt{\frac{4}{121}} = \sqrt{\left(\frac{2}{11}\right)^2} = \frac{2}{11}$

10. $\sqrt[3]{-125} = \sqrt[3]{(-5)^3} = -5$

11. $\sqrt[7]{-1} = -1$

12. $36^{\frac{1}{2}} = \sqrt{36} = 6$

13. $0.008^{\frac{1}{3}} = \sqrt[3]{0.008} = \sqrt[3]{(0.2)^3} = 0.2$

14. $160,000 = \sqrt[4]{160,000} = \sqrt[4]{16(10,000)} = \sqrt[4]{16}\ \sqrt[4]{10,000}$
    $= \sqrt[4]{2^4}\ \sqrt[4]{10^4} = 2(10) = 20$

15. $8000^{\frac{2}{3}} = \left(\sqrt[3]{8000}\right)^2 = 20^2 = 400$

16. $32^{\frac{4}{5}} = \left(32^{\frac{1}{5}}\right)^4 = \left(\sqrt[5]{32}\right)^4 = 2^4 = 16$

17. $\left(\frac{27}{8}\right)^{-\frac{2}{3}} = \left(\frac{8}{27}\right)^{\frac{2}{3}} = \left[\left(\frac{8}{27}\right)^{\frac{1}{3}}\right]^2$
    $= \left(\sqrt[3]{\frac{8}{27}}\right)^2 = \left(\frac{2}{3}\right)^2 = \frac{4}{9}$

18. $\left(\frac{16}{25}\right)^{-\frac{1}{2}} = \left(\frac{25}{16}\right)^{\frac{1}{2}} = \sqrt{\frac{25}{16}} = \frac{5}{4}$

19. $\sqrt{81} + \sqrt{1600} = 9 + 40 = 49$

20. $\sqrt{81 + 1600} = \sqrt{1681} = 41$

21. $\sqrt{v^2 + 10v + 25} = \sqrt{(v + 5)^2} = |v + 5| = v + 5$

   Note: $|v + 5| = v + 5$ only because it was given that all variables, including v, are positive. This means that $v + 5$ is also positive.

22. $\sqrt{4v^2 + 12v + 9} = \sqrt{(2v + 3)^2} = |2v + 3| = 2v + 3$

   $|2v + 3| = 2v + 3$ because $2v + 3 \geq 0$.

23. $\sqrt{45} = \sqrt{9 \cdot 5} = 3\sqrt{5}$

24. $\sqrt{48} = \sqrt{16 \cdot 3} = 4\sqrt{3}$

25. $\sqrt{2}\sqrt{3} = \sqrt{6}$

26. $\sqrt{3}\sqrt{15} = \sqrt{45} = \sqrt{9 \cdot 5} = 3\sqrt{5}$

   or: $\sqrt{3}\sqrt{15} = \sqrt{3}\sqrt{3}\sqrt{5} = 3\sqrt{5}$

27. $\sqrt{10}\sqrt{35} = \sqrt{10 \cdot 35} = \sqrt{2 \cdot 5 \cdot 5 \cdot 7} = 5\sqrt{14}$

28. $\sqrt{\frac{80}{5}} = \sqrt{16} = 4$

29. $\frac{\sqrt{75}}{\sqrt{12}} = \sqrt{\frac{75}{12}} = \sqrt{\frac{25}{4}} = \frac{5}{2}$

30. $\frac{12}{\sqrt{3}} = \frac{12}{\sqrt{3}} \cdot \frac{\sqrt{3}}{\sqrt{3}} = \frac{12\sqrt{3}}{3} = 4\sqrt{3}$

31. $13\sqrt{2} - \sqrt{2} = (13 - 1)\sqrt{2} = 12\sqrt{2}$

32. $3\sqrt{8} + 4\sqrt{50} = 3\sqrt{4 \cdot 2} + 4\sqrt{25 \cdot 2}$
$$= 3(2)\sqrt{2} + 4(5)\sqrt{2}$$
$$= 6\sqrt{2} + 20\sqrt{2}$$
$$= 26\sqrt{2}$$

Chapter Eleven Review

33. $7\sqrt{27} - 5\sqrt{3} + \sqrt{300}$

$= 7\sqrt{9 \cdot 3} - 5\sqrt{3} + \sqrt{100 \cdot 3}$

$= 7(3)\sqrt{3} - 5\sqrt{3} + 10\sqrt{3}$

$= 21\sqrt{3} - 5\sqrt{3} + 10\sqrt{3}$

$= 26\sqrt{3}$

34. $\sqrt{2}(5 - \sqrt{2})$

$= 5\sqrt{2} - (\sqrt{2})^2$

$= 5\sqrt{2} - 2$

35. $\sqrt{3}(\sqrt{6} + 2\sqrt{33})$

$= \sqrt{3}\sqrt{6} + 2\sqrt{3}\sqrt{33}$

$= \sqrt{3}\sqrt{3}\sqrt{2} + 2\sqrt{3}\sqrt{3}\sqrt{11}$

$= 3\sqrt{2} + 6\sqrt{11}$

36. $(\sqrt{2} - 1)(\sqrt{2} + 1)$ Use the form $a^2 - b^2 = (a + b)(a - b)$.

$= (\sqrt{2})^2 - 1^2$

$= 2 - 1$

$= 1$

37. $(5 + \sqrt{2})(3 - \sqrt{2})$ Use FOIL.

$= 15 - 5\sqrt{2} + 3\sqrt{2} - (\sqrt{2})^2$

$= 15 - 2\sqrt{2} - 2$

$= 13 - 2\sqrt{2}$

38. $(\sqrt{7} + \sqrt{11})(\sqrt{7} - \sqrt{11})$ Use the form $a^2 - b^2 = (a + b)(a - b)$.

$= (\sqrt{7})^2 - (\sqrt{11})^2$

$= 7 - 11$

$= -4$

39. $\dfrac{15}{\sqrt{7} + \sqrt{2}} \cdot \dfrac{\sqrt{7} - \sqrt{2}}{\sqrt{7} - \sqrt{2}} = \dfrac{15(\sqrt{7} - \sqrt{2})}{(\sqrt{7})^2 - (\sqrt{2})^2}$

$= \dfrac{15(\sqrt{7} - \sqrt{2})}{7 - 2} = \dfrac{15(\sqrt{7} - \sqrt{2})}{5} = 3(\sqrt{7} - \sqrt{2})$ or $3\sqrt{7} - 3\sqrt{2}$

40. $\sqrt{xy^2}\ \sqrt{x}$

$\qquad = \sqrt{x}\ \sqrt{y^2}\ \sqrt{x}$

$\qquad = xy \qquad\qquad$ (Since $x \geq 0$ and $y \geq 0$.)

41. $\dfrac{\sqrt{5x^3y^5}}{\sqrt{5xy}} = \sqrt{\dfrac{5x^3y^5}{5xy}} = \sqrt{x^2y^4} = xy^2$

42. $\sqrt{26v}\ \sqrt{65v}$

$\qquad = \sqrt{26v\ \cdot\ 65v}$

$\qquad = \sqrt{2(13)(5)(13)v^2}$

$\qquad = 13\sqrt{10}\ |v|$

$\qquad = 13\sqrt{10}\ v \quad$ or $\quad 13v\sqrt{10}$

43. $\sqrt{3x + 1} = 5 \qquad$ Square both sides.

$\quad 3x + 1 = 25$

$\quad 3x = 24$

$\qquad x = 8 \qquad$ *Be sure to check for extraneous solutions when solving radical equations.*

44. $\sqrt{3m + 5} = -1$

There can be no solution, because the principal square root is always nonnegative. There is no need to do a thing - there is no solution.

45. $\sqrt{v^2 - 3v + 6} = v \qquad$ Square both sides.

$\quad v^2 - 3v + 6 = v^2$

$\qquad -3v + 6 = 0$

$\qquad\qquad v = 2$

46. $\sqrt{y^2 - 6y} = -2y$

$\quad y^2 - 6y = (-2y)^2$

$$y^2 - 6y = 4y^2$$
$$-3y^2 - 6y = 0$$
$$3y^2 + 6y = 0$$
$$y^2 + 2y = 0$$
$$y(y + 2) = 0$$
$$y = 0, \ y = -2$$

47. $\sqrt{2v^2 - 17v + 17} - 7 = v$ Isolate the radical.

$$\sqrt{2v^2 - 17v + 17} = v + 7$$
$$2v^2 - 17v + 17 = (v + 7)^2$$
$$2v^2 - 17v + 17 = v^2 + 14v + 49$$
$$v^2 - 31v - 32 = 0$$
$$(v - 32)(v + 1) = 0$$
$$v = 32, \ v = -1$$

48. $\sqrt{y + 1} = \sqrt{y^2 + 3y - 2}$

$$y + 1 = y^2 + 3y - 2$$
$$y^2 + 2y - 3 = 0$$
$$(y + 3)(y - 1) = 0$$
$$y = -3 \quad y = 1$$

           Solution: $y = 1$

           $y = -3$ is extraneous

49. $\sqrt{9} + \sqrt{16} + \sqrt{144} = 3 + 4 + 12 = 19$

50. $\sqrt{9 + 16 + 144} = \sqrt{169} = 13$

51. $\sqrt{9} - \sqrt{16}\ \sqrt{144} = 3 - 4(12) = 3 - 48 = -45$

52. $(\sqrt{9} - \sqrt{16})(\sqrt{144} = (3 - 4)(12) = -12$

53. $x^{\frac{1}{4}} x^{\frac{2}{3}} = x^{\frac{1}{4} + \frac{2}{3}} = x^{\frac{3}{12} + \frac{8}{12}} = x^{\frac{11}{12}}$

54. $\dfrac{x^{\frac{2}{3}}}{x^{\frac{1}{4}}} = x^{\frac{2}{3} - \frac{1}{4}} = x^{\frac{8}{12} - \frac{3}{12}} = x^{\frac{5}{12}}$

55. $\left(x^{\frac{2}{3}}\right)^{\frac{1}{4}} = x^{\left(\frac{2}{3} \cdot \frac{1}{4}\right)} = x^{\frac{1}{6}}$ Notice that the exponents are multiplied.

56. $\left(x^{-\frac{3}{5}}\right)^{-\frac{5}{6}} = x^{\left(\frac{3}{5} \cdot \frac{5}{6}\right)} = x^{\frac{3}{6}} = x^{\frac{1}{2}}$

57. $\dfrac{x^{\frac{7}{12}}}{x^{-\frac{1}{4}}} = x^{\left(\frac{7}{12} + \frac{1}{4}\right)} = x^{\left(\frac{7}{12} + \frac{3}{12}\right)} = x^{\frac{10}{12}} = x^{\frac{5}{6}}$

58. $\left(x^{\frac{1}{2}}x^{\frac{1}{3}}\right)^{\frac{3}{5}}$

$= \left(x^{\frac{1}{2} + \frac{1}{3}}\right)^{\frac{3}{5}}$

$= \left(x^{\frac{5}{6}}\right)^{\frac{3}{5}}$

$= x^{\frac{5}{6} \cdot \frac{3}{5}} = x^{\frac{1}{2}}$

59. $\sqrt{109} \approx 10.44$ 60. $85^{\frac{1}{2}} = \sqrt{85} \approx 9.22$

61. $90^{\frac{1}{3}} = \sqrt[3]{90} \approx 4.48$

62. Answer : ≈ 36.41.

$400^{\frac{3}{5}}$ can be rewritten as $\left(\sqrt[5]{400}\right)^{3}$ or $\sqrt[5]{400^{3}}$.

Or, if your calculator has a y^{x} key, use 400 for y and 0.6 for x.

63. $\sqrt{121} - \sqrt{169} = 11 - 13 = -2$ 64. $\sqrt{64} - \sqrt[3]{64} = 8 - 4 = 4$

65. Let n be the number.
$\sqrt{2n + 5} = 7$
$2n + 5 = 49$
$2n = 44$
$n = 22$

Chapter Eleven Review

66. $2^2 + h^2 = 8^2$

$h^2 = 60$

$h = \sqrt{60} = 2\sqrt{15}$ m ≈ 7.75 m

67. Use the distance formula:

$$h = \sqrt{(-8 - 4)^2 + (8 + 1)^2}$$

$$= \sqrt{(-12)^2 + 9^2}$$

$$= \sqrt{144 + 81}$$

$$= \sqrt{225}$$

$$= 15$$

68a. $C = 6\sqrt{0 + 400} = 6\sqrt{400} = 6(20) = 120$ dollars

b. $C = 6\sqrt{129 + 400} = 6\sqrt{529} = 6(23) = 138$ dollars

c. $192 = 6\sqrt{n + 400}$

$32 = \sqrt{n + 400}$

$1024 = n + 400$

$n = 624$ (cables)

69. See answer section in the text.

70. $\sqrt{25 + 144} = \sqrt{169} = 13$

It is incorrect to write $\sqrt{25 + 144} = \sqrt{25} + \sqrt{144}$. There is no such rule.

71. See answer section in the text.

72. $\sqrt{16 - 1} = \sqrt{15}$

It is incorrect to write $\sqrt{16 - 1} = \sqrt{16} - \sqrt{1}$. There is no such rule.

Solutions To All Mastery Test Problems – Chapter Eleven

| 1a. | 28 | 1b. | 13 | 1c. | 10 | 1d. | 0.2 |
| 2a. | 4.24 | 2b. | 21.47 | 2c. | 45.27 | 2d. | 0.12 |

Chapter Eleven Mastery Test

3a. $\sqrt{33}$ **3b.** $x + 3$ **3c.** 1 **3d.** $\sqrt{2x}$

4a. $\sqrt{16(2)} = 4\sqrt{2}$ **4b.** $\sqrt{5(121)} = 11\sqrt{5}$

4c. $\sqrt{9} = 3$ **4d.** $\dfrac{9}{12} = \dfrac{3}{4}$

5a. $3\sqrt{29}$ **5b.** $-3\sqrt{31}$

5c. $\sqrt{4(5)} + 3\sqrt{100(5)}$
$= 2\sqrt{5} + 30\sqrt{5}$
$= 32\sqrt{5}$

5d. $2\sqrt{25(3)} - 4\sqrt{25(2)} + 7\sqrt{16(3)}$
$= 10\sqrt{3} - 20\sqrt{2} + 28\sqrt{3}$
$= 38\sqrt{3} - 20\sqrt{2}$

6a. $\sqrt{5}\sqrt{3} - \sqrt{5}\sqrt{7}$
$= \sqrt{15} - \sqrt{35}$

6b. $\sqrt{(15)(5)} + \sqrt{15(3)}$
$= 5\sqrt{3} + 3\sqrt{5}$

6c. $(\sqrt{17})^2 - (\sqrt{3})^2$
$= 17 - 3 = 14$

6d. $2(\sqrt{7})^2 - 8\sqrt{7}\sqrt{5} + 3\sqrt{5}\sqrt{7} - 12(\sqrt{5})^2$
$= 2(7) - 5\sqrt{35} - 12(5)$
$= -46 - 5\sqrt{35}$

7a. $\dfrac{24(\sqrt{6})}{\sqrt{6}\sqrt{6}} = \dfrac{24\sqrt{6}}{6}$

$= 4\sqrt{6}$

7b. $\dfrac{\sqrt{6}\sqrt{10}}{\sqrt{10}\sqrt{10}} = \dfrac{\sqrt{2(3)(2)(5)}}{10}$

$= \dfrac{2\sqrt{15}}{10} = \dfrac{\sqrt{15}}{5}$

8a. $\sqrt{4(14)}$
$= 2\sqrt{14}$

8b. $8\sqrt{7} + 2\sqrt{9(7)}$
$= 8\sqrt{7} + 6\sqrt{7} = 14\sqrt{7}$

8c. $\sqrt{5(11)(7)(11)} = 11\sqrt{35}$

8d. $\sqrt{\dfrac{5}{6}} = \dfrac{\sqrt{5}}{\sqrt{6}} = \dfrac{\sqrt{5}\sqrt{6}}{\sqrt{6}\sqrt{6}} = \dfrac{\sqrt{30}}{6}$

9a. $\sqrt{2x+1} = 3$
$2x + 1 = 9$
$2x = 8$
$x = 4$

9b. $\sqrt{3m+4} = 7$
$3m + 4 = 49$
$3m = 45$
$m = 15$

> *Don't forget to check for extraneous solutions!*

page 289

Chapter Eleven Mastery Test

9c.

$\sqrt{a^2 + 3a - 15} = a$

$a^2 + 3a - 15 = a^2$

$3a - 15 = 0$

$3(a - 5) = 0$

$a - 5 = 0$

$a = 5$

9d.

$\sqrt{v + 9} = v - 3$

$v + 9 = (v - 3)^2$

$v + 9 = v^2 - 6v + 9$

$v^2 - 7v = 0$

$v(v - 7) = 0$

$v = 0;\ v = 7$ *Only v = 7 checks.*

10a.   Let the two consecutive natural numbers be n and n + 1; their sum is n + (n + 1).

$\sqrt{n + (n + 1)} = n - 7$ (Notice that n must be the smaller number.)

$2n + 1 = n^2 - 14n + 49$ Both sides have been squared.

$n^2 - 16n + 48 = 0$

$(n - 12)(n - 4) = 0$

$n = 12;\ n = 4$ *Only n = 12 checks. Answer: The numbers are 12 and 13.*

10b.   Let the height of the kite be represented by h.

$h^2 + 25^2 = 75^2$

$h^2 + 625 = 5625$

$h^2 = 5000$

$h = \sqrt{5000} = \sqrt{2500(2)}$

$\quad = 50\sqrt{2} \approx 70.71$ (meters)

10c.

$90 = 6\sqrt{n + 80}$

$15 = \sqrt{n + 80}$

$15^2 = n + 80$

$225 = n + 80$

$n = 145$ (units)

11a. $d = \sqrt{(-10 - 2)^2 + (12 - 7)^2}$

$\quad = \sqrt{144 + 25} = \sqrt{169} = 13$

11b. $d = \sqrt{(1 - (-1))^2 + (-3 - (-1))^2}$

$\quad = \sqrt{4 + 4} = \sqrt{8} = 2\sqrt{2}$

12a. $\sqrt{10,000} = \sqrt{(100)^2} = 100$

12b. $\sqrt[4]{10,000} = \sqrt[4]{(10)^4} = 10$

12c. $[\,(32)^{\frac{1}{5}}\,]^3 = [\,\sqrt[5]{32}\,]^3 = 2^3 = 8$

12d. $\left(\frac{125}{8}\right)^{\frac{2}{3}} = \left(\sqrt[3]{\frac{125}{8}}\right)^2 = \left(\frac{5}{2}\right)^2 = \frac{25}{4}$

Chapter Eleven Mastery Test

Problem Set 12.1

1. See answer section in the text.

   Take the square root of both sides.

5. $v^2 - 13 = 0$

   $v^2 = 13$

   $v = -\sqrt{13}, \sqrt{13}$

9. $x^2 - 0.36 = 0$

   $x^2 = 0.36$

   $x = -\sqrt{0.36} \quad = -0.6$

   $x = \sqrt{0.36} \quad = 0.6$

13. $3m^2 + 5 = 17$

    $3m^2 = 12$

    $m^2 = 4$

    $m = -2, 2$

17. $4v^2 - 201 = 999$

    $4v^2 = 1200$

    $v^2 = 300$

    $v = \pm \sqrt{300} = \pm 10\sqrt{3}$

    answer: $-10\sqrt{3}$, $10\sqrt{3}$

21. $7b^2 + 2 = 10$

    $7b^2 = 8$

    $b^2 = \frac{8}{7}$

    $b = \pm \sqrt{\frac{8}{7}} = \pm \frac{2\sqrt{2}}{\sqrt{7}} = \pm \frac{2\sqrt{2}\sqrt{7}}{\sqrt{7}\sqrt{7}}$

    $= \pm \frac{2\sqrt{14}}{7}$

    Answer: $-\frac{2\sqrt{14}}{7}$, $\frac{2\sqrt{14}}{7}$

25. $(x - 3)^2 = 4$

    $x - 3 = 2$ or $x - 3 = -2$

    $x = 5$ $x = 1$

29. $(z + 6)^2 = 8$

    $z + 6 = \sqrt{8}$ or $z + 6 = -\sqrt{8}$

    $z + 6 = 2\sqrt{2}$ $z + 6 = -2\sqrt{2}$

    $z = -6 + \sqrt{2}$ $z = -6 - 2\sqrt{2}$

33. $(5y - 1)^2 = 20$

    $5y - 1 = \sqrt{20}$ or $5y - 1 = -\sqrt{20}$

    $5y - 1 = 2\sqrt{5}$ $5y - 1 = -2\sqrt{5}$

    $5y = 1 + 2\sqrt{5}$ $5y = 1 - 2\sqrt{5}$

    $y = \frac{1}{5} + \frac{2\sqrt{5}}{5}$ $y = \frac{1}{5} - \frac{2\sqrt{5}}{5}$

37.   We may estimate by solving $x^2 = 100$, which has solutions -10 and 10.   Answer: B

41.   Let n be the number.
       1 more than twice the number: $1 + 2n$
       the square of that quantity: $(1 + 2n)^2$

       $(1 + 2n)^2 = 169$
       $1 + 2n = \sqrt{169} = 13$ or $1 + 2n = -\sqrt{169} = -13$
       $2n = 12$ $2n = -14$
       $n = 6$ $n = -7$

45. $V = \pi r^2 h$
       $3200\pi = \pi r^2 (8)$ Divide both sides by 8π.
       $400 = r^2$
       $r = 20$ or $r = -20$ The radius is 20 cm. (We reject $r = -20$, since radius is nonnegative.)

49. $(2y + 1)^2 = 39.5641$

       $2y + 1 = \sqrt{39.5641}$ or $2y + 1 = -\sqrt{39.5641}$
       $2y + 1 = 6.29$ $2y + 1 = -6.29$
       $2y = 5.29$ $2y = -7.29$
       $y \approx 2.65$ $y \approx -3.65$

53. $(2x - 3)^2 = y$
       $2x - 3 = \sqrt{y}$ or $2x - 3 = -\sqrt{y}$
       $2x = \sqrt{y} + 3$ $2x = -\sqrt{y} + 3$

       $x = \dfrac{\sqrt{y}}{2} + \dfrac{3}{2}$ $x = -\dfrac{\sqrt{y}}{2} + \dfrac{3}{2}$

Problem Set 12.2

1. $x^2 - 8x + 16 = x^2 - 2(x)(4) + 4^2 = (x - 4)^2$

5. $v^2 + v + \frac{1}{4} = v^2 + 2(v)\left(\frac{1}{2}\right) + \left(\frac{1}{2}\right)^2 = (v + \frac{1}{2})^2$

9.   (1) Take half the coefficient of the x-term: $\dfrac{-14}{2} = -7$

       (2) Square that amount: $c = (-7)^2 = 49$

13. $\left(\dfrac{20}{2}\right)^2 = 10^2 = 100$ 17. $\left(\dfrac{-5}{2}\right)^2 = \dfrac{25}{4}$

21. $y^2 + 10y + 25 = \frac{1}{4}$ The left side is a perfect square trinomial.

       $(y + 5)^2 = \frac{1}{4}$

       $y + 5 = \frac{1}{2}$ or $y + 5 = -\frac{1}{2}$

       $y = -\frac{9}{2}$ $y = -\frac{11}{2}$

25. $v^2 + 6v = -5$ $\left(\frac{6}{2}\right)^2 = 3^2 = 9$ Add 9 to both sides.

       $v^2 + 6v + 9 = -5 + 9$

       $(v + 3)^2 = 4$

       $v + 3 = 2$ or $v + 3 = -2$

       $v = -1$ $v = -5$

29. $m^2 - \frac{3}{2}m = 1$

       $m^2 - \frac{3}{2}m + \left(-\frac{3}{4}\right)^2 = 1 + \left(-\frac{3}{4}\right)^2$

       $m^2 - \frac{3}{2}m + \left(-\frac{3}{4}\right)^2 = 1 + \frac{9}{16}$

       $\left(m - \frac{3}{4}\right)^2 = \frac{25}{16}$

       $m - \frac{3}{4} = \frac{5}{4}$ or $m - \frac{3}{4} = -\frac{5}{4}$

       $m = 2$ or $m = -\frac{1}{2}$

33. $2y^2 - 8y + 2 = 0$ Divide both sides by 2. The coefficient of y^2 must be 1 before you can complete the square.

       $y^2 - 4y + 1 = 0$ Move the constant to the right side.

       $y^2 - 4y = -1$

       $y^2 - 4y + 4 = -1 + 4$

       $(y - 2)^2 = 3$

       $y - 2 = \sqrt{3}$ or $y - 2 = -\sqrt{3}$

       $y = 2 + \sqrt{3}$ or $y = 2 - \sqrt{3}$

37. $4x^2 - 4x - 1 = 0$

       $4x^2 - 4x = 1$ Divide both sides by 4. To complete the square you need a coefficient of 1 for x^2.

       $x^2 - x = \frac{1}{4}$

       $x^2 - x + \frac{1}{4} = \frac{1}{4} + \frac{1}{4}$ $\left(\frac{-1}{2}\right)^2 = \frac{1}{4}$, which was added to both sides.

$$\left(x - \tfrac{1}{2}\right)^2 = \tfrac{1}{2}$$

$$x - \tfrac{1}{2} = \sqrt{\tfrac{1}{2}} \qquad \text{or} \qquad x - \tfrac{1}{2} = -\sqrt{\tfrac{1}{2}}$$

$$x - \tfrac{1}{2} = \frac{\sqrt{2}}{2} \qquad\qquad x - \tfrac{1}{2} = -\frac{\sqrt{2}}{2}$$

$$x = \tfrac{1}{2} + \frac{\sqrt{2}}{2} \qquad\qquad x = \tfrac{1}{2} - \frac{\sqrt{2}}{2}$$

41. $(y - 3)(y + 2) = 1$ Expand the left side.

$$y^2 - y - 6 = 1$$
$$y^2 - y = 7$$
$$y^2 - y + \tfrac{1}{4} = 7 + \tfrac{1}{4} \qquad\qquad \left(\frac{-1}{2}\right)^2 = \tfrac{1}{4}$$

$$\left(y - \tfrac{1}{2}\right)^2 = \frac{29}{4}$$

$$y - \tfrac{1}{2} = \sqrt{\frac{29}{4}} \qquad \text{or} \qquad y - \tfrac{1}{2} = -\sqrt{\frac{29}{4}}$$

$$y = \tfrac{1}{2} + \frac{\sqrt{29}}{2} \qquad\qquad y = \tfrac{1}{2} - \frac{\sqrt{29}}{2}$$

45. $\sqrt{35} \approx \sqrt{36} = 6$
$$-7 + \sqrt{35} \approx -7 + 6 = -1$$
$$-7 - \sqrt{35} \approx -7 - 6 = -13 \qquad\qquad \text{Answer: B}$$

49. $2v^2 = 4v - 1$
$$v^2 = 2v - \tfrac{1}{2}$$
$$v^2 - 2v = -\tfrac{1}{2}$$
$$v^2 - 2v + 1 = -\tfrac{1}{2} + 1$$
$$(v - 1)^2 = \tfrac{1}{2}$$
$$v - 1 = \sqrt{\tfrac{1}{2}} = \frac{\sqrt{2}}{2} \qquad v - 1 = -\frac{\sqrt{2}}{2}$$

$$v = 1 + \frac{\sqrt{2}}{2} \qquad\qquad v = 1 - \frac{\sqrt{2}}{2}$$

$$v \approx 1.71 \qquad\qquad v \approx 0.29$$

53. The width and length of the room make up the width and length of a right triangle with the wire being the hypotenuse. Let ℓ be the length of the wire.

$\ell^2 = 14^2 + 20^2 = 596$
$\ell = \sqrt{596} = 2\sqrt{149} \approx 24.4$ The length of the wire is about 24.4 feet.

Problem Set 12.3

1. $x^2 + 5x - 24 = 0.$ See answer section in the text.

5. $(x - 3)(x + 9) = -20$
    $x^2 + 6x - 27 = -20$
    $x^2 + 6x - 7 = 0$ See answer section in the text.

9. $a = 6, b = -5$ and $c = 1$

$$\frac{-(-5) \pm \sqrt{(-5)^2 - 4(6)(1)}}{2(6)} = \frac{5 \pm \sqrt{25 - 24}}{12} = \frac{5 \pm 1}{12}$$

$\dfrac{5 + 1}{12} = \dfrac{6}{12} = \dfrac{1}{2}$ $\dfrac{5 - 1}{12} = \dfrac{4}{12} = \dfrac{1}{3}$ Answer: $\dfrac{1}{2}, \dfrac{1}{3}$

13. $2m^2 - 7m + 4 = 0$ $a = 2, b = -7, c = 4$

$$\frac{-(-7) \pm \sqrt{(-7)^2 - 4(2)(4)}}{2(2)} = \frac{7 \pm \sqrt{49 - 32}}{4} = \frac{7 \pm \sqrt{17}}{4}$$

Answer: $\dfrac{7 + \sqrt{17}}{4}, \dfrac{7 - \sqrt{17}}{4}$

17. $a = 1, b = -4, c = 2$

$$\frac{-(-4) \pm \sqrt{(-4)^2 - 4(1)(2)}}{2(1)} = \frac{4 \pm \sqrt{16 - 8}}{2} = \frac{4 \pm \sqrt{8}}{2} = \frac{4 \pm 2\sqrt{2}}{2}$$

$$= \frac{2(2 \pm \sqrt{2})}{2} = 2 \pm \sqrt{2}$$ Answer: $2 + \sqrt{2} , 2 - \sqrt{2}$

21. $a = 4, b = 0, c = -7$

$$\frac{0 \pm \sqrt{0^2 - 4(4)(-7)}}{2(4)} = \frac{\pm \sqrt{16 \cdot 7}}{8} = \pm \frac{4\sqrt{7}}{8} = \pm \frac{\sqrt{7}}{2}$$

Answer: $\dfrac{\sqrt{7}}{2} , -\dfrac{\sqrt{7}}{2}$

25. $m^2 + 4m - 32 = 0$

$(m + 8)(m - 4) = 0$

$m + 8 = 0 \qquad m - 4 = 0$

$m = -8 \qquad m = 4$

Factoring is often the easiest, if it can be done. Consider factoring before trying methods that are more involved.

29. $x^2 - 4x - 3 = 0$ (does not factor) $\qquad a = 1, b = -4, c = -3$

$$\frac{-(-4) \pm \sqrt{(-4)^2 - 4(1)(-3)}}{2(1)} = \frac{4 \pm \sqrt{16 + 12}}{2} = \frac{4 \pm \sqrt{28}}{2} = \frac{4 \pm 2\sqrt{7}}{2}$$

$$= \frac{2(2 \pm \sqrt{7})}{2} = 2 \pm \sqrt{7}$$

Answer: $2 + \sqrt{7}$, $2 - \sqrt{7}$

33. $(w - 1)(w - 5) = -4$

$w^2 - 6w + 5 = -4$

$w^2 - 6w + 9 = 0$

$(w - 3)^2 = 0$

$w = 3$ (a double root)

Expand and collect like terms.

This is a perfect square trinomial.

37. $\dfrac{z^2}{144} = 0$ $\qquad$ Multiply both sides by 144.

$144\left(\dfrac{z^2}{144}\right) = 144(0)$

$z^2 = 0$

$z = 0$ (double root)

41. $x^2 + 2x + 2 = 0$

The directions call for the quadratic formula.
$a = 1, b = 2, c = 2$

$$x = \frac{-2 \pm \sqrt{2^2 - 4(1)(2)}}{2(1)} \qquad\qquad x = \frac{-b \pm \sqrt{b^2 - 4ac}}{2a}$$

Since $b^2 - 4ac = 2^2 - 4(1)(2) = -4$, there is no real solution.

45. $3x^2 + 3x - 1 = 0 \qquad a = 3, b = 3, c = -1$

$$\frac{-3 \pm \sqrt{3^2 - 4(3)(-1)}}{2(3)} = \frac{-3 \pm \sqrt{9 + 12}}{6} = \frac{-3 \pm \sqrt{21}}{6}$$

$$\frac{-3 + \sqrt{21}}{6} \approx = 0.26 \qquad\qquad \frac{-3 - \sqrt{21}}{6} \approx = -1.26$$

49. The area of the wasted metal is what's left after we remove a circle from a square.

area of the square − area of the circle = 124 cm^2

$$(2r)^2 \ - \ \pi r^2 = 124$$
$$4r^2 \ - \ \pi r^2 = 124$$
$$r^2(4 - \pi) = 124$$

$$r^2 = \frac{124}{4 - \pi}$$
$$r = \sqrt{\frac{124}{4 - \pi}} \approx 12.0 \qquad \text{The radius is approximately 12.0 cm.}$$

Problem Set 12.4

1. Let x be the smaller number and x + 8 be the larger number.
$$x\,(x + 8) = 5$$
$$x^2 + 8x = 5$$
$$x^2 + 8x - 5 = 0 \qquad \text{This does not factor, so the quadratic formula will be used.}$$

$$\frac{-8 \ \pm \ \sqrt{8^2 - 4(1)(-5)}}{2(1)} = \frac{-8 \pm \sqrt{64 + 20}}{2} = \frac{-8 \pm \sqrt{84}}{2} = \frac{-8 \pm 2\sqrt{21}}{2}$$

$$= \frac{2(-4 \pm \sqrt{21})}{2} = -4 \pm \sqrt{21}$$

Remember, these are two values for x, *not the two numbers referred to in the problem.*
For each value of x, there is a value of x + 8 :

$$x = -4 + \sqrt{21} \qquad\qquad x + 8 = (-4 + \sqrt{21}) + 8 \ = 4 + \sqrt{21}.$$
$$x = -4 - \sqrt{21} \qquad\qquad x + 8 = \ (-4 - \sqrt{21}) + 8 \ = 4 - \sqrt{21}.$$

5. Let x be the number.
its square: x^2
3 less than the square: $x^2 - 3$

$$x = x^2 - 3$$
$$x^2 - x - 3 = 0$$

$$\frac{-(-1) \ \pm \ \sqrt{(-1)^2 - 4(1)(-3)}}{2(1)} = \frac{1 \pm \sqrt{1 + 12}}{2} = \frac{1 \pm \sqrt{13}}{2}$$

$$\text{Answer: } \frac{1 + \sqrt{13}}{2} \ \text{or} \ \frac{1 - \sqrt{13}}{2}$$

9. Let the smaller number be x and the larger number be x + 1.

Their reciprocals: $\frac{1}{x}$, $\frac{1}{x+1}$

$\frac{1}{x} + \frac{1}{x+1} = 1$ Multiply both sides by x(x +1) to remove the fractions.

$x(x+1)\left(\frac{1}{x} + \frac{1}{x+1}\right) = x(x+1)(1)$

$(x+1) \quad + \quad x \quad = x(x+1)$

$2x + 1 = x^2 + x$

$x^2 - x - 1 = 0$

$\frac{-(-1) \pm \sqrt{(-1)^2 - 4(1)(-1)}}{2(1)} = \frac{1 \pm \sqrt{1+4}}{2} = \frac{1 \pm \sqrt{5}}{2} = \frac{1}{2} \pm \frac{\sqrt{5}}{2}$

Remember, these are two possibilities for x, *not the two numbers referred to in the problem.*
For each value of x, the other number is x + 1 :

$x = \frac{1}{2} + \frac{\sqrt{5}}{2}$ and $x + 1 = \left(\frac{1}{2} + \frac{\sqrt{5}}{2}\right) + 1 = \frac{3}{2} + \frac{\sqrt{5}}{2}$

$x = \frac{1}{2} - \frac{\sqrt{5}}{2}$ and $x + 1 = \left(\frac{1}{2} - \frac{\sqrt{5}}{2}\right) + 1 = \frac{3}{2} - \frac{\sqrt{5}}{2}$

13. Let s be the length of a side of the computer board. The area of a square is the square of the length of a side:

$s^2 = 70$
$s = \pm \sqrt{70}$ We reject $-\sqrt{70}$ because s represents length and cannot be negative.

Answer: $\sqrt{70}$ cm $\approx$ 8.37 cm

17. The area of a triangle is one-half the product of the base and altitude (height).
Let b = the measure of the base and b − 4 = the measure of the altitude.

$25 = \frac{1}{2}b(b-4)$ Multiply both sides by 2 to eliminate fractions.

$50 = b(b-4)$

$50 = b^2 - 4b$

$b^2 - 4b - 50 = 0$

$\frac{-(-4) \pm \sqrt{(-4)^2 - 4(1)(-50)}}{2(1)} = \frac{4 \pm \sqrt{16+200}}{2} = \frac{4 \pm \sqrt{216}}{2} = \frac{4 \pm \sqrt{36 \cdot 6}}{2}$

$\frac{4 \pm 6\sqrt{6}}{2} = \frac{2(2 \pm 3\sqrt{6})}{2} = 2 \pm 3\sqrt{6}$

Answer: The base has length $2 + 3\sqrt{6} \approx 9.35$ cm. $(2 - 3\sqrt{6}$ is negative)

21.   Let d be the distance up the wall to the top of the ladder.

$$d^2 + 5^2 = 16^2$$ Use the Pythagorean Theorem.

$$d^2 + 25 = 256$$

$$d^2 = 231$$

$$d = \sqrt{231} \approx 15.20 \text{ (feet)}$$

25.   Let r = the rate of the eastbound plane and $r + 50$ = rate of the southbound plane.

Notice that the triangle is a picture of their paths, and the lengths of the sides correspond to the *distances* involved. Since distance = rate · time, we can write expressions for the distance each plane has traveled. These distances will be the lengths of the legs of the triangle, and we can apply the Pythagorean Theorem.

distance for the eastbound plane = $(r \text{ mi/h})(\frac{1}{2} \text{ hr}) = \frac{r}{2}$ miles

distance for the southbound plane = $(r + 50 \text{ mi/h})(\frac{1}{2} \text{ hr}) = \frac{1}{2}(r + 50)$ miles

$$\left(\frac{r}{2}\right)^2 + \left(\frac{r + 50}{2}\right)^2 = 230^2$$ Apply the Pythagorean Theorem.

$$\frac{r^2}{4} + \frac{(r + 50)^2}{4} = 52,900$$ Multiply both sides by 4.

$$r^2 + (r + 50)^2 = 211,600$$

$$r^2 + r^2 + 100r + 2500 = 211,600$$

$$2r^2 + 100r - 209,100 = 0$$ Divide both sides by 2.

$$r^2 + 50r - 104,550 = 0$$

$$\frac{-50 \pm \sqrt{50^2 - 4(1)(-104,550)}}{2(1)} = \frac{-50 \pm \sqrt{2500 + 418,200}}{2} = \frac{-50 \pm \sqrt{420,700}}{2}$$

$$= \frac{-50 \pm \sqrt{100 \cdot 4207}}{2} = \frac{-50 \pm 10\sqrt{4207}}{2} = -25 \pm 5\sqrt{4207}$$

The positive value is approximated by 299.31.

Answer: the eastbound plane's rate is $-25 + 5\sqrt{4207} \approx 299.31$ mi/h, and
the southbound plane's rate is $(-25 + 5\sqrt{4207}) + 50 = 25 + 5\sqrt{4207} \approx 349.31$ mi/h.

29.   Let r = the rate of the slower bicycle and $r + 4$ = the rate of the faster bicycle.
Notice that 20 minutes = $\frac{1}{3}$ hour. *Be sure to use consistent units!*

page 299

section   12.4

	rate	×	time	=	distance
slower bicycle	r		$\frac{70}{r}$		70
faster bicycle	r + 4		$\frac{70}{r + 4}$		70

Since rate × time = distance, time = $\frac{distance}{rate}$.

The difference in their times is $\frac{1}{3}$ hour:

$\left(\text{larger time}\right) - \left(\text{smaller time}\right) = \frac{1}{3}$ This means that we can write:

$\left(\text{time for the } \textit{slower} \text{ bike}\right) - \left(\text{time for the } \textit{faster} \text{ bike}\right) = \frac{1}{3}$

$\frac{70}{r} - \frac{70}{r + 4} = \frac{1}{3}$ Multiply both sides by $3r(r + 4)$ to remove the fractions.

$3r(r + 4)\left(\frac{70}{r} - \frac{70}{r + 4}\right) = 3r(r + 4)\left(\frac{1}{3}\right)$

$3(70)(r + 4) - 3(70)r = r(r + 4)$

$210r + 840 - 210r = r^2 + 4r$

$r^2 + 4r - 840 = 0$ The quadratic formula will give: $r = -2 \pm 2\sqrt{211}$

Since this represents a rate, we reject the negative value.
The rate of the slower bicycle is $-2 + 2\sqrt{211} \approx 27.05$ km/h;
the rate of the faster bicycle is $(-2 + 2\sqrt{211}) + 4 = 2 + 2\sqrt{211} \approx 31.05$ km/h.

33.   t = time required for the faster belt
t + 6 = time required for the slower belt

$$\boxed{rate\ of\ work = \frac{1}{time\ required\ alone}}$$

	rate	×	time	=	work done
faster belt	$\frac{1}{t}$		4		$\frac{4}{t}$
slower belt	$\frac{1}{t + 6}$		4		$\frac{4}{t + 6}$

$\left(\text{work done by the faster belt}\right) + \left(\text{work done by the slower belt}\right) = 1$ job done

$\frac{4}{t} + \frac{4}{t + 6} = 1$ Multiply both sides by $t(t + 6)$ to remove the fractions.

$$t(t + 6)\left(\frac{4}{t} + \frac{4}{t + 6}\right) = t(t + 6)(1)$$

$$4(t + 6) + 4t = t(t + 6)$$

$$4t + 24 + 4t = t^2 + 6t$$

$$t^2 - 2t - 24 = 0$$

$$(t - 6)(t + 4) = 0$$

$$t - 6 = 0 \qquad t + 4 = 0$$

$$t = 6 \qquad t = -4 \qquad\qquad \text{(we reject } t = -4\text{)}$$

time required of the faster belt: 6 hours
time required of the slower belt: $6 + 6 = 12$ hours

Problem Set 12.5
Note: See answer section in the text for the graphs.

1.   All parabolas of the form $y = ax^2$ have vertex $(0,0)$.   Since $a = 2$, the parabola opens upward and is narrower than the graph of $y = x^2$.

5. $y = (x -(-5))^2$. We have $a = 1$, which indicates that the graph of this parabola will have the same shape as the graph of $y = x^2$. Since $h = -5$ and $k = 0$ it will be shifted left 5 units. The vertex will be at $(-5,0)$.

9. $y = 2(x - 4)^2$. We have $a = 2$, which indicates that the graph of this parabola will open upward and be narrower than the graph of $y = x^2$. Since $h = 4$ and $k = 0$, the parabola will be shifted to the right 4 units (compared to $y = x^2$) and so the vertex will be at $(4, 0)$.

13. $y = 1(x - 2)^2 + 1$, which gives $h = 2$, $k = 1$. This tells us that the graph of this parabola will be shifted right 2 units and up 1 unit compared to the graph of $y = x^2$.   The vertex will be at $(2, 1)$. Since $a = 1$, the shape will be the same as $y = x^2$.

17.   We have $a = \frac{1}{2}$, so the parabola opens upward and is wider than the graph of $y = x^2$. Since $h = 4$ and $k = 3$ it will be shifted to the right 4 units and up 3 units compared to the graph of $y = x^2$.   The vertex will be at $(4,3)$.

21.   These parabolas all have $a = 1$ and so all have the same shape. In each case $h = 0$, so there has been no horizontal shifting. The only difference between them is a vertical shift.
(b) is obtained by shifting the parabola $y = x^2$ down 2 units, because $k = -2$
(c) is obtained by shifting the parabola $y = x^2$ upward 2 units, because $k = 2$.

25.   Both parabolas have $a = 1$, so their graphs will have the same shape.
(a) is the basic parabola, opening upward with vertex $(0,0)$.
(b) has $h = 1$ and $k = 1$, so the graph of that parabola will be obtained by shifting
     the graph of $y = x^2$ to the right 1 unit and up 1 unit.

29. Both parabolas have a = 2, so their graphs will have the same shape. They are turned upward, and will appear more narrow than the graph of $y = x^2$ ($y = 2x^2$ grows twice as quickly as $y = x^2$).
(a) has vertex at $(0,0)$, with no shifting relative to the graph of $y = x^2$.
(b) has vertex $(0,1)$, and is shifted up 1 unit relative to the graph of $y = x^2$ or $y = 2x^2$.

33. $y = x^2 - 6x + 8$
$y = x^2 - 6x + 9 - 9 + 8$
$y = (x - 3)^2 - 1$ The vertex is $(3, -1)$.

37. Since h = 1 and k = −3, the vertex is at $(1, -3)$. The graph will be shifted to the right 1 unit and down 3 units compared to the graph of $y = x^2$. The value of a is 2, so the parabola will be narrower than the graph of $y = x^2$. It will open upward.

41. $0 = x^2 + 5x + 6$
$0 = (x + 2)(x + 3)$
$x + 2 = 0 \qquad x + 3 = 0$
$x = -2 \qquad x = -3 \qquad$ x-intercepts: $(-2, 0)$, $(-3, 0)$

45. We will use the form $y = a(x - h)^2 + k$.
Notice that the graph is shifted down four units relative to the graph of $y = x^2$. There is no horizontal shifting, so the equation is $y = ax^2 - 4$.
It can be dangerous to *assume* it is $y = x^2 - 4$; perhaps a is <u>not</u> 1.

If you need to find a, or need to verify what you <u>think</u> is a, look at the graph and identify coordinates of a point on the graph (other than the vertex.)
Which point you choose is up to you, but we will use $(2, 0)$. Substitute these coordinates into the equation and solve for a:

$0 = a(2)^2 - 4$
$0 = 4a - 4$
$a = 1 \qquad$ The complete equation is $y = x^2 - 4$.

49. $y = \frac{1}{4}x^2 - \frac{33}{4}x + 35$

$0 = \frac{1}{4}x^2 - \frac{33}{4}x + 35 \qquad$ Multiply both sides by 4 to clear the fractions.

$0 = x^2 - 33x + 140$

$0 = (x - 28)(x - 5)$

$x = 28 \qquad x = 5 \qquad$ Drill the holes at 5 inches and at 29 inches.

Problem Set 12.6

1. $\sqrt{-16} = \sqrt{(-1)(16)} = \sqrt{-1}\sqrt{16} = (i)(4) = 4i$

5. $\sqrt{-12} = \sqrt{(-1)(12)} = \sqrt{-1}\sqrt{12} = i\sqrt{12} = i(2\sqrt{3}) = 2i\sqrt{3}$

9.
$(5 - 7i) + (8 + 9i)$
$= 5 + 8 + (-7 + 9)i$
$= 13 + 2i$

13.
$-13 + 4 - 8i - 9i$
$= -13 + 4 + (-8 - 9)i$
$= -9 - 17i$

17. $(3 + 5i)(3 - 5i)$ Use the form $(a + b)(a - b) = a^2 - b^2$.
$= 9 - (5i)^2$
$= 9 - 25i^2$ $\boxed{i^2 = -1}$
$= 9 + 25 = 34$

21.
$\dfrac{6 - 4i}{1 + i} = \dfrac{(6 - 4i)(1 - i)}{(1 + i)(1 - i)}$

$= \dfrac{6 - 10i + 4i^2}{1 - i^2}$

$= \dfrac{6 - 10i - 4}{1 - (-1)}$

$= \dfrac{2 - 10i}{2}$

$= 1 - 5i$

25.
$\dfrac{5 + 3i}{i}$

$= \dfrac{(5 + 3i)(-i)}{i(-i)}$

$= \dfrac{-5i - 3i^2}{-i^2} = \dfrac{-5i + 3}{-(-1)}$

$= -5i + 3$

$= 3 - 5i$

29.
$(2 - i)^2 - 4(2 - i) + 5$
$= 4 - 4i + i^2 - 8 + 4i + 5$ $\boxed{i^2 = -1}$
$= 0$

33.
$x^2 = -36$
$x = \sqrt{-36}$ $x = -\sqrt{-36}$
$x = 6i$ $x = -6i$

37. $(2m + 1)^2 = -3$

$2m + 1 = i\sqrt{3}$ $2m + 1 = -i\sqrt{3}$

$2m = -1 + i\sqrt{3}$ $2m = -1 - i\sqrt{3}$

$m = -\dfrac{1}{2} + i\dfrac{\sqrt{3}}{2}$, $m = -\dfrac{1}{2} - i\dfrac{\sqrt{3}}{2}$

41. $m^2 - 4m + 8 = 0$ $x = \dfrac{-b \pm \sqrt{b^2 - 4ac}}{2a}$ $a = 1, b = -4, c = 8$

$x = \dfrac{-(-4) \pm \sqrt{(-4)^2 - 4(1)(8)}}{2(1)}$

$x = \dfrac{4 \pm \sqrt{16 - 32}}{2} = \dfrac{4 \pm \sqrt{-16}}{2} = \dfrac{4 \pm 4i}{2} = 2 \pm 2i$

$\boxed{\text{section} \quad 12.6}$

45. $3x^2 + 45 = 0$ Divide both sides by 3.

$x^2 + 15 = 0$ We will use extraction of roots.

$x^2 = -15$

$x = \sqrt{-15}$, $x = -\sqrt{-15}$

$x = i\sqrt{15}$, $x = -i\sqrt{15}$

49. $(2m - 1)(3m - 1) = m + 13$

$6m^2 - 5m + 1 = m + 13$

$6m^2 - 6m - 12 = 0$

$m^2 - m - 2 = 0$

$(m - 2)(m + 1) = 0$

$m - 2 = 0$ $m + 1 = 0$

$m = 2$ $m = -1$

53. It is handy to remember how powers of i work from the start:

$i = i$

$i^2 = -1$ $\boxed{i^2 = -1}$

$i^3 = -i$

$i^4 = 1$ After this, they repeat the pattern again. The number i raised to

                            an integer power will always be one of i, -1, $-i$, or 1.

$i^5 = i^1\, i^4 = i(1) = i$

$i^6 = i^2\, i^4 = i^2(1) = -1$

$i^7 = i^3\, i^4 = i^3(1) = -i$

$i^8 = i^4 i^4 = i^4(1) = 1$ and so forth.

57. $(1 + i)^3 = (1 + i)^2(1 + i) = (1 + 2i + i^2)(1 + i) = (2i)(1 + i) = 2i + 2i^2 = -2 + 2i$

Solutions to All Review Exercises – Chapter Twelve

1. First rewrite as $3x^2 - 7x + 8 = 0$.
$a = 3, b = -7, c = 8$

2. First rewrite as $2x^2 - 11 = 0$.
$a = 2, b = 0, c = -11$

3. First rewrite:
$$5x(x - 1) = x(x - 3)$$
$$5x^2 - 5x = x^2 - 3x$$
$$4x^2 - 2x = 0 \qquad a = 4, b = -2, c = 0$$

4. First rewrite:
$$(x + 4)(4x - 1) = 3$$
$$4x^2 - x + 16x - 4 = 3$$
$$4x^2 + 15x - 7 = 0 \qquad a = 4, b = 15, c = -7$$

COMMENT *In #5-#8, you can either substitute in the number and see if it yields a true statement, or you can solve the equation and see if the given number turns up as a solution.*

5. $$20\left(\frac{3}{4}\right)^2 - 7\left(\frac{3}{4}\right) = 6 \qquad ?$$

$$20\left(\frac{9}{16}\right) - \frac{21}{4} = 6$$

$$\frac{45}{4} - \frac{21}{4} = 6$$

$$\frac{24}{4} = 6 \quad \text{(true)} \qquad \frac{3}{4} \text{ is a solution.}$$

6. $$20\left(-\frac{2}{5}\right)^2 - 7\left(-\frac{2}{5}\right) = 6 \quad ?$$

$$20\left(\frac{4}{25}\right) + \frac{14}{5} = 6$$

$$\frac{16}{5} + \frac{14}{5} = 6$$

$$\frac{30}{5} = 6 \quad \text{(true)} \qquad -\frac{2}{5} \text{ is a solution.}$$

7. Instead of checking by substitution, you could also solve $n^2 = 8n - 13$ and see if $4 - \sqrt{3}$ is a solution. Either way is about the same amount of work, so we will proceed with substitution:
$$(4 - \sqrt{3})^2 = 8(4 - \sqrt{3}) - 13 \qquad ?$$
$$16 - 8\sqrt{3} + 3 = 32 - 8\sqrt{3} - 13$$
$$19 - 8\sqrt{3} = 19 - 8\sqrt{3} \qquad\qquad \text{(true)}$$

$$4 - \sqrt{3} \text{ is a solution.}$$

8. $$(4 + \sqrt{3})^2 = 8(4 + \sqrt{3}) - 13 \qquad ?$$
$$16 + 8\sqrt{3} + 3 = 32 + 8\sqrt{3} - 13$$
$$19 + 8\sqrt{3} = 19 + 8\sqrt{3} \qquad\qquad \text{(true)}$$

$$4 + \sqrt{3} \text{ is a solution.}$$

COMMENT *Compare exercises #7 and #8. Once you find that $4 - \sqrt{3}$ is a solution to $n^2 = 8n - 13$, we automatically know that $4 + \sqrt{3}$ is a solution, since it is the conjugate of $4 - \sqrt{3}$.*

Chapter Twelve Review

9. $y^2 = 19$ Take square root of both sides.

         $\sqrt{y^2} = \sqrt{19}$

         $|y| = \sqrt{19}$ Remember $\sqrt{x^2} = |x|$.

         $y = -\sqrt{19}, \sqrt{19}$

10. $(v - 4)^2 = 36$

         $\sqrt{(v-4)^2} = \sqrt{36}$ Remember $\sqrt{x^2} = |x|$

         $|v - 4| = 6$

         $v - 4 = 6$ or $v - 4 = -6$

         $v = 10$ or $v = -2$

11. $(2w + 1)^2 = 49$

         $\sqrt{(2w+1)^2} = \sqrt{49}$ Remember $\sqrt{x^2} = |x|$.

         $|2w + 1| = 7$

         $2w + 1 = 7$ or $2w + 1 = -7$

          $w = 3$ or $w = -4$

12. $-2(x + 7)^2 = -5$

         $(x + 7)^2 = \frac{5}{2}$

         $x + 7 = \sqrt{\frac{5}{2}}$ or $x + 7 = -\sqrt{\frac{5}{2}}$

         $x = -7 + \sqrt{\frac{5}{2}}$ or $x = -7 - \sqrt{\frac{5}{2}}$

         $x = -7 + \frac{\sqrt{10}}{2}$ or $x = -7 - \frac{\sqrt{10}}{2}$

13. $30m^2 + 11m - 30 = 0$ $a = 30, b = 11, c = -30$

         $m = \dfrac{-11 \pm \sqrt{11^2 - 4(30)(-30)}}{2(30)}$

         $m = \dfrac{-11 \pm \sqrt{3721}}{60} = \dfrac{-11 \pm 61}{60}$

         $m = \frac{5}{6}$ or $m = -\frac{6}{5}$

Chapter Twelve Review

14. $12v^2 - 95v = 8$ First put into standard form.

    $12v^2 - 95v - 8 = 0$ $a = 12, b = -95, c = -8$

$$v = \frac{95 \pm \sqrt{(-95)^2 - 4(12)(-8)}}{2(12)} = \frac{95 \pm \sqrt{9409}}{24}$$

$$v = \frac{95 \pm 97}{24}$$

$$v = 8 \qquad \text{or} \qquad v = -\frac{1}{12}$$

15. $y^2 = 2(y + 5)$ First put into standard form.

    $y^2 = 2y + 10$

    $y^2 - 2y - 10 = 0$ $a = 1, b = -2, c = -10$

$$y = \frac{2 \pm \sqrt{(-2)^2 - 4(1)(-10)}}{2}$$

$$y = 1 - \sqrt{11} \qquad \text{or} \qquad y = 1 + \sqrt{11}$$

16. $(x + 10)^2 = 4x + 41$ First put into standard form.

    $x^2 + 20x + 100 = 4x + 41$

    $x^2 + 16x + 59 = 0$ $a = 1, b = 16, c = 59$

$$x = \frac{-16 \pm \sqrt{16^2 - 4(1)(59)}}{2} = \frac{-16 \pm \sqrt{20}}{2}$$

$$x = -8 + \sqrt{5} \qquad \text{or} \qquad x = -8 + \sqrt{5}$$

COMMENTS ON SOLVING QUADRATIC EQUATIONS

(a) Factoring is usually easiest, if it can be done.

(b) If the problem is set up so that each side is a perfect square, try extraction of roots.

(c) The quadratic formula usually takes longest, so don't use it unless you must.

(d) If you are having trouble factoring (maybe there are lots of possibilities), try this:

> *If $b^2 - 4ac$ is a perfect square, then $ax^2 + bx + c$ can be factored using rational numbers. Keep trying. If $b^2 - 4ac$ is not a perfect square, then $ax^2 + bx + c$ cannot be factored using rational numbers. In that case, you should use some other method, such as the quadratic formula.*

Chapter Twelve Review

17. $15t^2 + 11t - 12 = 0$ We will factor.

$(5t - 3)(3t + 4) = 0$ Use trial and error.

$t = \frac{3}{5}$ or $t = -\frac{4}{3}$

18. $16m^2 - 24m + 9 = 0$ This will factor.

$(4m - 3)^2 = 0$

$4m - 3 = 0$

$m = \frac{3}{4}$ (a double root)

19. $(5n + 2)^2 = 9$ Use extraction of roots.

$5n + 2 = 3$ or $5n + 2 = -3$

$5n = 1$ $5n = -5$

$n = \frac{1}{5}$ $n = -1$

20. $(2n - 5)^2 = 12$ Use extraction of roots.

$2n - 5 = \sqrt{12}$ or $2n - 5 = -\sqrt{12}$

$2n - 5 = 2\sqrt{3}$ $2n - 5 = -2\sqrt{3}$

$2n = 5 + 2\sqrt{3}$ $2n = 5 - 2\sqrt{3}$

$n = \frac{5}{2} + \sqrt{3}$ or $n = \frac{5}{2} - \sqrt{3}$

21. $v^2 = 2(4v - 7)$ First rewrite in standard form.

$v^2 - 8v + 14 = 0$ This won't factor. Use the quadratic formula.

$$v = \frac{8 \pm \sqrt{(-8)^2 - 4(1)(14)}}{2}$$ We leave the details to you.

$v = 4 - \sqrt{2}, v = 4 + \sqrt{2}$

22. $v(v - 8) = -4$ Rewrite in standard form.

$v^2 - 8v = -4$

$v^2 - 8v + 4 = 0$ This won't factor.

(we leave the details to you)   Apply the quadratic formula.

Answer: $v = 4 - 2\sqrt{3}, v = 4 + 2\sqrt{3}$

Chapter Twelve Review

23. $2w^2 + 3w = 0$ Use factoring.

$w(2w + 3) = 0$

$w = 0$ or $w = -\dfrac{3}{2}$

24. $(y + \pi)^2 = 0$ Simply take the square root of both sides.

$y + \pi = 0$

$y = -\pi$ (double root)

25. $\dfrac{x^2}{4} + \dfrac{x}{2} = 1$ Multiply both sides by 4 to remove fractions.

$x^2 + 2x = 4$ Rewrite in standard form.

$x^2 + 2x - 4 = 0$ This doesn't factor.

Apply the quadratic formula.

$x = -1 - \sqrt{5},\ x = -1 + \sqrt{5}$ We leave the details to you.

26. $9z^2 - 49 = 0$ Factor the left side.

$(3z + 7)(3z - 7) = 0$

$z = -\dfrac{7}{3},\ z = \dfrac{7}{3}$

Alternate Solution:

$9z^2 - 49 = 0$

$9z^2 = 49$

$z^2 = \dfrac{49}{9}$

$z = \dfrac{7}{3},\ z = -\dfrac{7}{3}$

27. $m^2 + 6m + c$ perfect square trinomial: $m^2 + 2bm + b^2$

The coefficient of the m term is 6, so $2b = 6$.

$2b = 6$

$b = 3$ and $b^2 = 9$.

Therefore $c = 9$ and we have $m^2 + 6m + 9$.

> *The rule here is to take one-half of the coefficient of the first-degree term, and then square it.*

page 309

Chapter Twelve Review

28. $n^2 - 8n + c$

$$c = \left(-\frac{8}{2}\right)^2 = (-4)^2 = 16$$ See comment with #27.

29. $x^2 - 5x + c$

$$c = \left(-\frac{5}{2}\right)^2 = \frac{25}{4}$$ See comment with #27.

30. $y^2 + \frac{2}{7}y + c$

$$c = \left(\frac{1}{2} \cdot \frac{2}{7}\right)^2 = \left(\frac{1}{7}\right)^2 = \frac{1}{49}$$ See comment with #27.

31. $m^2 - 6m = -8$

$m^2 - 6m + 9 = -8 + 9$

$(m - 3)^2 = 1$

$m - 3 = 1$ .... or .... $m - 3 = -1$

$m = 4$ .... or .... $m = 2$

| COMMENT | *Why Use Completing the Square?*
When allowed our choice of methods we usually don't complete the square, since the other methods are often faster, or at least we are more practiced with them. The reason for learning this method is not just for solving equations, but because it is an important algebraic technique that is needed in later courses.

32. $n^2 = 4n - 3$ Put all but the constant term on the left.

$n^2 - 4n = -3$

$n^2 - 4n + 4 = -3 + 4$

$(n - 2)^2 = 1$

$n - 2 = 1$ .... or .... $n - 2 = -1$

$n = 3$ .... or .... $n = 1$

33. $(y + 1)(y - 1)(2y - 3) = 0$

$y + 1 = 0$ .... $y - 1 = 0$ .... $2y - 3 = 0$

$y = -1$ .... $y = 1$ .... $y = \frac{3}{2}$

Chapter Twelve Review

34. $x^3 - 4x = 0$

$x(x^2 - 4) = 0$

$x(x + 2)(x - 2) = 0$

$x = 0 \quad x = -2 \quad x = 2$

35. $\dfrac{3 + \sqrt{15}}{3} \approx \dfrac{3 + \sqrt{16}}{3} = \dfrac{3 + 4}{3} = \dfrac{7}{3} = 2\dfrac{1}{3}$ 　　　　Answer: C

36. $\dfrac{3 - \sqrt{15}}{3} \approx \dfrac{3 - \sqrt{16}}{3} = \dfrac{3 - 4}{3} = -\dfrac{1}{3}$ 　　　　Answer: B

37. $d = 4.9t^2$ 　　　　　Given $d = 30$, find t.

$30 = 4.9t^2$

$t^2 = \dfrac{30}{4.9} = \dfrac{300}{49}$

$t = \sqrt{\dfrac{300}{49}} = \sqrt{\dfrac{3(100)}{7^2}} = \dfrac{10\sqrt{3}}{7} \approx 2.47$ sec

38. 　w　　　　　　width

　3w + 2　　　length

$w(3w + 2) = 56$

$3w^2 + 2w = 56$

$3w^2 + 2w - 56 = 0$ 　　　　We will factor.

$(3w + 14)(w - 4) = 0$

$w = -\dfrac{14}{3} \qquad w = 4$ 　　　　We reject $w = -\dfrac{14}{3}$.

The width is 4 meters and the length is $3(4) + 2 = 14$ meters.

As a check, notice that $4(14) = 56$.

39. 　h　　　　　　height

　3h　　　　　width

Use the Pythagorean Theorem:

Chapter Twelve Review

$$h^2 + (3h)^2 = 6.5^2$$
$$h^2 + 9h^2 = 6.5^2$$
$$10h^2 = 42.25$$
$$h^2 = 4.225$$
$$h = \sqrt{4.225} \approx 2.06 \text{ m}$$

To obtain an exact answer with no decimals, consider this:

$$h^2 = 4.225$$
$$1000h^2 = 4225 \qquad \text{Divide by 25.}$$
$$40h^2 = 169$$
$$h^2 = \frac{169}{40}$$

$$h = \frac{13}{\sqrt{40}} = \frac{13}{2\sqrt{10}} = \frac{13\sqrt{10}}{20}.$$

40. Let t be the number of hours needed for the faster belt. Then the slower belt would require $t + 3$ hours. In two hours, 1 complete job is done.

	R	·	T	=	W
Slower belt	$\dfrac{1}{t + 3}$		2		$\dfrac{2}{t + 3}$
Faster belt	$\dfrac{1}{t}$		2		$\dfrac{2}{t}$

$$\frac{2}{t + 3} + \frac{2}{t} = 1 \qquad \text{Multiply by } t(t + 3)$$

$$2t + 2(t + 3) = t(t + 3)$$
$$2t + 2t + 6 = t^2 + 3t$$
$$t^2 - t - 6 = 0 \qquad \text{This factors.}$$
$$(t - 3)(t + 2) = 0$$
$$t = 3 \qquad\quad t = -2 \qquad \text{We reject } t = -2.$$

The faster belt will take 3 hours and the slower belt 6 hours.

Chapter Twelve Review

41. $y = -3x^2$ We have $h = 0$ and $k = 0$, so there is no shifting relative to the graph of $y = x^2$; the vertex is $(0,0)$.

                          $a = -3$, so the parabola opens downward and is narrower than the parabola $y = x^2$.

42. $y = x^2 + 3$ Since $a = 1 > 0$, the parabola opens upward. $h = 0$ and $k = 3$, so the vertex is $(0,3)$.

                          You can also conclude that the vertex is $(0,3)$ by noticing that this graph is shifted upward 3 units relative to the graph of $y = x^2$, which has vertex $(0,0)$.

43. $y = -(x + 3)^2$

There is more than one way to do this. Here are two:

(1)   Since $y = -(x + 3)^2$ is written in the form $y = a(x - h)^2 + k$, we note:

        $h = -3$ and $k = 0$, so the vertex is $(-3,0)$.

        $a = -1$, which is negative, so the parabola opens downward.

(2)   Another way to deal with this is to first consider the parabola $y = (x + 3)^2$. It is shifted to the left 3 units relative to the graph of $y = x^2$, so its vertex is $(-3,0)$.

        Now think about $y = -(x + 3)^2$. All the y-values of this equation are just opposite in sign of the y values for $y = (x + 3)^2$, which means the graph of $y = -(x + 3)^2$ has the same shape, but opens downward. This does not alter the vertex; it is still $(-3,0)$.

44. $y = (x - 2)^2 + 1$

(1)   Since $a = 1 > 0$, the parabola opens upward. $h = 2$ and $k = 1$, so the vertex is $(2,1)$.

(2)   Another approach:

        This graph is shifted right 2 units and up 1 unit compared to the graph of $y = x^2$. The vertex is $(2,1)$ and it opens upward, as $y = x^2$ does.

45.   They are the same.

        If an equation is multiplied on both sides by a non-zero constant, you obtain an equivalent equation; that is, one that has the same solution set as the original. If you begin with either equation and multiply both sides by -1, you obtain the other, and so they have the same solution set.

Chapter Twelve Review

46. $y = 2x^2$ This parabola opens upward and there is no shifting relative to $y = x^2$.
It has vertex (0,0) and passes through (1,2), and so must be D.

47. $y = -2x^2$ This parabola opens downward and there is no shifting relative to $y = x^2$.
It has vertex (0,0) and passes through $(1, -2)$, and so must be E.

48. $y = \frac{1}{2}x^2$ This parabola opens upward and there is no shifting relative to $y = x^2$.
It has vertex (0,0) and passes through $(1, \frac{1}{2})$ and so must be H.

49. $y = (x - 2)^2$ This parabola is shifted right 2 units compared to $y = x^2$. The vertex
is (2,0) and it opens upward. The match is A.

50. $y = (x + 2)^2$ This parabola is shifted left 2 units compared to $y = x^2$. The vertex is
$(-2, 0)$ and it opens upward. The match is C.

51. $y = x^2 - 2$ This parabola is shifted down 2 units compared to $y = x^2$. The vertex is
$(0, -2)$ and it opens upward. The match is B.

52. $y = x^2 + 2$ This parabola is shifted up 2 units compared to $y = x^2$. The vertex is
$(0, 2)$ and it opens upward. The match is F.

53. $y = (x+1)^2+1$ This parabola is shifted up 1 unit and left 1 unit compared to $y = x^2$. The
vertex is $(-1, 1)$ and it opens upward. The match is I.

54. $y = (x-1)^2-1$ This parabola is shifted right 1 unit and down 1 unit compared to $y = x^2$.
The vertex is $(1, -1)$ and it opens upward. The match is G.

55. $(11 - 7i) + (3 - 5i) = 11 + 3 - 7i - 5i = 14 - 12i$

56. $(11 - 7i) - (3 - 5i) = 11 - 3 - 7i - (-5i) = 8 - 7i + 5i = 8 - 2i$

57. $(11 - 7i)(3 - 5i)$ Use FOIL.
$= 33 - 55i - 21i + 35i^2$
$= 33 - 76i - 35$ $\boxed{i^2 = -1}$
$= -2 - 76i$

Chapter Twelve Review

58. $\dfrac{11 - 7i}{3 - 5i} = \dfrac{11 - 7i}{3 - 5i} \cdot \dfrac{3 + 5i}{3 + 5i} = \dfrac{(11 - 7i)(3 + 5i)}{(3 - 5i)(3 + 5i)} = \dfrac{33 + 55i - 21i - 35i^2}{3^2 - (5i)^2}$

$= \dfrac{33 + 34i + 35}{9 - 25i^2} = \dfrac{68 + 34i}{9 + 25} = \dfrac{34(2 + i)}{34} = 2 + i$

59. $(2x + 5)^2 = -1$ This is ideal for extraction of roots.

$2x + 5 = \sqrt{-1}$ or $2x + 5 = -\sqrt{-1}$

$2x = -5 + \sqrt{-1}$ or $2x = -5 - \sqrt{-1}$

$x = -\dfrac{5}{2} + \dfrac{1}{2}i$ or $x = -\dfrac{5}{2} - \dfrac{1}{2}i$

The answer may also be written as: $x = -\dfrac{5}{2} + \dfrac{i}{2}$ or $x = -\dfrac{5}{2} - \dfrac{i}{2}$.

60. $x^2 + 2x + 2 = 0$ This does not factor, so apply the quadratic formula. $a = 1$, $b = 2$ and $c = 2$

$x = \dfrac{-2 \pm \sqrt{2^2 - 4(1)(2)}}{2}$

$x = \dfrac{-2 \pm \sqrt{-4}}{2} = \dfrac{-2 + 2i}{2} = -1 \pm i$ Answer: $x = -1 - i$ or $-1 + i$.

Solutions To All Mastery Test Problems – Chapter Twelve

1a. $m = -10, 10$

1b. $n = -\sqrt{11}, \sqrt{11}$

1c. $7x^2 = 112$
$x^2 = 16$
$x = -4, 4$

1d.

$2y - 3 = 5$	$2y - 3 = -5$
$2y = 8$	$2y = -2$
$y = 4$	$y = -1$

2a. $x^2 + 2x + 1 = 3 + 1$
$(x + 1)^2 = 4$

$x + 1 = 2$	$x + 1 = -2$
$x = 1$	$x = -3$

2b. $y^2 - 3y - 4 = 24$
$y^2 - 3y = 28$
$y^2 - 3y + \dfrac{9}{4} = 28 + \dfrac{9}{4}$
$(y - \dfrac{3}{2})^2 = \dfrac{121}{4}$

$$y - \frac{3}{2} = \frac{11}{2} \qquad\qquad y - \frac{3}{2} = -\frac{11}{2}$$

$$y = 7 \qquad\qquad\qquad y = -4$$

2c. $v^2 + 8v + 16 = -14 + 16$

$(v + 4)^2 = 2$

$v + 4 = \sqrt{2} \qquad\qquad v + 4 = -\sqrt{2}$

$v = -4 + \sqrt{2} \qquad\quad v = -4 - \sqrt{2}$

2d. $m^2 + \frac{3}{10}m = \frac{1}{10}$

$$m^2 + \frac{3}{10}m + \left(\frac{3}{20}\right)^2 = \frac{1}{10} + \frac{9}{400}$$

$$(m + \frac{3}{20})^2 = \frac{49}{400}$$

$$m + \frac{3}{20} = \frac{7}{20} \qquad \text{or} \quad m + \frac{3}{20} = -\frac{7}{20}$$

$$m = \frac{4}{20} = \frac{1}{5} \qquad\qquad m = -\frac{10}{20} = -\frac{1}{2}$$

3a. $\dfrac{-3 \pm \sqrt{3^2 - 4(1)(-54)}}{2}$

$$= \frac{-3 \pm \sqrt{225}}{2} = \frac{-3 \pm 15}{2}$$

$$= 6 \text{ or } -9$$

3b. $\dfrac{-(-1) \pm \sqrt{(-1)^2 - 4(6)(-77)}}{2(6)}$

$$= \frac{1 \pm \sqrt{1849}}{12} = \frac{1 \pm 43}{12}$$

$$= \frac{11}{3} \text{ or } -\frac{7}{2}$$

3c. $x^2 - 10x + 9 = x - 12$

$x^2 - 11x + 21 = 0$

$$\frac{-(-11) \pm \sqrt{(-11)^2 - 4(1)(21)}}{2}$$

$$= \frac{11 \pm \sqrt{37}}{2}$$

Answer: $\frac{11}{2} + \frac{\sqrt{37}}{2}, \frac{11}{2} - \frac{\sqrt{37}}{2}$

3d. $2y^2 - 2y = 1$

$2y^2 - 2y - 1 = 0$

$$\frac{-(-2) \pm \sqrt{(-2)^2 - 4(2)(-1)}}{2(2)}$$

$$= \frac{2 \pm \sqrt{12}}{4} = \frac{2 \pm 2\sqrt{3}}{4} = \frac{2(1 \pm \sqrt{3})}{2(2)}$$

Answer: $\frac{1}{2} - \frac{\sqrt{3}}{2}, \frac{1}{2} + \frac{\sqrt{3}}{2}$

4a.   Let the two consecutive integers be n and n + 1.

$n^2 + (n + 1)^2 = 113$

$n^2 + n^2 + 2n + 1 = 113$

$2n^2 + 2n - 112 = 0$ *Save yourself some work!* Divide both sides by 2.

Chapter Twelve Mastery Test

$n^2 + n - 56 = 0$

$(n + 8)(n - 7) = 0$ Use factoring when you can.

$n = -8$ $n = 7$ Remember to also find n + 1. Answer: -8 and -7 or 7 and 8

4b.   Let the height of the tower be h, and the length of the guy-wire be 2h.

Geometrically, the guy-wire represents the hypotenuse of a right triangle.

$h^2 + 50^2 = (2h)^2$

$h^2 + 2500 = 4h^2$

$3h^2 = 2500$

$h^2 = \frac{2500}{3}$ $h = \sqrt{\frac{2500}{3}} = \frac{\sqrt{2500}}{\sqrt{3}} = \frac{50\sqrt{3}}{3} \approx 28.87$ meters.

Notice that we reject the negative square root because h represents length.

4c. $V = \pi r^2 h$ V = volume of the cylinder; r = the radius of the end; h = the height.

$1.6 = \pi r^2 (2)$

$r^2 = \frac{1.6}{2\pi} = \frac{16}{20\pi} = \frac{4}{5\pi}$ $r = \sqrt{\frac{4}{5\pi}} = \frac{2}{\sqrt{5\pi}} = \frac{2\sqrt{5\pi}}{5\pi} \approx 0.50$ meters

4d.   Let r be the rate at which the first CD earns interest.

	Principal · Rate	=	Interest Earnings
first CD	$\frac{280}{r}$	r	280
second CD	$\frac{135}{r+0.01}$	r+0.01	135

$Prt = I$ (with t=1 in this problem), which allows us to calculate $P = \frac{I}{r}$.

The sum of the principal amounts is $5000:

$\frac{280}{r} + \frac{135}{r + 0.01} = 5000$ Multiply both sides by r(r + 0.01).

$280(r + 0.01) + 135r = 5000(r)(r + 0.01)$

$280r + 2.8 + 135r = 5000r^2 + 50r$

$5000r^2 - 365r - 2.8 = 0$

Chapter Twelve Mastery Test

$$r = \frac{365 \pm \sqrt{(365)^2 - 4(5000)(-2.8)}}{2(5000)}$$ Only one of these values is positive: $\frac{365 + 435}{10,000} = 0.08$

Answer: The first CD earns interest at a rate of 8%, and the second at 9%.

5.   See answer section in the text for graphs.

6a.   2i   6b. $3i\sqrt{2}$ 6c. $3 + 3i$ 6d. $-2 + 5i$

7a. $6 - 2i$ 7b. $-2 + 8i$

7c. $8 - 10i + 12i - 15i^2$ 7d. $\frac{i(2 - 3i)}{(2 + 3i)(2 - 3i)} = \frac{2i - 3i^2}{4 - 9i^2}$

       $= 8 + 2i - 15(-1)$

       $= 23 + 2i$ $= \frac{3 + 2i}{4 + 9} = \frac{3 + 2i}{13} = \frac{3}{13} + \frac{2}{13}i$

8a. $3x-1 = \sqrt{-16}$ $3x-1 = -\sqrt{-16}$ 8b. $\frac{-(-6) \pm \sqrt{(-6)^2 - 4(1)(13)}}{2}$

       $3x - 1 = 4i$ $3x - 1 = -4i$

       $3x = 1 + 4i$ $3x = 1 - 4i$ $\frac{6 \pm \sqrt{-16}}{2} = \frac{6 \pm 4i}{2} = 3 \pm 2i$

       $x = \frac{1}{3} + \frac{4}{3}i$ $x = \frac{1}{3} - \frac{4}{3}i$ Answer: $3 + 2i$, $3 - 2i$

Chapter Twelve Mastery Test

1. $\sqrt{81} = 9$

2. $\sqrt{0.01} = \sqrt{\frac{1}{100}} = \frac{1}{10} = 0.1$

3. $\sqrt{1,000,000} = \sqrt{10^6} = 10^3 = 1000$

4. $\sqrt[3]{125} = \sqrt[3]{5^3} = 5$

5. $\sqrt{169 - 144} = \sqrt{25} = 5$

6. $\sqrt{169} - \sqrt{144} = 13 - 12 = 1$

7. $\sqrt[3]{1,000,000} = \sqrt[3]{10^6} = 10^2 = 100$ (Contrast this with #3.)

8. $\sqrt{125} = \sqrt{25(5)} = 5\sqrt{5}$ (Contrast this with #4.)

9. $\sqrt{18} = \sqrt{9(2)} = 3\sqrt{2}$

10. $5\sqrt{2} + 6\sqrt{2} = 11\sqrt{2}$

11. $4\sqrt{3} - 9\sqrt{3} = -5\sqrt{3}$

12. $(2\sqrt{3})(4\sqrt{5}) = 8\sqrt{15}$

13. $\sqrt{(-3)^2} = \sqrt{9} = 3$ This illustrates the rule $\sqrt{x^2} = |x|$.

14. $\sqrt{6}(\sqrt{2} - \sqrt{3}) = \sqrt{6}\sqrt{2} - \sqrt{6}\sqrt{3} = \sqrt{3}\sqrt{2}\sqrt{2} - \sqrt{2}\sqrt{3}\sqrt{3} = 2\sqrt{3} - 3\sqrt{2}$

15. $(\sqrt{11} - \sqrt{2})(\sqrt{11} + \sqrt{2})$ Use the form $a^2 - b^2 = (a + b)(a - b)$.
    $= (\sqrt{11})^2 - (\sqrt{2})^2$
    $= 11 - 2$
    $= 9$

16. $\sqrt{\frac{25}{36}} = \sqrt{\left(\frac{5}{6}\right)^2} = \frac{5}{6}$

17. $\frac{12}{\sqrt{3}} = \frac{12}{\sqrt{3}}\frac{\sqrt{3}}{\sqrt{3}} = \frac{12\sqrt{3}}{3} = 4\sqrt{3}$

18. $5\sqrt{8} - 3\sqrt{50} = 5\sqrt{4(2)} - 3\sqrt{25(2)} = 10\sqrt{2} - 15\sqrt{2} = -5\sqrt{2}$

19. $100^{\frac{1}{2}} = \sqrt{100} = 10$

20. $64^{\frac{1}{2}} = \sqrt{64} = 8$

21. $64^{\frac{1}{3}} = \sqrt[3]{64} = 4$

22. $1^{\frac{1}{3}} = \sqrt[3]{1} = 1$

23. $\left(\frac{1}{3}\right)^1 = \frac{1}{3}$ Any number raised to the first power is just that number.

Cumulative Review

24. $8^{\frac{2}{3}} = (8^{\frac{1}{3}})^2 = (\sqrt[3]{8})^2 = 2^2 = 4$

25. $\sqrt{-36} = \sqrt{36}\sqrt{-1} = 6i$

26. $2i(3 - 4i) = 6i - 8i^2 = 8 + 6i$

27. $2i - (3 - 4i) = -3 + 6i$

28. $(3 + 2i)(3 - 2i) = 9 - (2i)^2$
    $= 9 - 4i^2 = 13$

29. $\dfrac{15}{2 - i} \cdot \dfrac{2 + i}{2 + i} = \dfrac{15(2 + i)}{4 - i^2} = \dfrac{15(2 + i)}{5} = 3(2 + i) = 6 + 3i$

30. $(1 + i)^2 = 1^2 + 2i + i^2 = 2i$ Use the form $(a + b)^2 = a^2 + 2ab + b^2$.

31. $-\sqrt{4} + \sqrt{-9} = -\sqrt{2^2} + \sqrt{9}\sqrt{-1} = -2 + 3i$

32. $\sqrt{-64} + \sqrt[3]{-64} = \sqrt{64}\sqrt{-1} + \sqrt[3]{(-4)^3} = 8i + (-4) = -4 + 8i$

33. $x^2 - 11x + 10 = 0$ This factors.

    $(x - 10)(x - 1) = 0$

    $x = 10$ $x = 1$

34. $x^2 = 100$ Use extraction of roots.

    $x = -10, \ 10$

35. $x^2 + 3x + 1 = 0$ Use the quadratic formula: $\dfrac{-b \pm \sqrt{b^2 - 4ac}}{2a}$

    $\dfrac{-3 \pm \sqrt{3^2 - 4(1)(1)}}{2(1)} = \dfrac{-3 \pm \sqrt{5}}{2}$ or $-\dfrac{3}{2} \pm \dfrac{\sqrt{5}}{2}$

36. $x^2 + 3x + 5 = 0$ Use the quadratic formula: $\dfrac{-b \pm \sqrt{b^2 - 4ac}}{2a}$

    $\dfrac{-3 \pm \sqrt{3^2 - 4(1)(5)}}{2(1)} = \dfrac{-3 \pm \sqrt{9 - 20}}{2} = \dfrac{-3 \pm \sqrt{-11}}{2} = \dfrac{-3 \pm i\sqrt{11}}{2}$ or $-\dfrac{3}{2} \pm i\dfrac{\sqrt{11}}{2}$

37. $\sqrt{3x + 1} = 5$ Square both sides.

    $3x + 1 = 25$

    $3x = 24$

    $x = 8$ *(Be sure to check for extraneous solutions when using this method.)*

38. $\sqrt{2m - 1} + 2 = m$ You will need to isolate the radical to one side before squaring.

    $\sqrt{2m - 1} = m - 2$ Now square both sides.

    $2m - 1 = (m - 2)^2$

$2m - 1 = m^2 - 4m + 4$

$m^2 - 6m + 5 = 0$ This factors, and factoring is easier than using the quadratic formula.

$(m - 5)(m - 1) = 0$

$m = 5, \ m = 1$ Be sure to check. You will find that 1 is extraneous.

Only $m = 5$ is a solution.

39.   Use the distance formula: $\sqrt{(x_2 - x_1)^2 + (y_2 - y_1)^2}$

$\sqrt{(8 - 5)^2 + (0 - (-4))^2} = \sqrt{3^2 + 4^2} = \sqrt{9 + 16} = \sqrt{25} = 5$

40.   They will be parallel. The reason for this is that an inconsistent system of equations is one that has no solution, which means that the lines have no point of intersection. Only parallel lines never meet.

41. $x = 2y + 3$ Since x is solved for, we will use substitution.

$2x + 5y = -3$

$2(2y + 3) + 5y = -3$

$9y + 6 = -3$

$9y = -9$

$y = -1$ Now find x. Technically, you can use either of the original equations, but we find the first easiest in this problem.

$x = 2(-1) + 3$

$x = 1$ The solution is the ordered pair $(1, -1)$.

42. $x - 5y = -2$ Since there is 5y in both, addition of the equations will eliminate y.

$\underline{2x + 5y = 26}$

$3x = 24$

$x = 8$ Now substitute into either of the original equations to find y.

$8 - 5y = -2$

$-5y = -10$

$y = 2$ The solution is $(8, 2)$.

Cumulative Review

43.　　$2x - 3y = 2$　　　　　We will use addition-subtraction.

　　　　$5x + 2y = 14$

$$10x - 15y = 10$$
$$-(10x + 4y = 28)$$
$$-19y = -18$$
$$y = \frac{18}{19}$$

Substitute into the first or second equation to find x. We will leave the details to you!   Answer: $\left(\frac{46}{19}, \frac{18}{19} \right)$

44.　　$3x + 6y = 9$　　　　　You should notice that each can be simplified. We do that first.
　　　　$2x + 4y = 6$

　　　　$x + 2y = 3$　　　　　We find that they are really the same equation, hence, the same line.
　　　　$x + 2y = 3$　　　　　The solution is the set of all points on the line, so we say there are are infinitely many solutions.

45.

	value per coin	number of coins	total value
nickels	0.05	n	0.05n
dimes	0.10	d	0.10d

The system is:

$0.05n + 0.10d = 2.40$　　　　Add the amounts of money in nickels and dimes to get the grand total.

$n + d = 30$　　　　The total number of coins is 30.

We leave the solving to you, but suggest that you multiply the first equation by 100 to remove the decimals.
Answer: There are 12 nickels and 18 dimes.

46.　　Let r_1 be the rate of train without speeding up.

	Rate	Time	Distance
going slower	r_1	$\frac{200}{r_1}$	200
going faster	r_2	$\frac{200}{r_2}$	200

The problem indicates that the difference in the times is 1 hour. This means we must subtract the two times, but in what order? We must do this so that the result is positive:

Cumulative Review

larger time − smaller time = 1 hour

Think about this: the larger time is *the time required when the train is going slower.* So we set it up with that time first:

$$\frac{200}{r_1} - \frac{200}{r_2} = 1$$

The second equation in the system is:
$r_2 = r_1 + 10$ The faster rate is 10 mi/h more than the slower rate.

We leave the solving of the system to you. It is reasonable to use substitution, since r_2 is solved for in the second equation.
Answer: The original rate of the train is 40 miles per hour.

47. See answer section in the text for graphs. We suggest that you graph each of the dashed lines by finding the intercepts. It is helpful to mark each half-plane with different colors or markings, so you can easily see where they overlap.

48. The inequalities x ≥ 0 and y ≥ 0 simply indicate the first quadrant (plus axes).

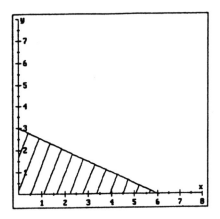

49. See answer section in the text.

50.

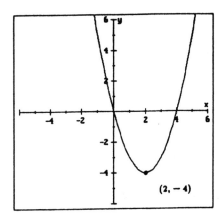

(2, −4)

See textbook for graphs for odd-numbered exercises.

Cumulative Review